This is now the third edition of a text that is already well established as one of the standard undergraduate books on the subject of elementary particle physics. Professor Hughes has updated the whole text in line with current particle nomenclature and has added material to cover important new developments. There is also a completely new major chapter on particle physics and cosmology, an exciting subject which has become an area of increasing importance in recent years.

In this field much can be learnt from the way the subject developed, and so, where this helps its understanding, an historical treatment is used. At all stages the theoretical development is closely linked to the relevant experimental measurements, providing a sound foundation to what might otherwise appear as a rather abstract subject.

This lucid text will fill the need for an introductory undergraduate book aimed at students taking a first course in this fascinating area of modern physics.

Elementary particles

Elementary particles

THIRD EDITION

I. S. HUGHES
University of Glasgow

The right of the
University of Cambridge
to print and sell
all manner of books
was granted by
Henry VIII in 1534.
The University has printed
and published continuously
since 1584.

CAMBRIDGE UNIVERSITY PRESS

Cambridge

New York   Port Chester

Melbourne   Sydney

Published by the Press Syndicate of the University of Cambridge
The Pitt Building, Trumpington Street, Cambridge CB2 1RP
40 West 20th Street, New York, NY 10011–4211, USA
10 Stamford Road, Oakleigh, Melbourne 3166, Australia

First edition © I. S. Hughes, 1972
Second edition © Cambridge University Press, 1985
Third edition © Cambridge University Press, 1991

First published by Penguin Books Ltd, 1972
Second edition published by Cambridge University Press, 1985
Reprinted 1987, 1988
Third edition published by Cambridge University Press, 1991

Printed in Great Britain at the University Press, Cambridge

British Library cataloguing in publication data
Hughes, I. S. (Ian Simpson)
Elementary particles. – 3rd ed.
1. Elementary particles
I. Title
539.721

Library of Congress cataloguing in publication data
Hughes, I. S. (Ian Simpson). 1930–
Elementary particles/I. S. Hughes. – 3rd ed.
       p.   cm.
Includes bibliographical references and index.
ISBN 0-521-40402-9. – ISBN 0-521-40739-7 (pbk.)
1. Particles (Nuclear physics) 2. Cosmology. 3. Nuclear
astrophysics. I. Title.
QC793.2.H83   1991   91–4492 CIP
539.7′2–dc20

ISBN 0 521 40402 9 hardback
ISBN 0 521 40739 7 paperback

There are therefore Agents in Nature able to make the Particles of Bodies stick together by very strong attractions. And it is the Business of experimental Philosophy to find them out.
Newton, *Opticks*.

My purpose is now to lead you into the Pallace where you shall have a clear and delightful view of all those various objects, and scattered excellencies, that lye up and down upon the face of the creation, which are only seen by those that go down into the Seas, and by no other.
Daniel Pell, Πέλαγοσ.

Contents

Preface to the third edition

Since the publication of the second edition of this book there have been no advances in the subject of the magnitude of the discovery of the Z and W bosons but there have been developments which justify an updated version of the text. From a wider perspective the links between particle physics and cosmology have become closer in a number of ways, in particular as we see the earliest moments after the Big Bang as the ultimate high energy laboratory for testing grand unification on the one hand and grand unification as an essential element in understanding the origins of our universe on the other. Equally the Dark Matter problem is another area where the symbiotic nature of particle physics and cosmology may make the essential contribution to our understanding. Since the 1987A supernova neutrinos are seen to have yet another aspect to their fascinating characteristics as both fundamental components of matter and tools by which we can learn about the cosmos. Without pretending in any way to offer a text in cosmology I have felt it worthwhile to include in this new edition a chapter dealing with these topics.

At a more detailed level I have updated the text to conform to the new (more logical) particle nomenclature as well as improving the treatment of some topics and correcting a few errors in the earlier version.

Ian Hughes, May 1990.

Preface to the first and second editions

This book is intended for undergraduates or others coming to the subject of particle physics for the first time. For this reason the only prior knowledge assumed is of the elements of quantum theory and statistical mechanics.

The story of the development of particle physics in the years since the Second World War has been one of almost continuous excitement. Much of this has been due to an unceasing interplay of experiment and theory in the best classical tradition. Few years have passed without a remarkable advance in theory or experiment, such as the discovery of the antiproton; of the strange particles; the Gell-Mann–Nishijima scheme; parity non-conservation; the difference between electron and muon neutrinos; the strongly-decaying resonances; the SU(3) symmetry scheme and the omega particle; evidence for quarks and for gluons; neutral currents; charm and beauty; electromagnetic-weak unification; the discovery of the W and Z bosons and a good many others.

This rapid progress has been a consequence of, and a justification for, parallel progress in technology and instrumentation. In the first chapter of the book I have outlined the principal techniques used in this work. I hope that this will enable the student to understand how the many experiments referred to in later parts of the book have actually been carried out, since I believe that such an understanding is essential to a proper appreciation of the subject. While I have not adopted a strictly historical approach I have felt it desirable to discuss the way in which many of the problems were originally seen and subsequently solved; as, for instance, the puzzle of the muons when first observed, the τ–θ problem and others, since the solution of the problems is itself often very instructive and aids an understanding of the phenomena.

In a book at this level many of the theoretical aspects of the subject cannot be treated in a rigorous way. I have chosen to emphasise those aspects derivable from conservation laws since they underlie much of the later theory and themselves afford an important degree of understanding.

The book has been very substantially modified compared with the first edition published in 1972. The seventies has been the decade of the leptons and the weak interaction, and new chapters (7 and 11) have been added to give fuller treatment to this side of the subject. On the other hand, the discovery of new quark flavours has deepened and extended our understanding of the quark–gluon structure of matter and this material is treated in new chapters 12 and 13. The giant step forward of the theory of electro–weak unification and its experimental verification (chapter 11) has led to increased confidence that all the forces of nature will eventually be understood in a unified way and this subject is treated briefly in a new chapter (14). In order to accommodate the new material without too greatly increasing the length of the book, some topics which are now seen to be of less importance (and results which have turned out to be wrong!) have been removed from the earlier text.

Particle physics is a very active subject in both its theoretical and experimental aspects, and in the technology of accelerators and detectors. I have tried to bring the discussion as near as possible to current work and in so doing I take the risk that some of the most recent results and ideas may prove in time to be wrong. I believe that this risk is justified by the attempt to show that the subject is very much alive and is continually generating the most fundamental and challenging problems.

I hope that this book presents the subject in sufficient depth to give the student an understanding of its fundamental nature, its fascination and its recent startling progress. In an introductory text, however, many subjects have to be dealt with superficially or not at all and a number of theoretical results presented on trust. In a subject as active as particle physics it is not entirely straightforward to recommend books for further reading, since much of the most useful material is contained in publications such as the reports of summer schools, like the annual series organised by CERN, the notes for which are published as CERN reports. However, as more advanced texts covering most of the material, I recommend *Introduction to High Energy Physics* by D. H. Perkins (Addison–Wesley, 1982) and *Quarks and Leptons* by F. Halzen and A. D. Martin (Wiley, 1984). A good introduction to group theory, particularly as applied in particle physics, is provided in *Lie Groups for Pedestrians* by H. J. Lipkin (North Holland, 1965). I have been indebted to these texts, among others, in preparing the revised version of this book.

Acknowledgements

I am grateful to Dr W. T. Morton for comments and suggestions on the original version of the book, to Dr C. D. Froggatt and Dr J. G. Lynch for similar help with the second edition and to Professor J. C. Brown, Dr A. T. Davies, Dr I. Ten Have, Dr J. F. L. Simmons and Dr K. M. Smith for advice and comments on the new chapter on Particle Physics and Cosmology and on some of the other new sections of the third edition. It is a pleasure to record my thanks to my wife Isobel for much encouragement and for editorial assistance particularly on the second edition.

I am indebted to the authors who have provided me with figures or allowed me to reproduce figures from their papers, to the CERN photographic service and to DESY for provision of material, and to the following journals for permission to reproduce diagrams originally published therein: *Nature*; *Philosophical Magazine*; *Physical Review*; *Physical Review Letters*; *Physics Letters* and *Nuclear Physics*.

Ian S. Hughes, August 1990.

1
Accelerators, beams and detectors

1.1 Introduction

An important part of the study of particle physics is an understanding of experimental tools – the accelerators, beams and detectors by means of which particles are accelerated, their trajectories controlled and their properties measured. There exist a limited number of types of accelerators and detectors in common use or which have in the past proved crucial to the progress of the subject. No more technical detail is included here than is essential to an understanding of the uses of these techniques in the study of particle physics. In the chapters which follow we shall assume that these techniques are familiar to the student, so that it will generally not be necessary to describe in detail the technique used in particular experiments.

1.2 Particle accelerators and beams
1.2.1 Introduction

Particle accelerators and their associated external beam lines are key elements in most particle physics experiments.

Charged particles are accelerated by passing across a region of potential difference which in practice is normally a cavity fed with radiofrequency power and phased such that the particle is *accelerated* as it passes through. Since practicable fields and dimensions are such that a single passage through the cavity can produce only a rather small acceleration, the particle must either pass through many such cavities or pass many times through the same group of cavities by guidance around a cyclic path.

In the *linear accelerator* a linear RF structure is fed with RF power from a bank of klystrons to produce a wave travelling down the structure with a velocity equal to the particle velocity so that the particle remains always on the accelerating phase of the wave throughout its flight. The largest operational linear accelerator is the two-mile-long accelerator at Stanford Linear Accelerator Centre (SLAC) which accelerates electrons to 50 GeV.

In *cyclic accelerators* a magnetic field is used to guide the particles around a cyclic path such that they repeatedly pass across the accelerating gap. In the *cyclotron*, which was an important machine in the early days of the subject, protons or heavier charged particles moved in a vacuum box containing two hollow D-shaped cavities between which could be applied an alternating potential difference (fig. 1.1). The magnetic guide field normal to the plane of the Ds caused particles of constant momentum to travel in a circular orbit of radius R, given by

$$R = \frac{pc}{Be} \quad p = \text{momentum}$$

$$e = \text{charge}$$
$$B = \text{magnetic field.}$$

The frequency of the potential difference applied to the gap between the Ds must be such that particles are accelerated each time they cross the gap.

Fig. 1.1. Principle of operation of the cyclotron.

The angular frequency is easily seen to be given by

$$\omega = \frac{Be}{mc} \quad m = \text{particle mass}$$

which is independent of the particle momentum as long as the mass is constant. For high energies, of course, the mass increases due to the relativistic effect and a constant frequency is no longer adequate.

In the *synchrotron*, particles are maintained at *constant radius* in a ring-shaped vacuum chamber contained in a magnetic field. The magnet is thus also in the form of a ring and need not cover the whole circular area as in the cyclotron. Since the radius of the orbit is constant, the magnetic field must increase to hold the particles in the same orbit as their momentum increases. The circulating frequency at any moment is then given by

$$\omega_0 = \frac{Bec}{E} \quad E = \text{total particle energy.}$$

Acceleration is achieved by having the particles pass through suitably-phased RF cavities at one or more positions in the ring. In practice, the particles are always bunched in the accelerator, although several bunches may be present at the same time. In all high-energy machines the particles are first accelerated in a linear accelerator before injection to the synchrotron. For electrons, which become relativistic at very low energy, the velocity and thus the accelerating frequency is essentially constant. For protons this is not so.

1.2.2   *Colliding beams and available energy in the centre of mass*
When a particle of rest mass m and total energy E collides with another particle of the same mass at rest the energy available in the centre of mass of the two particles is (see appendix A) E' given by

$$E'^2 = s = 2m^2 + 2mE.$$

Thus at high energies where $E \gg m$ the energy available in the centre of mass increases only as the square root of the particle energy E, with much of the total energy going to increase of the velocity of the complete centre-of-mass system (cms). In order to obtain the very high energies in the cms necessary, for instance, to make very heavy particles like the W and Z, the energy for a *fixed-target* machine thus becomes very great. For two particles of equal mass and equal and opposite momentum, however, the centre-of-mass frame is the same as the laboratory frame and the energy available is simply $2E$. Thus for two 50 GeV particles colliding head-on we have 100 GeV available in the centre of mass, whereas to achieve this result in a fixed-target collision would require an accelerated proton to have an energy ~ 5000 GeV.

Practical limits on attainable magnetic fields (presently up to ~ 6–7 T with superconducting magnets) and on ring radius have led to colliding-beam machines as the favoured way to attain the highest energies. For particles of opposite electric charge, such as electrons and positrons or protons and antiprotons, both beams can be accelerated as bunches circulating in opposite directions in the same vacuum chamber and colliding at the intersection positions where experiments are placed. For particles of the same charge, separate but intersecting rings are necessary.

In order to achieve an adequate rate of interactions the number of incident particles and the target density in a fixed-target machine must be sufficiently high. The number of interactions is approximately (for a 'thin' target)

$$N_I = N_P \cdot N_T \cdot n \cdot \sigma$$

N_I = no. of interactions/second

N_P = no. of incident particles/pulse

N_T = no. of target particles/unit area

σ = interaction cross-section

n = no. of pulses/second

$$N_T = \frac{N_A}{A} \rho \cdot t \cdot N$$

N_A = Avogadro's number

A = atomic weight

ρ = target density

t = target thickness

N = no. of target particles/atom.

We are often interested in processes with very small cross-sections $\sim$ few nanobarns (10^{-33} cm^2). A typical fixed-target experiment using a 1 m-long liquid-hydrogen target bombarded with 10^7 particles per pulse every 10 seconds will yield $\sim 4 \times 10^{-4}$ interactions per second for each nanobarn of cross-section with $N_P \cdot N_T \cdot n \sim 4 \times 10^{31}$. The full circulating beam $\sim 10^{13}$ particles per pulse will yield ~ 40 interactions s^{-1} nb^{-1}, but such fluxes are seldom usable in experiments.

For two colliding beams the reaction rate is customarily written in terms of the 'luminosity' L. Thus the number of interactions per unit time is

$$N_I = L\sigma.$$

The luminosity is then simply

$$L = \frac{n_1 n_2}{a} \cdot b \cdot f$$

$n_1, n_2 =$ no. of particles/bunch in each beam

$a =$ cross-sectional area of beams at intersection

(total overlap assumed)

$b =$ no. of bunches/beam

$f =$ revolution frequency.

Luminosities $\sim 10^{30}$–10^{31} cm^{-2} s^{-1} have been achieved or are anticipated in colliding beam machines yielding raw (unselected) rates $N_I \sim 10^{-3}$–10^{-2} interactions per nanobarn per second. Thus, perhaps surprisingly, accelerator technology has produced beams of intensity such that colliding-beam machines can produce highly-useful interaction rates.

1.2.3 *Beam stability and accelerator magnet configurations*

In a high-energy accelerator such as the CERN Super Proton Synchrotron (SPS), the protons will orbit the ring 10^5–10^6 times covering a total distance of over a million kilometres. In order to retain the beam over the complete cycle it is essential that the structure of the guiding magnetic lattice be such that the beam is not allowed progressively to defocus and that instabilities are controlled. In particular, cyclic instabilities which would add to the deviation of particles from the stable orbit progressively on each orbit must be avoided. The focusing properties of the magnet system are thus crucial parameters in the machine design.

Particles may stray from the perfect situation by deviating radially or perpendicularly from the stable orbit – transverse or betatron oscillations, and also by deviation from the ideal phase with respect to the RF acceleration – synchrotron oscillations.

To control radial and vertical deviations from the equilibrium orbit requires a non-uniform magnetic field which will focus the particle beam as does a lens in an optical system. *Quadrupole* magnets are the most commonly used magnetic focusing devices. A cross-section through a quadrupole magnet is shown in fig. 1.2. On-axis particles are unaffected by such a magnet. In the plane AB, off-axis particles are deflected back towards the axis so that in this plane the quadrupole acts as a convergent lens. In the orthogonal plane, however, particles are deflected off-axis and the equivalent lens is divergent. Convergence in both planes can be achieved by a combination of two or more quadrupole magnets.

In earlier generations of machines it was customary to combine the functions of bending and focusing the particles in the same magnets by shaping the poles to produce the appropriate field shape. Latterly, the practice has been to separate the bending and focusing functions by using simple dipoles for bending and separate quadrupoles (plus some more complex magnets) for focusing.

In earlier machines the betatron oscillations were of magnitude such as to require large vacuum chambers with consequent need for large-aperture magnets of high cost. For stable radial betatron oscillations the condition is that the vertical (i.e. normal to the orbit plane) component of the magnetic field should vary as r^{-n} with $0 < n < 1$. The frequencies of the vertical and horizontal oscillations are then given respectively by

$$v_V = \omega_0 n^{\frac{1}{2}} \quad \text{and} \quad v_H = \omega_0 (1-n)^{\frac{1}{2}}$$

where ω_0 is the circulating frequency as already defined. The amplitude of

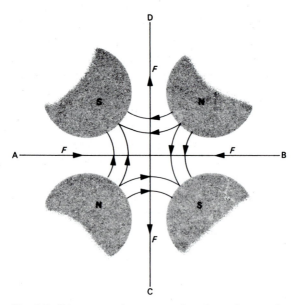

Fig. 1.2. Diagrammatic cross-section through a quadrupole magnet. The field is as shown. Negative particles passing normally into the plane of the paper from above will suffer forces F as shown. Thus the quadrupole focuses such particles in the plane AB and defocuses in the plane CD.

the vertical oscillation is, on the other hand, proportional to $n^{-\frac{1}{2}}$ so that a high value of n is required to minimise the size of the vacuum chamber and thus the magnet aperture. The *strong-focusing* idea proposed by Courant, Livingston and Snyder (1952) has proved to provide the best solution to the focusing problem. A high value of n is used to minimise the aperture but the sign of n is reversed in alternate magnets. Thus a typical configuration for a unit of a separated function machine is of the FODO variety:

Focusing (quadrupole) – No focus (dipole) – Defocusing

(quadrupole) – No focus (dipole)

Focusing in one plane is accompanied by defocusing in the other but with *overall* focusing in both planes.

The synchrotron oscillations arise from the fact that a particle arriving early can be arranged to see a higher accelerating field than that for the synchronous particle. Such an early particle takes up a slightly larger radius orbit thus taking longer to circulate back to the cavity. Particles arriving late experience the opposite effect, so that particles execute oscillations about the synchronous position for the bunch.

An important development critical to the success of $\bar{p}p$ colliders has been the invention and implementation of the idea of *stochastic cooling* (Van der Meer, 1972). Antiprotons produced in collisions of protons with a fixed target will have a substantial spread in angles and energies. In order to accelerate the antiprotons with minimum loss, however, the spread in energy and direction must be small. The useful flux of antiprotons can be maximised if they can be 'cooled', i.e. made as uniform as possible in all components of momentum. The stochastic cooling method is to store the antiprotons in a ring at relatively low energy and to use pick-up electrodes to sense the particle position relative to the mean. Signals from particles off the mean are then sent across the ring and are used to correct the off-mean particles. The beam may be thus cooled before being injected into the main accelerator.

1.2.4   *Synchrotron radiation*

A charged particle which suffers acceleration emits electro-magnetic radiation. Thus particles moving in a circular orbit in an accelerator will lose energy by such radiation. For a particle travelling in a circle the energy loss per turn due to synchrotron radiation is given by

$$\Delta E = \frac{4\pi}{3} e^2 \beta^3 \frac{1}{R} \left(\frac{E}{mc^2} \right)^4$$

$$= \frac{4\pi}{3} \left(\frac{e^2}{mc^2} \right) mc^2 \beta^3 \frac{1}{R} \left(\frac{E}{mc^2} \right)^4$$

$$= \frac{4\pi}{3} \left(\frac{r_e}{R} \right) mc^2 \beta^3 \left(\frac{E}{mc^2} \right)^4$$

$$= 8.856 \times 10^{-2} \left(\frac{E^4}{R} \right)$$

e = charge
β = velocity/c
E = total energy
m = mass
R = orbit radius
r_e = classical electron radius

where ΔE is in MeV, E is in GeV and R is in metres.

The m^{-4} factor means that the effect is much more important for electrons than for protons and, in practice, up to energies presently attained, it is only for electrons that synchrotron radiation is an important consideration in the accelerator design. Due to the E^4 factor the effect very rapidly becomes serious at higher energies, although some easement of the problem can be achieved by increasing R. In the LEP e^+–e^- collider at CERN the average radius is 4.2 km and at 55 GeV the synchrotron radiation loss is 260 MeV per turn (orbit is not quite circular). For energies much higher than 150–200 GeV it seems likely that in a circular electron accelerator synchrotron radiation would be such as to require unacceptable RF power compensation and that linear accelerators will be the only possibility.

1.2.5   *An example of an operating accelerator – the CERN Super Proton Synchrotron (SPS) and the LEP e^+–e^- collider*

The CERN SPS, which started operating in 1976, is a proton accelerator with a maximum energy of 450 GeV which is also operated as a proton-antiproton collider at energies of 315 GeV in each beam. The layout of the machine and its injectors and experimental halls is shown in fig. 1.3(a) and some of the most important parameters of the accelerator are given in table 1.1.

The main SPS ring is housed in a tunnel of diameter 4.14 m bored in the molasse rock. There are six straight sections for injection, extraction and RF acceleration. The average bending radius is $\sim$ 1.48 km and bending is provided by 744 dipole magnets with a peak field of 2 T. The accelerator uses a strong focusing, separated-function magnet lattice of the type: focus–bend–defocus–bend

{(quad)–(4 dipoles)–(quad)–(4 dipoles)}

(FODO) repeated 108 times round the ring. The vacuum chamber is of stainless steel with elliptical cross-section. Acceleration is achieved with

Fig. 1.3(a). Layout of the SPS accelerator and its associated facilities at CERN.

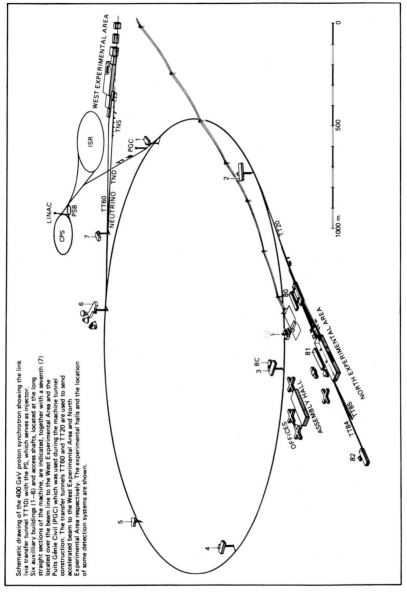

Schematic drawing of the 400 GeV proton synchrotron showing the link (via transfer tunnel TT10) with the PS, which serves as injector. Six auxilliary buildings (1–6) and access shafts, located at the long straight sections of the machine, are indicated, together with a seventh (7) located over the beam line to the West Experimental Area and the Puits Génie Civil (PGC) which was used during the machine tunnel construction. The transfer tunnels TT60 and TT20 are used to send accelerated beam to the West Experimental Area and North Experimental Area respectively. The experimental halls and the location of some detection systems are shown.

two sets of RF cavities operating at about 200 MHz. Since protons are injected into the main ring at 10 GeV the required frequency swing during acceleration is only 0.5%.

The protons are initially accelerated by a radio frequency quadrupole then transferred to a proton linear accelerator which takes them to 50 MeV. Further acceleration to ~ 800 MeV in the proton booster rings and 10 GeV in the proton synchrotron is needed before injection to the SPS ring.

In fixed-target operation protons can be extracted in either 'fast' (few µs) or 'slow' (2s) spills. The first is suited to bubble chamber operation where the beam should be confined to a limited time at or near the bottom of the pressure pulse providing a well-controlled time for bubble growth while the second is suited to counter experiments where a long beam spill allows the maximum number of interactions with minimum dead time for the detection system. Two experimental halls are fed with extracted beams of a remarkable variety of energies and spills on a repetition rate at maximum energy of about ten seconds. Circulating beam currents have been steadily improved throughout the life of the machine and 2×10^{13} protons per pulse is regularly achieved.

The SPS has also been used with great success as a proton–antiproton colliding ring. The antiprotons are produced in the CPS (CERN Proton Synchrotron) by protons accelerated to 26 GeV and focused onto a tungsten target. Antiprotons having energies around the production maximum of 3.5 GeV are collected and transferred to an 'antiproton-

Table 1.1. *The CERN SPS parameters*

Peak energy	450 GeV
Machine diameter	2.2 km
Injection energy	10 GeV
Magnet field (peak energy)	1.8 T
Magnet field (injection)	0.045 T
Total number of bending magnets	744
Apertures	39×129 mm² 52×92 mm²
Number of quadrupoles	216
Total peak voltage (bending magnets)	24.3 kV
Peak current (bending magnets)	4900 A
Number of RF cavities	2
Interaction length of cavity	20.196 m
Maximum power per cavity	500 kW
Frequency swing	0.44%
Design pressure	3×10^{-7} torr

accumulator' ring which has a large aperture and fixed field. In order to achieve the required antiproton flux it is necessary to accept particles having a certain spread in vector momentum. In the accumulator this spread is reduced by a process of 'stochastic cooling' developed at CERN by S. Van der Meer. Signals from pick-up electrodes in the ring measure the deviations in antiproton momentum from the mean value. These signals are sent across the accumulator ring to generate correction signals applied to the same antiprotons as they pass so that the spread in momentum is progressively reduced until the stack is sufficiently homogeneous for the subsequent stages of acceleration. The antiproton bunch is then ejected back into the CPS for acceleration to 26 GeV and then transferred to the SPS where three bunches of protons have already been injected and are coasting at 26 GeV in the opposite direction. Two further $\bar{p}$ bunches are injected and $\bar{p}$ and p are then simultaneously accelerated to 315 GeV/c. Since in the storage mode the bending magnets must be continuously powered they are not generally run at as high a field as in pulsed operation so that the operating energy is lower. Collisions between the $\bar{p}$ and p bunches take place in the six underground caves for experiments. Luminosities of 3×10^{30} cm^{-2} s^{-1} have been achieved with beam lifetime between fills $\sim$ 12 hours.

In summer of 1989 there was completed at CERN a large electron–positron collider, LEP. Since the e^+ and e^- are oppositely charged they can circulate in bunches travelling in opposite directions within the same vacuum chamber which consists of a ring with eight straight sections linked by eight arcs. The particles are guided by 3368 dipole magnets and focused by 808 quadrupoles the total path length around the ring being some 27 km. The ring is situated in a tunnel crossing the French–Swiss frontier and lying largely beneath the plain between the West end of the Lake of Geneva and the Jura mountains. In four of the straight sections underground halls house elaborate detector facilities while two of the remaining straights accommodate the RF accelerating cavities. Sufficient RF cavities have been initially commissioned to accelerate each beam to 55 GeV while at a later stage it is planned to increase the RF installation to reach 100 GeV in each beam. Luminosity of 10^{31} cm^{-2} s^{-1} appears to be well within reach with four bunches colliding every 22 µs in each of the experiments.

The injection system for LEP has to provide fills of positrons and electrons at intervals of a few hours depending on the beam lifetime in LEP where bunches are gradually degraded by scattering on residual gas even though the vacuum in the ring is very good ($\sim 10^{-9}$ torr). The injection chain involving four main elements is illustrated in fig. 1.3(b). The first element is a tandem linac (LIL) the first section of which produces pulses of electrons at 200 MeV (2.5 A for 10–20 ns) which generate positrons by striking a tungsten target. The second section of the linac accelerates the positrons to 600 MeV before they are injected into the electron–positron accumulator ring (EPA) where the bunches are stored. To produce electrons the tungsten target is removed and the energy gain in the second section of the linac is reduced to 400 MeV so that the electron bunches can also be stored at 600 MeV. A very substantial cost saving is achieved by exploiting the existing CERN PS and SPS machines for the next stages of acceleration. The PS raises the energy to 3.5 GeV following which the bunches are injected into the SPS and accelerated to 20 GeV before transfer into LEP for acceleration to the final collision

Fig. 1.3(b). The early acceleration stages and injection for the LEP e^+e^- collider at CERN.

energy. A number of important modifications were necessary to the PS and SPS machines to accommodate the electrons and positrons but a remarkable feature of the system is that the e^+/e^- are accelerated in the dead time of the 'normal' SPS 450 GeV cycles so that operation of the SPS for fixed target proton physics is scarcely affected. The injection system can deliver almost simultaneously 1.6×10^{12} positrons and electrons in 12 minutes when LEP requires to be filled.

The LEP machine was commissioned in summer 1989 and the first results on Z physics were obtained in September of the same year (see section 11.6.4)

1.2.6 *Particle beams for fixed-target experiments*

In carrying out fixed-target experiments in particle physics it is generally necessary to know the nature and momentum of the incident particles which interact with targets of hydrogen or complex nuclei. This may be achieved in electronic-detector experiments either by using a pure beam in which only the wanted particles are present in a narrow momentum bite, or by using a mixed-particle beam produced from a primary target bombarded by accelerated protons but *identifying* the particles before they pass into the target (for instance, by a DISC counter – see below) and studying only interactions produced by the wanted variety of incident particle. Bubble chambers cannot be triggered in the expansion, the cycle time is relatively long (~ 0.1–few seconds) and the number of particles acceptable per expansion is small compared with electronic experiments (20–30 compared with 10^6–10^7). For these reasons it is generally advantageous to use as pure a beam as possible in bubble-chamber experiments.

In a collision of accelerated protons of greater than a few GeV with a solid target either inside or outside the accelerator vacuum chamber, many kinds of particle will be produced. With positive charge we expect p, π^+, K^+, Σ^+ and, due to decay of these, μ^+ and e^+, with negative charge we will get π^-, $\bar{p}$, K^-, Σ^-, Ξ^-, Ω^-, μ^- and e^-, while neutrals will consist of π^0, K^0, n, Λ^0, Ξ^0, ν. Some of these particles have very short lifetimes at rest ($\sim 10^{-10}$ s) and at lower energies will decay close to the primary target. For high energies, however, the lifetime is relativistically dilated (see appendix A) and beams of even the short-lived hyperons such as Σ and Ξ have been used in experiments. Protons, neutrons, π^+ and π^0 are by far the most prolifically directly produced particles from the primary target. Experiments requiring pure or enriched beams of other particles need an efficient way of rejecting protons and pions.

Dipole bending magnets combined with collimators are used to select particle charge and momentum (fig. 1.4). Quadrupole magnets are used to focus the particle beam which may be many hundreds of metres in total length. While the bending magnets are equivalent to optical prisms the quadrupoles are equivalent to lenses.

1.2.7   *Particle separators*

Separation of particles of different mass by combinations of electric and magnetic fields has been practised since the early days of atomic physics. For particle beams of energies up to about 7 GeV, *electrostatic separation* has been commonly used, while in the range up to about 40 GeV *radiofrequency* separation is effective. In both cases the beam entering the separator system must be of well-defined momentum so that particles of different mass will have different velocities.

The electrostatic separator consists of a pair of parallel plates maintained at a high potential difference. Particles passing into the gap between the plates then experience a transverse force Ee where E is the field and e the charge. The angular deflection is

$$\theta = \frac{p_T}{p} = \frac{EeL}{p\beta}$$

p = beam momentum

p_T = transverse momentum

E = electric field strength

L = separator length

$\beta = v/c$, v = velocity, $c = 1$.

Thus for two masses of particle m_1 and m_2 with velocities defined by β_1 and β_2 the difference in the deflection is

$$\Delta\theta = \frac{EeL}{p}\left(\frac{1}{\beta_1} - \frac{1}{\beta_2}\right)$$

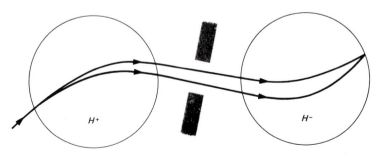

Fig. 1.4. The use of two magnets with opposite fields normal to the paper to produce momentum analysis using a collimator, with correction of dispersion.

Writing

$$\beta^2 = \frac{p^2}{m^2 + p^2}$$

we have

$$\frac{1}{\beta_1} - \frac{1}{\beta_2} = \left(1 + \frac{m_1^2}{p^2}\right)^{\frac{1}{2}} - \left(1 + \frac{m_2^2}{p^2}\right)^{\frac{1}{2}} \simeq \frac{1}{2}\left(\frac{m_1^2}{p^2} - \frac{m_2^2}{p^2}\right)$$

(relativistic approximation)

and

$$\Delta\theta \simeq \frac{EeL}{2p^3} \Delta(m^2).$$

The deflection of the wanted particles is generally compensated by vertical bending magnets at the end of the separators. The angular separation resulting from the separator is converted to a spatial separation at the 'mass slit' by a converging quadrupole lens focusing at the slit through which the wanted particles are allowed to pass.

For an electrostatic system the separation varies as the inverse cube of the momentum and it becomes impossible to achieve a satisfactory separation above about 7 GeV/c. For beams of higher momentum a method of velocity separation has been developed which depends on the time of flight of the particles between two radiofrequency cavities. The principle of operation is illustrated in fig. 1.5. A bunch of particles of well-defined momentum passes through the first RF separator $R1$, which is normally a cylindrical iris-loaded waveguide, and suffers a transverse deflection which will vary in magnitude according to the part of the RF cycle at which the bunch passes through the cavity. The deflected beam is focused on the second cavity $R2$ by a quadrupole system Q. If the distance L and the relative phase of the cavities is suitably adjusted the unwanted particles will receive a deflection exactly cancelling the original one whilst

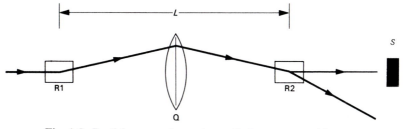

Fig. 1.5. Particle separation using radiofrequency cavities.

that of the wanted particles is doubled. The unwanted particles are stopped by a beam stopper S. The required condition is

$$\frac{L}{\beta_1 c} = \frac{1}{2f} \quad f = \text{radio frequency.}$$

For cancellation of the deflection of the unwanted particles

$$\frac{L}{\beta_2 c} = \frac{1}{f}$$

so that $fL = \beta_2 c$ and the condition is satisfied for one (or a discrete series, allowing for other multiples of $1/f$) momentum if f and L are fixed. For two contaminants, such as π^+ and p in a K^+ beam, it is still possible to achieve separation at certain momenta by arranging the length such that the two contaminants arrive at $R2$ at a phase interval of 2π. This fixes the momentum at which the system operates but flexibility can be achieved by using three RF cavities. For electronic experiments, requiring a long beam pulse, a highly-enriched beam has been operated up to 38 GeV/c using superconducting RF cavities.

1.3   Particle detectors

1.3.1   *General considerations*

For none of the detectors discussed here will we consider more of the technical detail than is essential to an understanding of the way in which they can be used in the study of particle physics.

Particle detectors are of several varieties. In general, the objectives of a complex detector are:

(a) to localise the particle trajectory in *space*. If this can be done in a region of known magnetic field then for a charged particle the momentum can be obtained from the relation

$$p = \frac{peB}{c}$$

where B is the magnetic flux density, p the radius of curvature, e the charge on the particle, c the speed of light and p the particle momentum. The higher the field and the smaller the distortions, the more precisely the sagitta and hence the momentum can be measured.

(b) to measure the particle energy. This may be done by means of calorimeters (see below) where a particle loses all its energy and is not available for further study or, for charged particles in certain circumstances, by measurement of the rate of energy loss by ionisation which varies as a function of energy as shown in fig. 1.6. In some circumstances it is possible to measure the velocity of a particle directly by time of flight or by Cerenkov radiation.

(c) to identify the particle. This may be done by finding the particle mass from measurements of its momentum *and* energy or momentum *and* velocity or in certain cases by the nature of its interactions with matter or by its decays. For instance, muons do not produce nuclear interactions and are thus distinguished from the strongly-interacting 'hadrons'. On the other hand, muons are so much heavier than electrons that they produce very little

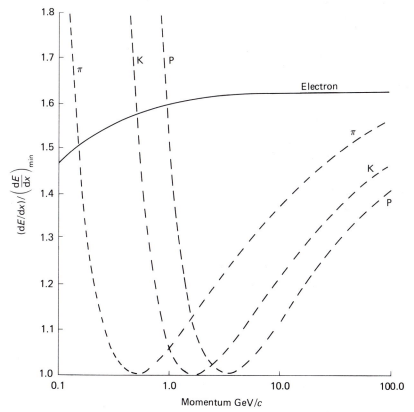

Fig. 1.6. Energy loss for ionising particles as a function of momentum (logarithmic scale). The high momentum increase corresponds to the 'relativistic rise'. The curves shown are for an 80% argon, 20% methane, gas mixture at one atmosphere but the behaviour is not sensitive to the material.

bremsstrahlung and do not give rise to electromagnetic showers, and this property distinguishes them from electrons.

(d) to localise the particle in *time*. This is not only necessary in the case of time of flight measurement to measure the velocity of particles as mentioned in (b) but more frequently because it is necessary to associate signals due to the same particle or particles from the same event in a number of different detectors against a big background of signals from unassociated particles. Signals within an appropriate time gate (with proper allowance for time of flight) can be used to eliminate unassociated particles.

It is generally true that we rely on processes of ionisation and excitation of atoms as the primary processes on which all particle detectors depend. Thus for detection of electrically neutral particles we require a first stage in which such particles lose at least part of their energy to produce charged particles which *can* be detected or in which the neutral particles decay into charged particles.

The direct loss of energy by ionisation for a charged particle is always very small. For instance the ionisation loss for a minimum ionising particle in Pb is ~ 1 MeV gm^{-1} $cm^2 \simeq 1.6 \times 10^{-13}$ Jg^{-1} cm^2, representing some 30 000 ion pairs gm^{-1} cm^2. Multiplying by the density to get the energy loss per cm yields a value of only 12.8 MeV cm^{-1} or 2×10^{-12} J cm^{-1}, a tiny value in terms of direct recording by an instrument even although the energy loss per cm in Pb is greater than in most materials. It is thus apparent that particle detectors must depend on some means of *amplifying* the primary signal to render it measurable.

In one class of detectors the amplification process depends on use of a detector material which is in a semi-stable condition. The energy released in the ion pairs then triggers off the instability along the track of ions left by the particle and so renders the track detectable. In the bubble chamber the unstable state is that of a superheated liquid, and we get boiling along the track which thus appears as a string of bubbles. In the cloud chamber used in the early days of nuclear and particle physics, the unstable state is that of a supersaturated vapour so that condensation of droplets is triggered off along the track. In the nuclear photographic emulsion the instability is chemical rather than physical and the ions result in development of the grains of silver bromide to provide a latent image of grains along the track.

An approach which has been used from the earliest days of nuclear and particle physics, and which is still being developed at the present time, is to detect the ionisation in gases or liquids by collection of the ions onto electrodes using an electric field and subsequently amplifying the signal electronically. An alternative in gas-filled detectors is to amplify the ionisation inside the detector by avalanche formation.

Another class of detectors exploits light emitted by particles either by Cerenkov radiation when the particle velocity is greater than the velocity of light in the medium through which the particle is passing, or by excitation of the atoms of the material which de-excite to give light – as in scintillation counters. In both cases the total light energy is small but a *photomultiplier* amplifies the signal to a usable level.

1.3.2 Bubble chambers

In the bubble chamber an unstable state of a liquid is created by superheating. The very small amount of energy deposited by a minimum ionising particle (20 MeV m^{-1} or 3.2×10^{-12} J m^{-1} in liquid hydrogen) is sufficient to trigger off the instability to produce boiling so that the tracks are visible as strings of small bubbles. The superheating of a bubble-

Table 1.2. *Properties of some commonly used bubble-chamber liquids*

Liquid	Operating temperature/K	Operating pressure/ kN m^{-2} (psi)	Density/ g cm^{-3}	Radiation length/ m	Index of refraction	Notes
H$_2$	28	538 (78)	0.06	11.45	1.09	Pure proton target Highly explosive
D$_2$	32	738 (107)	0.13	9.50	1.1	Simplest neutron target
He	3	3.4 (0.5)	0.14	9.00	1.03	Spin 0, I-spin 0 Hyperfragment source
C$_3$H$_8$ (propane)	333	2069 (300)	0.44	1.18	1.22	Highly inflammable
CF$_3$Br (Freon)	303	1862 (270)	1.5	0.11		Non-inflammable
Xenon	252	2551 (370)	2.18	0.035	1.18	Extremely expensive

chamber liquid is achieved by a sudden reduction of the liquid pressure, usually by a piston in contact with the liquid. The properties of some bubble-chamber liquids are summarised in table 1.2. The timing of the expansion is arranged so that the beam pulse from the accelerator enters the chamber when the pressure is at or near the minimum and the chamber is at its most sensitive. The flash is delayed for up to about a millisecond depending upon the bubble size desired in order to allow the bubbles to grow to a size suitable for photography. The cameras are wound on before the next expansion. The cycle is shown schematically in fig. 1.7. Bubble chambers have been operated at repetition rates up to 20 or 30 Hz which is sufficient to match the rate at which beam pulses can be supplied from proton accelerators, although not the higher-rate electron linacs.

Bubble chambers are almost always used with a high magnetic field orthogonal to the direction of the particle beam and parallel to the optical axis of the cameras. Superconducting magnets are frequently used with fields up to 2 or 3 T.

Holography has also been used for bubble-chamber photography. In a conventional optical system, use of a large aperture to improve the resolution reduces the depth of focus. A holographic system allows high resolution without loss of depth of focus. This feature not only yields the resolution necessary for the study of very-short-lived particles but allows the use of larger fluxes of particles through the chamber, since layers at successive depths can be studied without interfering with each other.

In the peak period for bubble-chamber physics during the years 1955–75 all the major accelerators had one or more bubble chambers in operation with typical dimensions of 2 m in length and depth and height of about 0.5 m. Even larger chambers were developed with the primary aim of studying neutrino interactions (see fig. 1.8, 3.5 m Big European Bubble

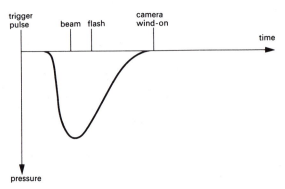

Fig. 1.7. Diagrammatic representation of a bubble chamber cycle.

Chamber at CERN). Smaller chambers with high resolution and repetition rate have also proved particularly useful as 'vertex detectors' in large *hybrid* systems involving a variety of detection devices. In these systems, although the chamber is expanded for every particle pulse, the lights are

Fig. 1.8. The Big European Bubble Chamber (BEBC) at CERN (courtesy of Photo CERN).

flashed and a photograph taken *only* when the external detectors indicate that an event of interest has occurred.

Summary for bubble chambers
Complete track reconstruction.
Resolution down to < 1 mm in suitable chambers for short-lived particles.
Good track separation and resolution of secondary vertices.
Mass identification from bubble density up to $\beta \sim 0.8$.
No trigger.
Not useful for colliding beams.
Scanning and measurement is time consuming.
Beam flux must be low for charged particles (< 15–100/expansion).
Repetition rate is low ($\lesssim 20$ Hz).

1.3.3   *Multiwire proportional chambers – MWPC*

In multiwire proportional chambers the ionisation produced by a charged particle in the gas of the chamber (usually argon $+ 10\%$ methane) is amplified by an ionisation avalanche in the gas as the ions move towards the electrodes under the influence of high electric fields.

A typical chamber geometry in its simplest form is shown in fig. 1.9. The wire spacing may be as close as 0.5 mm and typical potentials of 3–4 kV

(a)

Cathode strips

Anode wires

(b)

Fig. 1.9. Illustrative MWPC geometry. The cathode may be in the form of a continuous plane or strips or pads. The electric field configuration is illustrated in (b).

may be applied between anode and cathode. The high electric fields which produce the avalanche exist close to the fine anode wires – 10 to 50 μm diameter – where the field varies as $1/r$ and attains values of 10^4–10^5 V cm^{-1}.

A critical advance in development of such multi-wire chambers was made by Charpak (1968), who showed that the sense wires act as independent detectors so that the chamber provides a fine-grained planar detector. A negative signal on an anode wire arising from the avalanche close to it and the motion of the positive ions away from it is found only on that wire with positive signals on all the other electrodes. A development of such chambers, where the relative magnitude of the signals on cathode strips is measured, has been used to obtain positions with an accuracy much less than the strip separation and yield precisions $\lesssim 100$ μm.

In the truly *proportional* mode with output signals proportional to the primary ionisation, amplification up to $\times 10^4$ is possible, so that a few primary electrons yield a charge well within the input range of available amplifiers. Above amplification of $\sim \times 10^5$ a range of non-proportional conditions arises leading to saturated streamer and spark formation used in spark chambers, streamer chambers and flash tubes of more robust construction than MWPCs. The rise time of the avalanche pulse from each electron in a MWPC is ~ 100 ps and the decay time, which is dependent on the ion mobility, ~ 20 ns.

Successive layers of MWPCs with wires oriented in different directions can be used to measure the trajectories of charged particles. For a single particle, the signals in two orthogonally oriented layers give a unique space point. When more than one particle is present ambiguities arise (fig. 1.10) in associating the signals in different layers and it is customary to use modules with planes oriented along x, y and also tilted 'u' and 'v' axes. Such arrays of MWPCs are frequently used either within magnetic fields or on either side of a magnet to measure particle momentum. The usefulness of systems of such chambers which may contain many thousands of wires depends on the availability of relatively cheap transistor electronic technology.

1.3.4 *Drift chambers*

A drift chamber operates with many fewer wires than a MWPC by measuring the *time* for the ions to drift to the wire in a uniform electric field. If the drift velocity is known, this time can then yield the coordinate to high precision. A start signal from some other detector in the system is needed for the timing. The multi-wire electronics is thus exchanged for a limited number of timing channels.

A typical chamber electrode geometry is shown in fig. 1.11. The structure is designed to produce a uniform drift field ~ 1 kV cm^{-1} over most of the drift trajectory until the electrons reach the high-field avalanche region in the immediate vicinity of the anode wire. Typical drift velocities in argon–hydrocarbon mixtures are ~ 40 μm ns^{-1}, almost independent of the field intensity. For a timing precision of a few ns the equivalent space precision is thus ~ 100 μm. Drift distances of 10 cm or more are commonly used.

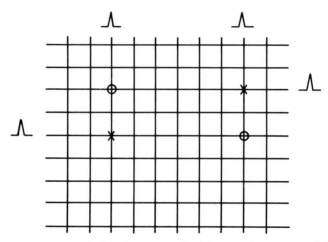

Fig. 1.10. Ambiguities in a MWPC with more than one particle and only two orthogonal wire planes. The pulses on the wires do not distinguish between real hits (X) and spurious combinations. Additional planes at, say, $\pm 45°$ ('u' and 'v' planes) remove the ambiguity.

Fig. 1.11. Drift chamber geometry. Electrons from ionisation along the particle trajectory drift to the anode wires. The field shaping wires are designed to maintain a uniform field as far as possible throughout the drift space.

An elaborate development of the drift chamber is the *time-projection chamber* (TPC) which consists of a gas volume subject to a uniform electric drift field along one dimension (fig. 1.12). On the plane orthogonal to the drift direction is a two-dimensional detector such as a multi-wire chamber with two-dimensional capability afforded for instance by wires and cathode pads. Along the drift direction the coordinate is measured by timing so that for a particle track in the gas a string of pulses is picked up at different positions on the end plate detector from which the trajectory can be uniquely reconstructed in three dimensions. Such detectors are becoming increasingly popular for studies of complex multi-particle interactions, particularly at colliding-beam machines. They can be placed in magnetic fields to give a measure of particle momentum while the ionisation can be sampled many times along the track to yield information on particle identification over a range of velocities, including the relativistic-rise region.

1.3.5 *Scintillation counters*

Many substances have been found to emit light, or *scintillate*, when a charged particle passes through them. Scintillators may be in the form of inorganic crystals such as sodium iodide, caesium iodide or bismuth germanate, organic crystals such as anthracene, organic liquids like toluene, or in the form of plastics (solid solutions).

Inorganic crystals are doped with a suitable activator such as thallium. A charged particle passing through the crystal produces free electrons and holes which are captured at the activator centre. The resultant excited state decays to produce light. The mechanism for organic scintillators is excitation of molecular states, the decay of which gives light in the ultra-violet range. This light must be converted to visible wavelengths by inclusion of suitable fluorescent dyes.

The light produced in the scintillator is detected by a photomultiplier. Electrons from the photomultiplier cathode are multiplied at each stage of a structure of successive dynodes to give an overall amplification of up to 10^8 for a 14-stage tube, to produce a pulse at the output which may be further amplified and used to drive scalars or to actuate coincidence or other circuits.

Although inorganic crystals such as sodium iodide are particularly useful for γ-ray detection, such crystals are expensive and most of the large scintillation counters used in high-energy experiments used to be made of organic plastic scintillator material. The plastic may be used in large slabs or in other shapes or in the form of a matrix of thin finger counters,

Graded
high-voltage
electrodes

Particle trajectory

0   − 45 kV   0

Ground planes

High-voltage
electrode

(a)

ALEPH   DALI_A8   ECM=91.31 Pch=40.4 Efl=81.6 Ewi=48.6
                   Nch=15   EV1=.939 EV2=.903 EV3=.129

(b)

Fig. 1.12(a). Schematic of a Time Projection Chamber (TPC).
Electrons drift to the two-dimensional position detectors (e.g. MWPC
with cathode pads) on the end plates which record the y and z
coordinates. The x coordinate is measured by drift time. The chamber
illustrated has the hv electrode in the centre with two-dimensional
detector planes at each end. (b). Computer reconstruction of an e^+–e^-
event with two particle jets in the ALEPH TPC at LEP. The discs
show the segmented end plate wire chamber detectors. The chamber
diameter is 4 m and the tracks are bent by a 1.5 T magnetic field.

Fig. 1.13. A large plastic scintillation counter with perspex light guides of equal length from all parts of the counter to the multiplier (courtesy of Photo CERN).

according to the application. It is often inconvenient or impossible to place the multiplier in contact with the actual scintillator, in which case lucite light pipes may be employed to link them. By this means, multipliers may be kept away from strong magnetic fields which affect their operation. An elaborate scintillation counter system is shown in fig. 1.13. An alternative technique is to use bars of wavelength-shifter material (acrylic doped with complex molecules such as quaterphenyl or **BBQ**) which shift the primary ultra-violet or blue scintillation light into the green. More recently large arrays of scintillating crystals such as bismuth germanate have been included in some large detector systems at colliding beam accelerators.

Properties of some commonly-used scintillation materials are given in table 1.3. A particularly important use of scintillation counters is in time-of-flight measurement. Coupled with a measurement of the momentum, such a measurement can distinguish between particles of different mass. The momentum p and the time of flight t are related by

$$p = m\beta\gamma = m\frac{v}{c}\left(1 - \frac{v^2}{c^2}\right)^{-\frac{1}{2}}$$

$$= \frac{mL}{tc}\left(1 - \frac{L^2}{t^2c^2}\right)^{-\frac{1}{2}}$$

where L is the flight path, m the mass and c the velocity of light. This relation yields curves of the form shown in fig. 1.14. In a small fast scintillator, measurement errors as low as 0.1 ns have been achieved using the rising edge of the pulse. From the figure it is then seen that if a 3 σ separation is required then πs and Ks can be distinguished up to about 1.0 GeV/c.

Table 1.3. *Properties of some typical scintillator materials*

	Light yield (photons/ keV)	Decay time (ns)	λ_{max} (nm)	Density (g cm^{-3})	Comments
NaI (Tl)	40	250	410	3.7	$X_0 = 2.6$ cm
BGO ($Bi_4Ge_3O_{12}$)	~ 5	300	480	7.1	$X_0 = 1.1$ cm
Anthracene	20	36	410	1.25	
Ne 102A	13	2.4	423	1.0	General-purpose plastic

1.3.6   *Cerenkov counters*

If a particle passes through a medium in which its velocity is greater than that of light in the medium it will emit electromagnetic radiation. This phenomenon was discovered by Cerenkov and bears his name. Cerenkov counters use one or more photomultipliers to detect the Cerenkov light and thus to record the passage of a particle. Compared with scintillation counters Cerenkov counters are less sensitive and in general more clumsy, but they possess the property of discrimination between particles of different velocity and thus, if the momentum is known, of different mass.

Basically, Cerenkov counters consist of a container filled with a gas or liquid or solid material transparent to the Cerenkov light, and for which the index of refraction μ is suitably chosen. For the emission of Cerenkov light we require

$$\mu > \frac{c}{v} = \frac{1}{\beta}$$

where v is the velocity of the particle to be detected. This simple condition underlies the operation of the *threshold Cerenkov detector*. Suppose that we wish to distinguish K^+-mesons at 1 GeV/c from protons and π^+-mesons at the same momentum. The values of β for protons, kaons and pions at this momentum are 0.73, 0.89 and 0.99 respectively. The corresponding values for μ are 1.37, 1.12 and 1.01. If we arrange two Cerenkov counters,

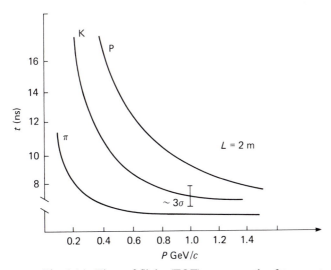

Fig. 1.14. Time of flight (TOF) over a path of two metres for π, K and p as a function of momentum.

one filled with water ($\mu = 1.33$) and one with carbon dioxide at pressure and temperature such that $\mu = 1.05$, then the protons will record in neither, the kaons only in the water counter and the pions in both.

The Cerenkov mechanism is related to the polarisation of the molecules along the track of the charged particle. The molecules very rapidly ($< 10^{-11}$ s) return to the ground state with emission of the Cerenkov light. The emitted light forms a coherent wavefront only when the Huygens condition is satisfied so that the light is emitted along the surface of a cone of angle given by

$$\theta = \cos^{-1}\left[\frac{1}{\beta\mu(v)}\right] \qquad \textbf{1.1}$$

where $\mu(v)$ is the index of refraction of the material for light of frequency v. The emitted light is plane-polarised such that the electric vector **E** is in the plane of the incident particle and the emitted photon.

The Cerenkov light is emitted as a continuous spectrum. The number of photons I emitted per unit path length per unit frequency by a single particle is given by

$$\frac{d^2 I}{dx\,dv} = \frac{4\pi^2 e^2}{hc^2}\left(1 - \frac{1}{\beta^2\mu^2(v)}\right)$$

$$= \frac{4\pi^2 e^2}{hc^2}\sin^2\theta$$

$$= \frac{2\pi}{137c}\sin^2\theta.$$

The rate of emission is low in that, for example, for a particle with $\beta \sim 1$ in water only about 200 photons cm^{-1} are emitted corresponding to $\sim 2.10^{-4}$ of the total energy loss and being $\sim \frac{1}{40}$ of the rate of photon production in a typical scintillator.

The angular condition **1.1** is exploited in *differential Cerenkov counters* which accept light over a narrow angular interval and hence over a narrow range of velocity. The resolution is limited by finite divergence in the particle directions (these counters are frequently used to identify particles in beams), multiple scattering in the radiator and chromatic dispersion. Special chromatically-corrected counters (DISC – Directional Isochronous Self-Collimating Counters) have been developed which achieve $\Delta v/v = 10^{-6}$–10^{-7}, equivalent to K/π separation up to ~ 500 GeV/c. With a ring diaphragm of fixed radius a gas-filled counter can be tuned over a range of velocities by variation of the gas pressure (fig. 1.15).

Differential counters such as that shown in fig. 1.15 require all the particles to be travelling in a closely-collimated parallel beam if they are to achieve the ultimate resolution. An important development designed to allow the use of such counters in the identification of particles emerging from an interaction over a wide range of angles is the *Ring Image Cerenkov* (RICH) counter. In such a device an optical system is required to focus the light emitted along the Cerenkov cone onto a detector in the form of a ring the radius of which is a measure of the cone angle and hence of the particle velocity. The principle of such a system is illustrated in fig. 1.16. Light from particles passing through the radiator is focused by a spherical

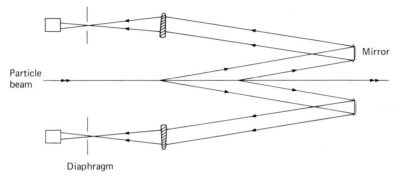

Fig. 1.15. Illustration of the principle of the differential Cerenkov counter.

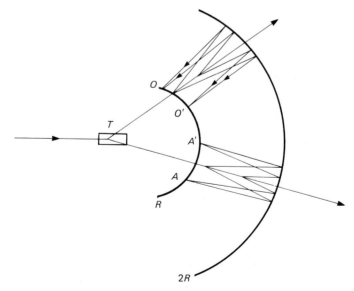

Fig. 1.16. Principle of the Ring Image Cerenkov Counter (RICH). Particles from the target *T* radiate Cerenkov light in the radiator between *R* and 2*R*. The mirror at 2*R* images the light into rings *OO'*, *AA'* on the detector placed on the surface at *R*.

mirror back onto a high-granularity detector, which may be a wire chamber or drift chamber. Special windows of quartz or CaF to allow passage of the higher-energy photons coupled with detector gases of low enough ionisation potential to be ionised by the transmitted photons are required. In very large counters of this kind the spherical mirror may by split into many sections. Results from a RICH counter are shown in fig. 1.17. It is noteworthy that in the minimum ionisation range Cerenkov counters afford the only method of particle identification. Properties of some materials commonly used in Cerenkov counters are given in table 1.4.

1.3.7   *Calorimeters*

For the measurement of the energy of very-high-energy particles it has become usual to employ devices known as '*calorimeters*'. In a calorimeter the particle loses practically *all* its energy by processes which include a stage of ionisation and ultimately result in heat. The measurement of the energy is achieved by measuring the energy deposited in the ionisation and excitation stage of energy loss. Since the particle loses all, or at least a large fraction, of its energy, it is no longer available for

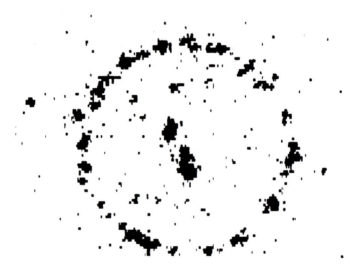

Fig. 1.17. Results from a ring-image Cerenkov counter. The picture records the signal from a single 5 GeV/c electron. The detector consists of a liquid radiator, a lead-glass plate as a pre-shower converter, a multistep parallel plate avalanche chamber and a CCD optical readout system (Charpak *et al.* 1989, photograph courtesy of A. Gougas, CERN).

further study, in contrast to the situation for energetic particles passing through, for instance, ionisation chambers or even bubble chambers. For this reason calorimeters are sometimes known as 'destructive' detectors.

The characteristics of a calorimeter depend on the nature of the dominant processes responsible for energy loss. For electrons and γ-rays the energy loss is dominated by electromagnetic interactions – bremsstrahlung, pair production and Compton scattering. The contribution of nuclear interactions in this case is small. For strongly-interacting particles such as nucleons and mesons the nuclear reactions are primarily responsible for the process of degradation of the energy. Neutrinos, which experience only the weak interaction, pass through any calorimeter with no signal. Muons do not produce electromagnetic showers due to their high mass and also pass through without interaction. Calorimeters thus fall into two categories – those designed to measure electron and γ-ray energies via electromagnetic processes and those designed to measure the energies of the strongly-interacting particles via nuclear interaction processes.

1.3.8 *Electromagnetic calorimeters*

An energetic electron passing into material will radiate photons due to the bremsstrahlung process. The photons will create further electrons by pair production (for energies greater than $2 \times$ electron mass) and will also transfer energy to electrons via Compton scattering. The second-generation electrons will in turn generate further photons and so on to produce an *electromagnetic shower* of photons, positive and negative

Table 1.4. *Properties of some Cerenkov radiators*

Radiator	μ	γ threshold	θ°_{max}	N_{max} (photons/ cm)	Comments
He	1.000033	123	0.47	0.03 ⎫	
Air	1.00029	41	1.4	0.26 ⎬ At STP	
CO_2	1.00043	34	1.7	0.37 ⎭	
H_2O	1.33	1.51	41	200	
Scintillator	1.58	1.30	51	270	cf. 10^4 photons cm^{-1} from scintillation
Lead glass (PbO/SiO_2, 55/45%)	1.67	1.25	53	290	

electrons (fig. 1.18). The shower particles continue to multiply until a maximum number of particles is reached when the average particle energy is no longer high enough to continue the multiplication process. Beyond this point the shower decays as the particles lose energy by ionisation and the photons by Compton scattering.

The energy loss process for electrons is characterised by *the radiation length*, X_0, defined by the relation

$$E(t) = E_0 \exp(-t/X_0)$$

where E_0 is the initial electron energy and $E(t)$ the average energy after passing through a thickness t of material. Thus the energy of an electron will on average drop by a factor of $1/e$ in a thickness X_0. The energy loss of a high-energy γ-ray by pair production is directly related to X_0. In a distance X_0 a high-energy γ will produce an electron–positron pair with probability 7/9.

X_0 is, in general, a function of electron energy and of the nature of the absorber material. However, for electron and photon energies above 1 GeV, X_0 is almost energy-independent.

The dependence on the atomic number Z and mass A is given to a good approximation by the expression

$$X_0(\text{g cm}^{-2}) \approx 716 \frac{A}{Z^2} \ln(183 Z^{-\frac{1}{3}}).$$

Thus to a first approximation the radiation length decreases as Z^{-2} so that calorimeters generally employ high-Z materials such as lead in order to minimise the overall size.

In the design of a calorimeter to measure electron or photon energy we are concerned to obtain maximum energy resolution and also to locate the position of the photon or electron as precisely as possible. For these reasons we need to understand the properties of the shower in terms of:

    (*i*) longitudinal extent;
    (*ii*) transverse spread;
    (*iii*) rate of energy deposition along length.

We have seen that the number of particles increases to a maximum and then decays when the energy of the particles drops below a critical value. We define the critical energy ε as the electron energy for which losses by radiation equal those by collision and ionisation. Typical values of ε are in the range 7 MeV (Pb) to 30 MeV (Argon). To a rough approximation, the number of particles plus photons doubles for each radiation length traversed up to the shower maximum at thickness t_{max} where the energy

Fig. 1.18. Electromagnetic shower in the BEBC filled with a mixture of liquid neon and liquid hydrogen. The magnetic field on the chamber is 3.5 T (Photo CERN).

66951

per particle equals ε. Thus if t_{max} corresponds to n_{max} radiation lengths then the number of particles plus photons present at t_{max} is $2^{n_{max}}$ and $\varepsilon(t_{max}) \sim (E_0/2^{n_{max}})$, so

$$t_{max} \text{ (in units of } X_0) = \ln(E_0/\varepsilon)/\ln 2$$

and t_{max} increases as the logarithm of the energy. A more exact treatment gives

$$t_{max} = \ln(E_0/\varepsilon) - 1.0.$$

Some measured distributions are given in fig. 1.19. The total longitudinal depth L beyond which there is no measurable increase in signal is then

$$L = t_{max} + x$$

where x measures the attenuation phase of the shower which is approximately exponential. For containment of 95% of the energy it is found that

$$L(95\%) \simeq (\ln E_0/\varepsilon - 1.0) + 12X_0 \simeq 20X_0 \quad \text{for } E = 50 \text{ GeV in Pb.}$$

The transverse spread of the shower, which determines the transverse dimensions of the calorimeter or the transverse dimensions of its cells if

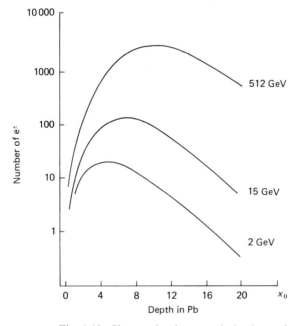

Fig. 1.19. Shower development in lead as a function of depth in radiation lengths. The 2 GeV and 15 GeV curves correspond to measured values while the curve for 512 GeV is calculated (Müller, 1972).

the shower position is to be measured, is primarily determined by the multiple scattering of the electrons which spreads the shower, since the bremsstrahlung and pair production processes at high energies are predominantly very forward-peaked and do not contribute significantly to the shower spread. The radial spread is determined by X_0 together with the angular deflection per radiation length at the critical energy. The Molière unit in multiple-scattering theory is defined in terms of these parameters as

$$R_{\mathrm{M}} = 21 \text{ MeV} \cdot \frac{X_0}{\varepsilon}$$

and in terms of this unit, all materials behave very similarly. About 90% of the shower energy is contained within R_{M} which is $\sim 3X_0$ for Pb.

The *energy resolution* depends primarily on the sampling fluctuations of the energy deposit in the active material. Many calorimeters (see below) are built of layers of absorbing material interleaved with layers of detector such as scintillator or ionisation detector and in such a case it is the statistical variations of electron path in the active material which determine the resolution. Thus if σ_E is the rms width of the energy spread then the percentage energy resolution

$$\frac{\sigma_E}{E} \sim \sqrt{\frac{2}{n_{\mathrm{e}}}}$$

where $n_{\mathrm{e}}/2$ is the number of samplings for electron pairs in the calorimeter. Then, since

$$n_{\mathrm{e}} \sim \frac{E_{\text{incident}}}{(\text{Energy loss per sampling gap})} = \frac{E}{\Delta E}$$

so that

$$\frac{\sigma_E}{E} \sim 0.05 \sqrt{\frac{\Delta E(\text{MeV})}{E(\text{GeV})}}.$$

This formula is approximately correct for $\Delta E \gtrsim 1$ MeV and a ratio R of (energy deposit in active material)/(energy deposit in passive material) within the limits $0.05 \lesssim R \lesssim 0.4$. Thus, in general, the resolution is proportional to $E^{-\frac{1}{2}}$. The only commonly used exception to this dependence is NaI where the resolution is limited by inhomogeneities rather than sampling fluctuations and is proportional to $E^{-\frac{1}{4}}$. It is thus customary to express the resolution as $x/\sqrt{E}\%$ where x varies from 5 for lead glass to 10–20 for Pb–MWPC sandwich construction.

Structures for calorimeters are shown diagrammatically in fig. 1.20 and properties of some commonly used materials are given in table 1.5.

Table 1.5. *Properties of some materials commonly used in calorimeters*

	X_0 (cm)	ε (MeV)	dE/dx (Mev cm^{-1})	λ_0 (cm)
NaI	2.6	12.5	4.84	41
Liquid argon	14.0	29.8	2.11	81
Fe	1.76	20.5	11.6	17
Pb	0.56	7.2	12.8	18
U	0.32	6.6	20.7	12

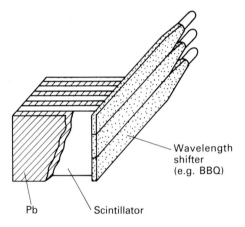

Wavelength shifter (e.g. BBQ)

Pb   Scintillator

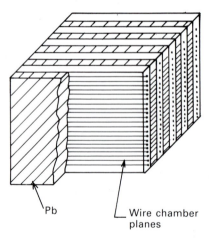

Pb   Wire chamber planes

Fig. 1.20. Some types of structure used in electromagnetic calorimeters.

1.3.9   *Hadron calorimeters*

The process of degradation and detection of hadron energies is more complex than for electromagnetic processes, depending on nuclear disintegrations and excitations.

The parameter corresponding to the radiation length X_0 in electromagnetic calorimeters is the *nuclear absorption length*

$$\lambda_0 = [(\text{nuclei/unit volume}) \cdot \sigma_I]^{-1}$$

$$= \frac{A}{N\rho\sigma_I}.$$

where A is the atomic number of the absorber, N is Avogadro's number, ρ is the density and σ_I the inelastic nucleon cross-section. σ_I and hence λ_0 vary with energy for energies in the region of the nucleon resonances, but above about 2 GeV, λ_0 is almost energy-independent.

Since the hadron calorimeter ultimately depends on energy deposited *within the instrument* by *charged particles* it is clear that not all the hadron energy will in general be 'visible'. This is so because a number of the interaction processes result in neutral particles such as neutrinos and slow neutrons which escape from the calorimeter completely, or in charged particles which escape from the calorimeter without depositing their total energy. This latter effect is important for muons and also for particles backscattered in the first layers of the instrument. The ratio of 'visible' or measured energy for hadrons to that for electrons averages ~ 0.7 in the energy range 1–20 GeV. A big fraction of the lost energy can be recovered by introducing layers of ^{238}U into the calorimeter. Neutrons in the 1–10 MeV range typical for nuclear break up, cause fission in the ^{238}U and their energy is thus converted into charged particle ionisation and measured. Such 'fission compensation' can bring the electron/hadron ratio close to unity.

The resolution of a hadron calorimeter is, in general, much poorer than for an electromagnetic calorimeter due to the much greater fluctuations in the development of the shower. Depending on the number of π^0-mesons produced in the early stages of the cascade, the shower may develop in either a predominantly electromagnetic or a hadronic mode contributing substantially to the variation in energy deposit.

For a calorimeter not employing fission compensation the resolution is approximately given by

$$\frac{\sigma_E}{E} \sim \frac{0.5}{\sqrt{E(\text{GeV})}} \sim \frac{0.1\sqrt{\Delta E(\text{MeV})}}{\sqrt{E(\text{GeV})}}$$

where σ_E is the rms spread in energy, E is the incident particle energy and ΔE the energy deposit per measuring gap.

The lateral shower size is such that $\sim 95\%$ of the energy is deposited within a cylinder of radius λ_0.

As for electromagnetic calorimeters, hadron calorimeters are usually constructed from a stack of alternate layers of absorbing material, such as iron or lead, and detection devices such as scintillator or proportional tubes. In detectors employing a magnetic field, the iron return yoke of the magnet may be instrumented in this way to provide a hadron calorimeter.

1.3.10   *Nuclear emulsions*

Nuclear emulsions were one of the most important tools in the early days of particle physics and have remained useful for some special purposes. These high-silver-concentration photographic plates can record the tracks of even minimum-ionising charged particles. They are small and continuously sensitive and were ideal for cosmic-ray work, particularly with high-flying balloons. Grain density measurements can yield velocities for non-relativistic tracks while multiple scattering was widely used to yield values of $p\beta$ which could be combined with the grain density to determine the mass. In most respects the nuclear emulsion has been overtaken by other techniques. In respect of spatial resolution, however, it remains the best technique available, and in some situations this feature outweighs the other shortcomings. The grain size in emulsions is 1–2 μm and space resolution of a few microns is possible in favourable cases. This feature may be particularly valuable in the study of very-short-lifetime particles. Emulsions have recently been used in hybrid systems along with bubble chambers and electronic detectors which indicate the location of potentially interesting reactions in the emulsion.

1.3.11   *Semiconductor counters*

Semiconductor counters have long been widely used in low-energy nuclear physics where their very high resolution for X-rays and low-energy charged particles has proved of great importance. In recent years such counters have been increasingly used in the form of very-fine-grain – 'microstrip' – devices to achieve high spatial resolution close to an interaction.

One or more semiconductor junctions is used in reverse bias to create a depleted zone without carriers which can act as an ionisation chamber. A very small ionisation signal can then be detected against a background of the remaining small leakage current.

In the 'silicon strip' device the electrodes are in the form of very fine strips ($\sim 10\,\mu m$). A stack of such planes gives a three-dimensional detector. Such detectors have been used as 'active targets'. An increase of track signals from one layer to the next is evidence for decay of a neutral particle to charged particles. Used in this way, or even placed close to the interaction target, these devices may detect decays of particles of very short life such as charmed or beauty mesons.

Track positions may be determined by measuring the centre of gravity of the strip signals. Resolutions of a few microns have been achieved in detectors with p-implanted strips $10\,\mu m$ wide at $10\,\mu m$ separation on a $280\,\mu m$-thick crystal of silicon. For detectors covering more than a few mm^2, however, the very large number of channels and the difficulty of readout generally lead to a pitch $\sim 100\,\mu m$. At this pitch a resolution of $8\,\mu m$ has been achieved.

1.3.12 *Transition radiation*

Electromagnetic radiation is emitted whenever charged particles cross an interface between materials having different dielectric properties. Although not widely used in particle detection and identification to date, this phenomenon may prove to be useful for identification of ultra-relativistic particles beyond the reach of other detectors.

For a macroscopically varying dielectric constant the particle produces a time-dependent electric field which results in a transient polarisation of the medium and the polarisation current gives the transition radiation. Detectors are designed in the form of a stack of layers of two different materials so that transition radiation is produced at all the interfaces. In such a stack we also have the possibility of multiple foil interference and the performance of the detector depends on the geometry and materials employed.

The total energy radiated is proportional to $\gamma\left((1-\beta^2)^{-\frac{1}{2}}\right)$ for the particle. The angular distribution is sharply forward-peaked with a sharp maximum for a single foil at $\theta \sim \gamma^{-1}$. Practical detectors exploit the X-ray emission using Li foils to optimise the ratio of detected ionisation + transition radiation to ionisation which is found to vary as $Z^{-3.5}$. To date, transition radiation detectors have been primarily used for identification of electrons and very high energy pions with $\gamma \gtrsim 2000$ in cosmic-rays in which circumstances the intensity of the radiation is relatively high. Detectors for accelerator experiments are under development and may be expected to become increasingly important at higher energies.

1.3.13   *Electronics and computers*

Every particle physics experiment is dependent on electronics and computers. Electronic detectors produce signals which require fast amplification, discrimination and often analogue to digital conversion (ADC) or time to digital conversion (TDC). Signals will frequently be combined or compared in simple coincidence units or in more complex matrices which can be programmed to accept only desired combinations.

In most experiments, conditions are imposed on the signals to select a particular class of events from a much more abundant background. Selection at levels of one event in 10^5 are not uncommon. For instance, it may be required to select only events with two or more oppositely-charged K-mesons in appropriate geometries where they could arise from ϕ-meson decay. An array of Cerenkov and scintillation counters with appropriate trigger electronics demanding signals in the allowable combinations of counters would be necessary. On-line computers to control the data collection, sample and display events and data, and even to make complex trigger decisions, are used in all the more elaborate experiments. Data for the selected events is normally written on magnetic tape. Depending on the complexity of the detector, each event may involve from a few hundred to more than 10^5 words of data. Since it is seldom possible to make the trigger conditions tight without losing genuine events it is therefore necessary to carry out further off-line filtering using more sophisticated algorithms. Subsequent stages in off-line analysis will generally include a pattern recognition stage to identify particle trajectories, geometrical reconstruction to obtain accurate positions, angles and curvatures, and sometimes also kinematical fitting to calculate missing energies and momenta. Calorimeters require shower identification and analysis programmes. Large experiments routinely record 10^6–10^7 events on tape so that the off-line computing task is formidable.

The technologies of electronics, on- and off-line computing and data analysis are in themselves big subjects with extensive literature and many specialists. The interested reader is referred to detailed texts and reviews for more information (e.g. Bologna and Vincelli (1983)).

2

Pions and muons

2.1   Introduction

We shall see later (chapter 10) that it is now clear that particles such as protons, neutrons and π-mesons (pions), once thought to be 'elementary' are, in fact, made from more basic constituents, the quarks. Thus the force between, say, two neutrons is basically the resultant of the forces between two groups, each of three quarks. In this respect it is approximately analogous to the Van der Waals force between two molecules, each of which consists of several atoms. As in the case of atomic or molecular problems, so for particles it is in many instances, however, very useful to treat the forces between particles such as nucleons rather than the forces between their constituent quarks. Although many of the ideas concerning particle interactions were developed before quarks were identified as the constituents of nucleons and mesons, these ideas are, in general, not invalidated by the discovery of the quark structure and, indeed, the subject is perhaps best understood by means of this semi-historical approach.

2.2   Prediction of the π-mesons by Yukawa

The prediction of the existence of mesons, by Yukawa in 1935, and their subsequent discovery in the cosmic radiation, present one of the most striking examples of the interaction of theory and experiment in modern physics.

We can understand Yukawa's argument in a qualitative way as follows. It had become well established that electromagnetic forces could be well understood in terms of a field of which the quanta were photons. In this theory of quantum electrodynamics the forces between two charged particles are attributed to the exchange of photons between them. Thus, for instance, in the centre-of-mass system (see appendix A.2) the Coulomb scattering of two electrons is described by a diagram like fig. 2.1(a), where energy and momentum have been exchanged by the exchange of a photon. This diagram is similar to a so-called Feynman diagram, which we shall have cause to use quite frequently. The more usual convention is to represent the process simply as in fig. 2.1(b). Let us first note the broad qualitative features of nuclear and electromagnetic forces.

Evidence is available from static nuclear properties and from scattering experiments. Our purpose here is not to treat these subjects in detail, but rather to discuss briefly those aspects which are relevant to the size of the nucleus and the strength of the nuclear forces.

It is immediately clear that the nuclear forces are attractive and sufficiently strong to overcome the Coulomb repulsion. It is also clear, since the nuclear radius R obeys the relation

$$R = r_0 A^{\frac{1}{3}}$$

(where r_0 is a constant and A is the mass number), that there is some form of 'repulsive core' which prevents the collapse of all the nucleons to the range of the nuclear force. We note also that the binding energy of nucleons in nuclei is about eight million electron volts throughout the greater part of the periodic table, i.e. the nuclear force is 'saturated'. This behaviour is similar to that observed for the intermolecular forces in liquids and solids. It is in contrast with the unsaturated Coulomb force, responsible for the binding of electrons in atoms, which are all of about the same size throughout the periodic table. The saturation is characteristic

Fig. 2.1. Photon exchange in electron–electron scattering (a) and Feynman diagram for the process (b).

of so-called exchange forces. We may describe such a force in terms not of a simple potential $V(r)$ but rather of the product $P_{12} V(r)$, where P is the so-called permutation operator. Thus, if the wave function of the pair of interacting particles is $\psi(1, 2)$, then the effect of P is to exchange certain of the properties of the particles. Thus

$$P_{12} \psi(1, 2) = \psi(2, 1).$$

The exchange may be one of charge, spin or position, or of any combination of these properties. The suggestion that the inter-nucleon force was due to some kind of exchange was first put forward by Heisenberg as early as 1932, although the nature of the exchange 'quantum' was not clear.

Before discussing further the magnitude of the range and strength of the forces, we may look at what evidence there is for their exchange nature from scattering experiments. Let us consider the scattering of neutrons on protons since in this case the complication of the Coulomb force is absent, as is the problem associated with having two identical particles. If we consider the scattering of neutrons of moderately high energy (say, 100 MeV), then in the case of no exchange force the scattered neutrons will be peaked in the forward direction, in the cms, since we will have contributions to the scattering from states having orbital angular momentum greater than zero. For exchange forces, on the other hand, the identity of the particles will be exchanged in the interaction and we will have protons peaked in the forward direction. In fact, the experimental results show peaks at both 0° and 180°, indicating the existence of both exchange and non-exchange forces.

An important question is whether the forces between the three possible pairings of nucleons, n–n, p–p, and n–p, are the same, apart from the Coulomb forces between the protons. Information on this point can be obtained from a comparison of the masses and energy levels of so-called mirror nuclei. Such pairs of nuclei differ only in that a neutron in one is replaced by a proton in the other. Thus, if the n–n and p–p forces are the same, with the exception of the Coulomb force, the masses and energy levels of, for instance, ^7_3Li and ^7_4Be or $^{11}_5\text{B}$ and $^{11}_6\text{C}$ should be the same when corrections have been made to take into account (a) the Coulomb effects and (b) the neutron–proton mass difference. Such a comparison, in fact, confirms the equality of the n–n and p–p forces. This property of nuclear forces is known as charge symmetry. If these forces are also equal to the n–p force, then we have charge independence. This is more difficult to establish from the static nuclear properties, but is supported by a comparison of isobars with even A in which the charge differs by two units. We shall return to this subject when we come to consider the property known as isotopic spin.

We now consider some quantitative evidence as to the range and strength of the nuclear force.

Ranges from nuclear size. We can get some indication of the range of the nuclear force if we recall that many experiments have established that the nuclear radius follows closely the relation

$$R = r_0 A^{\frac{1}{3}}$$

where

$$r_o = 1.5 \, \text{fm} \quad (1 \, \text{fm} = 1 \, \text{fermi} = 10^{-13} \, \text{cm}).$$

If we imagine that in nuclei the inter-nucleon distance is of the order of the range of the nuclear force, and that the nucleons are uniformly distributed so that each nucleon occupies a volume equal to πr_0^3, then the range of the force is also of the order of r_0. Clearly such a consideration involves several crude approximations and assumptions and only gives an order-of-magnitude value.

Alpha-particle scattering on nuclei. The original results on the scattering of alpha particles by nuclei, which established the nuclear atom, were well fitted on the assumption that the scattering was due to the Coulomb force between the alpha particle and the nucleus, considered as point charges. However, it is clear that the effects, both of the nuclear size and of the nuclear forces, will be important if the closest distance of approach becomes sufficiently small. This is indeed found to be true. For scattering of alpha particles by gold, for instance, the cross-section falls dramatically below the value to be expected for Coulomb scattering for an energy greater than about 20 MeV. Various techniques have been used to attempt to fit the data in terms of the nuclear radius and the range of the nuclear force as parameters. Again, the value for the range of the force is found to be of the order of 1.5 fm.

Neutron–proton scattering. The study of this phenomenon would appear to be particularly suitable for obtaining the range, since Coulomb effects are absent and we are dealing with a two-body situation. The simplest approach is to assume that the scattering can be described in terms of a square-well potential. In fact, changes in the potential shape have only small effects on the results. The parameters to be determined are then the potential depth V_0 and the well diameter d, where we consider neutron energies such that only s-state scattering is possible (less than ~ 5 MeV). Although we know the binding energy of the deuteron, and can measure the neutron–proton scattering cross-section, we cannot determine V_0 and d separately, because the scattering takes place in both the singlet and triplet spin states, so that, in fact, there are four parameters rather than two: $V_{0,t}$, d_t, $V_{0,s}$ and d_s. The result of fitting the data in terms of these four parameters is

$$V_{0,t} \simeq 35 \text{ MeV} \quad d_t \simeq 2 \text{ fm}$$
$$V_{0,s} \simeq 12 \text{ MeV} \quad d_s \simeq 3 \text{ fm}.$$

Low-energy proton–proton scattering also yields values

$$V_{0,s} \simeq 13 \text{ MeV} \quad d_s \simeq 2.6 \text{ fm}.$$

Deuteron binding energy. The application of the Schrödinger equation to the case of the deuteron, which, on the basis of the value of the magnetic moment, can be taken to be largely in the triplet s-state, yields a relationship between the potential and the range of the form

$$V_0 d^2 \simeq \frac{\pi^2 \hbar^2}{4M}$$

where M is the nucleon mass. Taking $V_0 \simeq 35$ MeV, we obtain $d \simeq 1.8$ fm.

Summarising, we can say that any satisfactory theory of nuclear forces had to account for a force which was very strong (well depth ~ 35 MeV), of short range (~ 2 fm) and which acted equally between proton–proton, neutron–neutron and proton–neutron. The force had also the characteristics, at least in part, of an exchange interaction, and the success of such a theory for electromagnetic forces made its pursuit particularly attractive also for the nuclear interaction.

Yukawa developed this idea in his famous paper of 1935. He first noted that a zero-mass quantum, such as the photon, will not give a force of sufficiently short range. In addition, since the quanta of the field must certainly be permitted to travel at high velocity, it is essential that the theory be relativistically correct. We can illustrate the results of the Yukawa treatment in the following way.

The relativistic relationship between total energy, momentum and mass for the field quantum of mass m is

$$E^2 - p^2 c^2 - m^2 c^4 = 0.$$

We can now use the usual quantum mechanical substitutions

$$E \rightarrow i\hbar\frac{\delta}{\delta t} \quad \text{and} \quad p \rightarrow i\hbar\nabla$$

to produce an operator equation

$$-\hbar^2\frac{\delta^2}{\delta t^2} + \hbar^2\nabla^2 c^2 - m^2 c^4 = 0.$$

If we now represent the force between nucleons by a potential $\phi(r,t)$, which may be regarded as a field variable, then we have

$$\left[\nabla^2 - \frac{1}{c^2}\frac{\delta^2}{\delta t^2} - \frac{m^2 c^2}{\hbar^2}\right]\phi = 0$$

as our 'wave equation'.

The time-independent part of the equation is

$$\left[\nabla^2 - \frac{m^2 c^2}{\hbar^2}\right]\phi = 0 \qquad\qquad \textbf{2.1}$$

to be compared with the equation for a static electric field,

$$\nabla^2\phi = 0$$

with solution

$$\phi = \frac{e}{r}.$$

The solution of the differential equation **2.1** is

$$\phi = \frac{g}{r}e^{-mcr/\hbar}$$

as can be checked by substitution, where g is a constant with the dimensions of electric charge, and is known as the *coupling constant* for the interaction. Thus the energy of a second nucleon in this field ϕ generated by the first is $g\phi$ and the interaction between the two nucleons is given in terms of the coupling constant by

$$\frac{g^2}{r}e^{-mcr/\hbar}$$

If we write the form of the potential as proportional to e^{-ar}/r and expand the exponential, we see that to first order the force becomes zero at $r = r_0 = 1/a$. Thus the range of the force, r, is $\sim \hbar/mc$. If we substitute 2 fm for our r_0 we find $m \simeq 200$ MeV/c^2 for the mass of the field quantum. We note that as $r \to 0$, $\phi \to \infty$ and the form of this 'Yukawa potential' is shown in fig. 2.2. Although this is certainly not a square well, we can use the value of V_0 obtained from the consideration above to obtain a rough value for the coupling constant. This yields $g^2 \simeq \hbar c$. It is clear that the potential cannot be infinitely deep, so that the form which we have derived cannot apply at very small values of r. However, we may hope that it is at least approximately true at the ranges with which we shall be initially concerned.

If we consider the various inter-nucleon forces in terms of the exchange of the Yukawa quanta, which we will refer to as π-mesons or pions, in anticipation of their later discovery, we obtain the following results. For proton–proton and neutron–neutron scattering, exchange of a neutral pion is required, unless exchange of two charged mesons is allowed (fig. 2.3(a)). For neutron–proton scattering, however, we may have exchange of both neutral and charged pions (figs. 2.3(b), (c), (d)). The equality of the n–n, n–p and p–p forces indicated that all are due to the same type of exchange, so that we must suppose that neutral, as well as charged, mesons should exist. This extension of the original Yukawa proposal was made in 1938 by Kemmer.

Finally, in predicting the properties of the pion, Yukawa pointed out that it was probably unstable. This proposal arose from attempts to explain nuclear β-decay, and to account for the fact that the mesons had not so far been observed. His explanation of the β-decay, in terms of the steps

$$n \to p + (\text{meson})^-$$
$$(\text{meson})^- \to e^- + \nu,$$

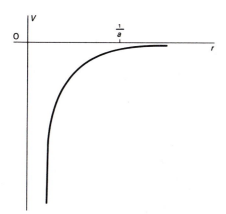

Fig. 2.2. Form of the Yukawa potential.

is now known to be wrong, as applied to the pion (although correctly describing β-decay in terms of the *W*!) though the meson lifetime obtained on this hypothesis was by chance approximately correct.

The picture of the nucleon which is implied by the Yukawa model is of a particle continually emitting and re-absorbing pions, so that it is effectively surrounded by a pion cloud. A single nucleon cannot emit a pion with conservation of energy, except within the limits allowed by the uncertainty principle. Thus a violation of energy conservation by an amount ΔE can exist only for a time Δt given by $\Delta t \, \Delta E \simeq \hbar$. If $\Delta E \simeq M_\pi$ (~ 140 MeV) then $\Delta t \simeq 4 \times 10^{-24}$ s. For velocity c this gives the extent of the pion cloud as $\sim 10^{-13}$ cm.

A further interesting application of this picture of the meson–nucleon relationship is to the problem of the anomalous magnetic moment. Dirac showed that it was a consequence of his relativistic wave equation that the electron should have a magnetic moment given by

$$\mu_e = \frac{-e\hbar}{2m_e c}.$$

The Dirac theory would then give

$$\mu_p = \frac{e\hbar}{2m_p c} \quad \text{and} \quad \mu_n = 0$$

for a proton and a neutron respectively.

In fact, measurement yields,

$$\mu_p = 2.78\mu_N \quad \text{and} \quad \mu_n = -1.93\mu_N$$

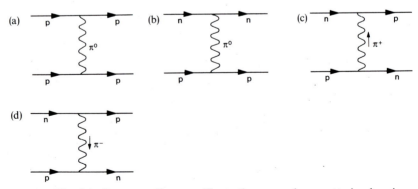

Fig. 2.3. Feynman diagrams illustrating pp and np scattering by pion exchange.

where

$$\mu_N = \frac{eh}{2m_pc}$$

the nuclear magneton.

The differences between the predicted Dirac values and the measured values are known as the anomalous magnetic moments. The meson picture affords a qualitative explanation of the observed magnetic moments. Since we assume that for part of the time the proton exists as a neutron plus a π^+-meson,

$$p \rightleftharpoons n + \pi^+$$

we must consider the effective magnetic moment for the $n-\pi^+$ system. We will assume that the pion has zero spin (see section 2.7) so that if the relative orbital angular momentum of neutron and π^+ is Lh, then

$$\mu = \frac{ehL}{2m_\pi c}$$

and for the p-state this gives $\mu = 7\mu_N$. The true μ_p is given by

$$\mu_p = x\mu_N + 7(1-x)\mu_N$$

where x is the probability that the proton exists in the so-called bare state, with no external pion. Using the measured value for μ_p we find $x = 0.7$. Considering the inverse process, where the neutron exists for part of the time as a proton plus a π^--meson, and again taking the pion spin to be zero, we obtain the effective magnetic moment for the $p-\pi^-$ combination by adding algebraically the true bare proton moment and that generated by the 'orbiting' pion. If we assume that x is here also equal to 0.7, and note that the orbital moment and the bare proton moment must be in opposite directions for the p-state in order to obtain spin $\frac{1}{2}$, we find that $\mu_n = -2.4\mu_N$ in qualitative agreement with the observed value.

In fact, of course, the hypothesis that nuclear forces are due to the exchange of π-mesons does not explain all their features, such as, for instance, the spin dependence. We now know that there exist a number of other strongly interacting mesons which must certainly also play a similar role to the pion in nuclear forces. However, the pion is the lightest of such mesons and must therefore account for the most important contribution at the outer edge of the potential. At the moment the fact that the situation has turned out to be much more complicated than envisaged by Yukawa is not important to us; the basic idea remains valid, and it was this idea which both stimulated the later experiments and enabled them to be interpreted in terms of the quantum of the nuclear force.

2.3   Discovery of pions and muons

In the late 1930s, following the suggestion of Yukawa, the only source of particles of sufficiently high energy to produce the mesons was the cosmic radiation. Evidence for the existence of mesons was forthcoming from both counter studies of the absorption of the soft and hard components of the cosmic rays and, even more strikingly, from cloud-chamber photographs.

In discussing the cloud-chamber evidence we first consider the possible information concerning a particle mass which can be obtained from measurements on a track in a cloud chamber. Measurements may be made of ρ, the radius of curvature in a magnetic field; of R, the range, if the particle comes to rest in the chamber; of the drop density and of the change in curvature of the track in passing through absorbing material, normally in the form of a plate across the chamber. We recall that ρ is a function of the particle momentum, R is a function of the particle energy, and the drop density or rate of energy loss, of the particle velocity. Thus a measurement of ρ and R, or of either of these quantities along with drop density or change in curvature on passing through a plate, will, in principle, yield a value of the particle mass. In practice, the precision obtained may be highly dependent on the energy of the particle and also on such experimental factors as possible turbulence in the cloud chamber, or non-uniform sensitivity.

The first clear evidence for the existence of a particle with mass intermediate between the electron and the proton was obtained by Anderson and Neddermeyer (1936), in cloud-chamber studies of cosmic rays at mountain altitude. In fig. 2.4 is shown one of the two photographs which they obtained showing the tracks of such a particle. Several particles are ejected, from the same point in a lead plate across the chamber, by an incident neutral particle. One of the tracks is densely ionising and stops in the chamber after a range of 40 mm. Its radius of curvature in the 632 kA m^{-1} field is 65 mm. A proton having this range would have an energy of 1.5 MeV and a radius of curvature of 0.2 m. As can be seen from some of the small spirals in the chamber, electrons having a range of this order have much smaller radii of curvature and are minimum-ionising tracks. In fact, if the measured range and curvature are used the mass of the particle is found to be about 180 MeV/c^2. A number of other such observations demonstrated the existence of such particles, having both positive and negative charges. The errors in the masses obtained showed wide variation, but indicated an average mass of about $100 \text{ MeV}/c^2$.

Concerning the meson decay, a cloud-chamber photograph was obtained in 1940, by Williams and Roberts, which showed a positive particle of mass about $120 \text{ MeV}/c^2$ decaying into an electron. The evidence concerning the *lifetime* of the meson came, in the early work, from measurements with Geiger counters on the absorption of cosmic rays. Indeed, these measurements also provided independent, indirect evidence on the actual existence of the particles. It was well established that the cosmic radiation consisted of two components known as the hard, or highly penetrating, component and the soft component, which was absorbed almost completely by 0.10–0.15 m of lead. The soft component was responsible for most of what were known as cosmic-ray 'showers', and was observed to consist of electrons, positrons and γ-rays. The hard component was observed to be absorbed roughly in proportion to the mass of the absorber, unlike the soft component in which absorption per atom was approximately proportional to the square of the atomic

Fig. 2.4. A stereoscopic pair of cloud-chamber photographs taken byAnderson and Neddermeyer. A neutral cosmic-ray particle caused an interaction in the lead plate across the chamber. A densely-ionising track emerging in the upper half of the photograph has a range of 40 mm and a radius of curvature of 65 mm. These data indicate a mass of about $180 \text{ MeV}/c^2$.

number. On the assumption that the hard component consists of very high energy electrons and positrons, it is possible to calculate the energy which these particles must have if they are to penetrate through the complete atmospheric layer. Such a calculation yields values of the energy which appear to be impossibly high, and for which there is no evidence in the energy spectrum of electrons and positrons observed at accessible altitudes. Initially it was thought that the explanation of this difficulty must lie in a breakdown of the energy-loss formula for particles of very high energy. However, as the formula was tested experimentally at increasing energies, it became clear that there is no evidence for such a breakdown. On the other hand, not more than a very small fraction of the particles making up the hard component can be protons, since all attempts to detect protons in the cosmic rays agreed that, at least near the Earth's surface, the proportion of protons in the penetrating component is not more than about 10 %. Thus the most plausible explanation of the nature of the hard component is that it is due to particles having a mass much greater than electrons, and for which the bremsstrahlung energy loss is thus very much less.

Detailed studies of the absorption of the hard component indicated that the absorption was not strictly proportional to the mass of the absorber, but also depended upon its thickness. Experiments of this nature used Geiger counters surrounded by sufficient lead to eliminate the soft component. In such an experiment the absorption of the hard component in air was measured by making counts at different altitudes. A mass of some other absorber, such as carbon, equivalent to the air layer between the two altitudes, could then be inserted above the counters. It was found that the air apparently absorbed more effectively than the carbon. This effect finds a natural explanation if we assume that the particles undergo radioactive decay, and quantitative comparison of the rates of absorption yielded a mean lifetime of approximately two microseconds.

A direct determination of the meson lifetime was made by Rasetti. The mesons were selected by a fourfold system of Geiger counters and allowed to pass into a 10 cm-thick iron absorber. Mesons which stopped in the iron, as indicated by an anticoincidence of a further set of counters, were found in about half of all cases to be associated with a charged particle emerging from the iron block, after a delay of the order of a few microseconds or less. If this delayed particle is taken to be the decay electron then the lifetime of the meson may be obtained from the distribution in delay times, and again a figure of about two microseconds was obtained. The fact that only half the incident mesons were observed to decay may be explained by assuming that negative and positive particles are present in equal numbers, that the positive particles suffer Coulomb repulsion from the nucleus, so that they are free to decay, but that the negative particles are rapidly absorbed.

Initially these mesons were taken to be the particles predicted by Yukawa since they roughly corresponded, in both mass and lifetime, to his predictions. However, increasing doubts arose concerning the interaction cross-section of these particles with nuclei. For the meson to play the appropriate role in nucleon forces it is essential that it has 'strong interactions' (see chapter 4) with nucleons. Even in the early experiments a difficulty was apparent in the assumption that mesons could pass through great distances in the atmosphere without interaction. If we assume that their cross-section for interaction is of the usual order of magnitude for 'strong' interactions ($\sim 10^{-26}$ cm^2), then we expect to find very few mesons remaining at ground level. More direct evidence was obtained, in 1947, in an experiment by Conversi, Pancini and Piccioni, who studied the absorption of negative mesons brought to rest in carbon and in iron. Broadly speaking, in carbon all the mesons were observed to decay, while in iron they were all absorbed without decay. Further observation proved that the capture rate for these mesons at rest was proportional to the atomic number, Z. At $Z = 12$ approximately half of the mesons decayed while half were captured, indicating an average time for capture of about one microsecond. Such a long capture time may be seen to be in strong disagreement with the hypothesis that the meson–nucleon interaction is 'strong'. The process of capture for the negative mesons is that the slow meson falls into an electron orbit and rapidly cascades downwards into the lowest state. We note that, due to its greater mass, the Bohr orbit for the meson is about two hundred times smaller than the corresponding orbit for an electron. At $Z = 12$ we can calculate that the meson will spend about 10^{-3} of its time actually in the nucleus. If we assume an interaction cross-section per nucleon of about ten millibarns, and a velocity of the order of c for the meson, then we obtain by crude classical arithmetic a reaction time of about 10^{-23} s for a strong interaction, differing by a factor of 10^{17} from the observed value and indicating that the absorption is a 'weak' process for which the time would be expected to have about the observed value.

These difficulties led Bethe and Marshak, in 1947, to suggest that there must exist another meson corresponding to the Yukawa particle. Conclusive evidence in support of this suggestion was obtained in the same year by Lattes, Muirhead, Ochialini and Powell, who observed the actual decay of one meson into another in a nuclear emulsion. Subsequent work revealed many examples of events like that shown in fig. 2.5 in emulsions flown for periods at very high altitudes by balloon. The masses of the particles in the three-particle decay chain of the figure were found to be ~ 140 MeV/c^2 and ~ 100 MeV/c^2, with the final particle identifiable as an electron.

We note here the difference in the measurements used for mass determination in a nuclear emulsion and in a cloud chamber. In the emulsion it has not normally been practicable to use a magnetic field to determine particle momentum. We may still obtain the energy from the range, and the velocity from the ionisation, of which the grain density in the track is a function. In the emulsion, a useful parameter is the multiple scattering of the track due to Coulomb scattering on emulsion nuclei. This may be measured by evaluating the deviation of the track from a straight line for a series of segments along its length. The mean value of such

Fig. 2.5. A $\pi \rightarrow \mu \rightarrow e$ decay in nuclear emulsion. The pion enters the picture at bottom centre, slows down and stops at top right. The muon travels to bottom left before decaying to an electron. The scale is in microns.

deviations is a function of the product of momentum and velocity. The parent meson now known as the π-meson, or pion, appears to fill the role of the Yukawa particle, and is not normally observed in cosmic rays at ground level. It is observed to decay into the lighter μ-meson, which forms the major part of the hard component of the cosmic radiation, and which in turn decays into an electron of the appropriate charge. Each of these decay processes must be accompanied by one or more neutral particles which will be discussed below. Although the term meson was originally used to indicate a particle intermediate in mass between the nucleons and the electron, modern terminology reserves this name for strongly-interacting particles with 'baryon number' zero (q̄q pairs, see sections 3.8 and 10.7). The muon has no strong interaction and is properly referred to as a lepton (see section 3.10), although the name μ-meson is still commonly used.

2.4   Properties of charged pions and muons

2.4.1   *Meson production*

As we have seen, certain properties were determined for the π- and μ-mesons, at least approximately, in the early cosmic-ray work, but the most accurate determinations have depended on experiments using mesons produced by high-energy accelerators. The Yukawa picture would suggest that it should be possible to produce pions in nucleon–nucleon collisions if the bombarding energy is high enough. We may picture the incident nucleon interacting with a pion in the 'cloud' and actually knocking it free (fig. 2.6). The threshold for this process is most easily worked out as follows (see also appendix A). We use the invariance of the quantity

$$E^2 - p^2 = m^2$$

where E, p and m are the total energy, momentum and mass of a particle, or systems of particles, and where we have taken $c = 1$. If we indicate

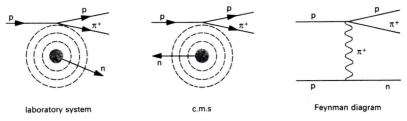

laboratory system   c.m.s   Feynman diagram

Fig. 2.6. Illustration of positive pion production in the process pp → pnπ⁺.

centre-of-mass (cms) quantities by dashed symbols, then, since in the cms $p' = 0$, we have

$$E^2 - p^2 = E'^2.$$

In the cms the threshold for

$$p + p \rightarrow pn\pi^+$$
$$p + n \rightarrow pp\pi^-$$

or similar processes occurs when

$$E' = 2m_n + m_\pi$$

where m_n and m_π are the nucleon and pion masses respectively. Also if we denote the incident proton kinetic energy by the symbol T then we have for the initial system

$$E^2 = m_n^2 + 2m_n T + T^2$$

and thus

$$2m_n T + T^2 = p^2.$$

This gives for the minimum kinetic energy necessary to produce a π-meson,

$$T = m_\pi \left[2 + \frac{m_\pi}{2m_n} \right]$$
$$= 285 \text{ MeV}.$$

Thus a cyclotron capable of accelerating to energies of the order of 300 MeV or greater should produce π-mesons when the protons strike stationary nucleons. In fact, the threshold for meson production for protons striking a cyclotron target is only about 180 MeV due to the great increase in the cms energy when head-on collisions occur, in the nucleus, between an incident proton and a moving nucleon.

2.4.2 *Pion mass*

A good determination of the pion mass was made by a magnetic analysis of the particles coming from the cyclotron target, to give precise values of their momentum, following which the particles were brought to rest in nuclear emulsions (see fig. 2.7). The energy can then be very precisely established from the range, and a comparison made between protons, pions and muons. The results of this experiment yielded mass values of 139.6 MeV/c^2 for positive and negative pions and 105.6 MeV/c^2 for positive muons.

A particularly accurate value for the mass of both pions and muons may be obtained from measurements of mesic X-rays. We have already discussed the absorption process for slow negative mesons, which proceeds via absorption of the meson into an outer atomic orbit followed by cascade into the deeper orbits and eventual nuclear interaction. The result of the transitions between atomic orbits is the emission of photons in the X-ray region. The X-ray energies may in many cases be measured to within ten or twenty electron volts. One may then use the Bohr formula, or the relativistic equivalent, to obtain the meson mass. This yields, for instance, a muon mass of 105.65839 ± 0.00006 MeV/c^2.

2.4.3 *Decay of charged pions and muons*

We have already seen that the charged π-meson decays into a μ-meson plus at least one additional neutral particle required for energy and momentum conservation. The energy of the μ-meson for π-mesons decaying at rest, or when evaluated in the cms system for the moving π-meson, is found to be always the same, 4.1 MeV, indicating that only one additional particle can be present in the decay. Since the masses of the π- and μ-mesons are known, it is possible to calculate the rest mass of the neutral particle; this is found to be consistent with zero. The only known particles of zero rest mass are the photon and the neutrino. Since the direction of the neutral particle is known, lying opposite to the μ-meson in the cms of the decay, it is possible to search for conversion electrons which may be present if the neutral object is a photon. Such an experiment was done in nuclear emulsions by O'Ceallaigh in 1950, when he examined the inferred neutral paths for a number of decays. No electron pairs were observed in a length for which the probability that the gamma ray should

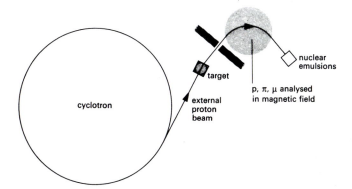

Fig. 2.7. Particles emitted from an external target are momentum analysed in a magnetic field. The range is then measured in nuclear emulsions to give a precise value of the mass.

not convert was 4×10^{-3}, indicating that the neutral was indeed a neutrino. This conclusion has been verified in later work. Thus we have

$$\pi^\pm \to \mu^\pm + \nu.$$

For the π^+ meson, the lifetime may be obtained by a calculation of the proper times for tracks observed in emulsions. More precise determinations have been made using a beam of π-mesons from an accelerator. The beam is stopped in a scintillator in which the pulse from the decay muon may also be observed. The latter method has yielded a value of $(2.603 \pm 0.002) \times 10^{-8}$ s. Negative π-mesons, when brought to rest are always absorbed before they can decay. However, the lifetime for either charge may be determined by observations of decay in flight, using a beam of π-mesons. For the decay of the μ-meson the energy of the observed decay electron is not unique, but is distributed through a spectrum of values, so that we infer that more than one neutral particle is present. The spectrum indicated that the total mass of the neutrals is again consistent with zero, and no γ-rays are observed to be associated with the decay. The simplest assumption consistent with these observations is for the decay process

$$\mu^\pm \to e^\pm + \nu + \bar{\nu}.$$

By methods similar to those described for the pion, the muon lifetime has been found to be $(2.19703 \pm 0.00004) \times 10^{-6}$ s. As we shall see in chapter 4, both of these decay processes are characteristic, in their products and lifetimes, of 'weak' interactions.

Other decay modes are also possible for the charged pion such as

$$\pi \to \begin{cases} e\nu & \text{branching fraction } (1.23 \pm 0.02) \times 10^{-4} \\ \mu\nu\gamma & (1.24 \pm 0.25) \times 10^{-4} \\ e\nu\gamma & (5.6 \pm 0.7) \times 10^{-8} \\ \pi^0 e\nu & (1.02 \pm 0.03) \times 10^{-8}. \end{cases}$$

2.5   The neutral π-mesons

We have already seen that the charge independence of nuclear forces demanded the existence of neutral, as well as charged, mesons. For such a neutral particle the most probable decay consistent with the conservation laws is into two γ-rays. Such a decay is an electromagnetic process and, as we shall see later, electromagnetic interactions are much stronger than the weak processes; we would thus expect this mode to dominate over possible weak modes, and to have a lifetime considerably shorter than that for the weak decay processes of the charged mesons.

A number of workers had suggested that the π^0-meson might be the source of the soft component of the cosmic radiation, its decay γ-rays giving rise to the showers of photons and electrons. In particular it had been noted that there was a correlation between hard and soft showers. More direct evidence came in 1950 from two experiments, one by Bjorklund *et al.* studying high-energy photons from a cyclotron target, and the other by Carlson, Hooper and King, using cosmic-ray 'stars' (interactions with secondary prongs in nuclear emulsions). In the experiment of Bjorklund *et al.*, γ-rays arising from the target of the Berkeley cyclotron were allowed to pass through holes in the shielding wall and to enter a magnetic spectrometer. The pairs which they produced in a thin sheet of tantalum were bent in the magnetic field, and detected by proportional counters in coincidence. γ-ray spectra were measured for protons incident on the cyclotron target at various energies between 175 and 340 MeV, and for angles of emission from the target of 0° and 180°. A number of possible sources of these γ-rays, which ranged in energy up to 200 MeV for 350 MeV incident protons, were considered. The spectra at 0° and 180° differed only by a Doppler shift, and became identical when transformed into a coordinate system moving with a velocity of $0.32c$. This corresponds to the expected cms velocity for a collision between a 340 MeV proton and a nuclear nucleon, moving in the opposite direction, with an energy of about 25 MeV and a velocity of about $0.2c$. This observation, and also the high energy of the γ-rays, makes it impossible that they should be due to any nuclear excitation. The absolute cross-section for γ-ray production in the energy range studied is of the order of 10^{-27} cm². For bremsstrahlung by protons of this energy we would expect a cross-section of only 10^{-29} cm². Also the spectrum shape, and the rather rapid increase in γ-ray production with increasing incident proton energy, are in complete disagreement with what is to be expected for bremsstrahlung, which cannot therefore explain the origin of the photons.

On the other hand, the hypothesis that the photons arise from the decay into two γ-rays of a neutral meson is found to give a good account of all the features of the observations. A mass of about 150 MeV/c^2 is required for the neutral pion. From this experiment its lifetime can only be said to be less than 10^{-11} s.

The data from the nuclear emulsion work of Carlson, Hooper and King were even more compelling. In this experiment the region around 'stars' formed by cosmic radiation when the emulsions were flown by balloon at a height of 21 km was examined for electron–positron pairs. The energy of the tracks of the pairs was determined by measurements of their multiple scattering. The direction of the parent γ-ray could be fixed to within 0.2°. An analysis of the γ-ray spectrum deduced from these measurements was then made on the assumption that these γ-rays arose from the decay of π^0-mesons. The fit to the observed spectrum was very good.

This experiment yielded a mass for the neutral π-meson of 150 ± 10 MeV$/c^2$. In confirmation of the interpretation of the photons as arising from π^0-decay, the π^0-spectrum was found to be the same as for charged pions emitted from these stars.

In this experiment it was also possible to set a somewhat lower limit on the lifetime of the neutral pion. We have seen that it was possible to fix the direction of a decay γ-ray, which produced an electron–positron pair, within 0.2°. If the π^0 has travelled a short distance from the star before decaying, the line of flight of the γ-ray will not pass through the star (see fig. 2.8). The distribution of the shortest distance between the line of flight and the star vertex is then a function of the π^0-energy spectrum, which is known, and of the π^0-lifetime. We should note that only the effect of the relativistic time dilatation allows a particle with a lifetime as short as that of the π^0 to travel any measurable distance. In fact, this experiment could only place an upper limit on the π^0-lifetime, which was, however, as short as 5×10^{-14} s.

Since this early work the existence of the π^0-meson has been confirmed in a multitude of experiments, perhaps the most striking of which are those where π^0 are produced in reactions in bubble chambers containing liquids of short radiation length (see section 1.3.2). Liquid xenon, for instance, has a radiation length of only 35 mm while other liquids, less expensive and more commonly used, have radiation lengths in the 0.1–0.3 m range. In such liquids the probability of both decay γ-rays producing pairs is quite large.

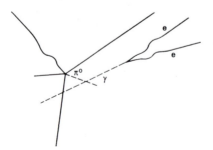

Fig. 2.8. In a cosmic ray star the line of flight of the γ-ray producing an electron pair may not pass through the vertex due to the finite lifetime (lengthened by time dilatation) of the π^0.

2.6 **Best determination of the mass and lifetime of the neutral π-meson**

The methods of the original experiment, described above, depend on the measurement of the absolute energy of one of the decay γ-rays. A more precise value may be obtained for the mass in relation to the mass of the π^--meson if

$$m_{\pi^-} - m_{\pi^0} > m_n - m_p.$$

In this case both the following reactions are possible:

$$\pi^- p \rightarrow n\gamma \qquad\qquad\qquad\qquad\qquad\qquad \textbf{2.2}$$
$$\pi^- p \rightarrow n\pi^0, \quad \pi^0 \rightarrow \gamma\gamma. \qquad\qquad\qquad\qquad \textbf{2.3}$$

We note that the γ-rays from reaction **2.2** for mono-energetic π^--absorption (in practice, absorption at rest) will themselves be mono-energetic, while for reaction **2.3** we expect a spectrum of γ-ray energies. If the mass of the π^0 is $2m_0$ and its velocity is $\beta_0 (c = 1)$, then a Lorentz transformation to the laboratory system gives for the energy of each γ-ray

$$E_\gamma = \frac{m_0 + m_0\beta_0 \cos\theta}{(1-\beta_0^2)^{\frac{1}{2}}}$$

where θ is the angle between the direction of motion of the π^0 and the γ-ray, in the π^0 system. The energy dependence is seen to be linear in $\cos\theta$, so that we expect a uniform distribution of γ-ray energies between $E_\gamma(\text{max})$ and $E_\gamma(\text{min})$, where the limiting values occur for γ-rays emitted along the direction, and opposite to, the π^0-motion. We can then write

$$\Delta E_\gamma = E_\gamma(\text{max}) - E_\gamma(\text{min}) = \frac{2m_0\beta_0}{(1-\beta_0^2)^{\frac{1}{2}}} = P_{\pi^0}.$$

Since the momentum of the neutron is equal to that of the π^0 we can thus obtain the kinetic energy of the neutron. Then writing

$$m_{\pi^-} - m_{\pi^0} = m_n - m_p + T_n + T_{\pi^0}$$

where T_n and T_{π^0} are the neutron and π^0 kinetic energies, and writing T_{π^0} as

$$[m_{\pi^0}^2 + P_{\pi^0}^2]^{\frac{1}{2}} - m_{\pi^0}$$

the values of T_n and P_{π^0} obtained from the γ-ray spectrum can be used to give $m_{\pi^-} - m_{\pi^0} = 5.4 \pm 1 \text{ MeV}/c^2$.

Concerning the π^0-lifetime, more accurate measurements have been made using a sample of decays into one γ-ray plus an electron pair (Dalitz pair). This decay mode occurs in only 1.2% of all cases, but the decay point of the π^0 is defined by the origin of the pair. The distances involved, however, are so small that the precision is still poor.

The most precise value for the π^0-lifetime has been determined indirectly by means of the Primakoff effect. This name is given to the process described by the following diagram (fig. 2.9(a)) where the π^0 is produced by interaction of an incident γ-ray with a virtual γ-ray of the Coulomb field of a nucleus. It is clear that the interaction described by the function η shown in the diagram is the same as that involved in the π^0-decay (fig. 2.9(b)). Thus it is possible that a measurement of the Primakoff effect *cross-section* may yield data from which the *lifetime* may be obtained. There are certain experimental and theoretical difficulties associated with this procedure. First, the cross-section for the Primakoff effect is expected to be quite small (of the order of 1 mb) and very forward-peaked. Moreover, in addition to production due to the Primakoff effect, there will also be present both coherent and incoherent photoproduction due to other nuclear effects. The Primakoff effect is identified by its characteristic angular behaviour, and the observed differential cross-section is then fitted as a sum of Primakoff and nuclear processes.

In the experiment of Bellettini *et al.* (1965) the π^0-photoproduction in lead, at 1 GeV γ-ray energy, was studied around the forward direction. The decay γ-rays were detected by Cerenkov counters in coincidence, each counter subtending a small solid angle of 4×10^{-3} steradians. Scintillation counters in front of the Cerenkov counters were placed in anticoincidence. Ten different counting channels were mounted on a high-precision frame and operated simultaneously. From the expected distributions, shown in fig. 2.10, for the Primakoff and nuclear production, it is clear that the important angular region is entirely below about 7°. Pulse heights from the coincidence counters were displayed on an oscilloscope and more than one million pulses were photographed. π^0-decays were unambiguously identified from the γ-ray energies and the requirement for a close coincidence in time. The observed angular distributions could be well fitted only by a combination of the Primakoff and nuclear production, and provided a measure of the Primakoff effect cross-section. The effect of the nuclear production was calculated with certain free parameters such as the nuclear radius. The fit to the data yielded values of these parameters as well as the value of the Primakoff cross-section, and thus of the π^0-lifetime. The Primakoff cross-section is proportional to the reciprocal of the lifetime, and the best fit to all the data yielded a value of $(0.73 \pm 0.1) \times 10^{-16}$ s. The value obtained for the quantity r_0, in the formula

$R = r_0 A^{\frac{1}{3}}$ for the nuclear radius, was 1.02 ± 0.07 fm, in satisfactory agreement with the values obtained for this quantity by other methods.

2.7   The spin of the charged meson

We will discuss the determination of the pion spin both for the reasons of the intrinsic importance of the quantity and also because of the interest of the method of analysis. We shall work always in the cms.

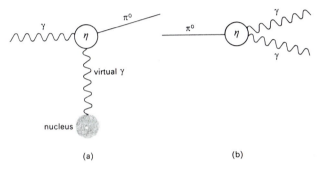

Fig. 2.9. (a) Illustration of the Primakoff effect where a π^0 is produced by the interaction of an incident γ-ray with the electromagnetic field of the nucleus. (b) Illustration of how the π^0-decay arises from the same kind of interaction.

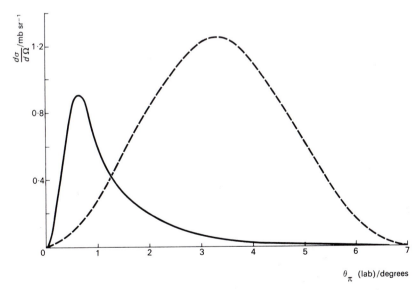

Fig. 2.10. Calculated angular distributions for production of π^0-mesons by the Primakoff process (solid line) and nuclear production (dashed line) (Belletini *et al.*, 1965).

For the π^+-particle the spin is determined by applying the principle of detailed balance to the reaction

$$\text{pp} \rightarrow \pi^+ d \qquad\qquad \textbf{2.4}$$

(which, at the energies under discussion, occurs approximately as frequently as the reaction

$$\text{pp} \rightarrow \pi^+ \text{pn})$$

and the inverse reaction

$$\pi^+ d \rightarrow \text{pp}. \qquad\qquad \textbf{2.5}$$

The principle of detailed balance then states that the number of transitions per unit time for the two reactions **2.4** and **2.5**, at the same centre-of-mass energy, will be equal. This number of transitions is given by the expression

$$\frac{2\pi}{\hbar} |M|^2 \frac{dN}{dE}$$

(this is often known as Golden Rule Number One – it can be proved by first order perturbation theory). M is the matrix element describing the interactions and dN/dE is the energy density of the final states. Often M can be taken to be independent of the momenta of the final particles; the argument here will not depend on any such special assumptions, but only on the result of the detailed balance theorem which implies also that

$$M_{\pi^+ d \rightarrow \text{pp}} = M_{\text{pp} \rightarrow \pi^+ d}.$$

Normally we are not interested in selecting a particular spin for the final state, and indeed most experiments will not measure this quantity. Thus, in evaluating the transition probability T we will sum over all the final spin states and write

$$T = \frac{2\pi}{\hbar} \sum_f |M_f|^2 \frac{dN}{dE},$$

where the summation is over all final states f.

This gives the total transition rate from all initial spin states, but each reaction proceeds from a particular spin configuration so that we must use an average of the transition probabilities from all the initial states, which will be all equally probable. For each of the initial particles of spin s there are $(2s+1)$ spin states, so that for the two initial particles having spins s_1 and s_2 we have altogether $(2s_1+1)(2s_2+1)$ initial states. Then the average over all states for the transition probability is

$$\frac{2\pi}{\hbar} [(2s_1+1)(2s_2+1)]^{-1} \sum_f |M_f|^2 \frac{dN}{dE}.$$

The cross-section is related to the transition probability in the following way:

Transition probability
= cross-section (σ)
× number of particles/unit volume
× relative velocity of interacting particles (v)

so that

$$\sigma = \frac{2\pi}{h}[(2s_1+1)(2s_2+1)]^{-1}\sum_f |M_f|^2 \frac{\mathrm{d}N}{\mathrm{d}E}$$

(unit volume/no. of particles) v^{-1}.

For a single particle in a volume V we have

$$\sigma = \frac{2\pi}{h}[(2s_1+1)(2s_2+1)]^{-1}\sum_f |M_f|^2 \frac{\mathrm{d}N}{\mathrm{d}E}\cdot V\cdot v_i^{-1},$$

where v_i is the relative velocity in the initial state.

Now we must consider the $\mathrm{d}N/\mathrm{d}E$ factor. Here we refer to a result familiar from statistical mechanics, that the number of states N for a particle with momentum in the interval between p and $p+\delta p$, in a box of volume V, is

$$\delta N = 4\pi p^2 \,\delta p \frac{V}{(2\pi h)^3},$$

so that

$$\frac{\mathrm{d}N}{\mathrm{d}p} = \frac{4\pi p^2 V}{(2\pi h)^3}.$$

But if ε_1 and ε_2 are the total particle energies, then

$$\frac{\mathrm{d}E}{\mathrm{d}p} = \frac{\mathrm{d}\varepsilon_1}{\mathrm{d}p}+\frac{\mathrm{d}\varepsilon_2}{\mathrm{d}p} = v_f \quad \left(\text{since } p^2 = \varepsilon^2 - m^2, \text{ so } \frac{\mathrm{d}\varepsilon}{\mathrm{d}p} = \frac{p}{\varepsilon} = v\right),$$

where v_f is the relative velocity of the particles in the final state. Thus

$$\frac{\mathrm{d}N}{\mathrm{d}E} = \frac{4\pi p^2 V}{(2\pi h)^3 v_f}.$$

Substituting, we obtain an expression for

$$\sigma = \frac{2\pi}{h}[(2s_1+1)(2s_2+1)]^{-1}\sum_f |M_f|^2 \frac{4\pi p^2 V^2}{(2\pi h)^3 v_i v_f}.$$

Now, at the same centre-of-mass energy, using the detailed balancing result that the matrix elements in both directions are equal, we write

$$M_{\pi^+\mathrm{d}\to pp} = M_{pp\to\pi^+\mathrm{d}}.$$

Thus we have

$$\sigma_{\pi^+d\to pp}(2s_{\pi^+}+1)(2s_d+1)v_{\pi^+d}v_{pp}(p_p)^{-2}$$
$$= \sigma_{pp\to\pi^+d}(2s_p+1)^2 v_{\pi^+d}v_{pp}(p_{\pi^+})^{-2},$$

so that

$$\sigma_{\pi^+d\to pp}(2s_\pi+1)\cdot 3 = \sigma_{pp\to\pi^+d}\cdot 4\cdot\left(\frac{p_p}{p_{\pi^+}}\right)^2$$

and

$$\sigma_{pp\to\pi^+d} = \sigma_{\pi^+d\to pp}\frac{3}{4}(2s_\pi+1)\left(\frac{p_{\pi^+}}{p_p}\right)^2$$

at any angle θ. Integrating over all angles, a factor of 2 must be introduced to take account of the indistinguishability of the two protons in the final state in the π^+d absorption reaction giving

$$\sigma_{pp\to\pi^+d} = 2\sigma_{\pi^+d\to pp}\frac{3}{4}(2s_\pi+1)\left(\frac{p_{\pi^+}}{p_p}\right)^2.$$

This last equation can now be used to determine the spin of the π^+.

We note first that even rather approximate values of the two cross-sections will yield the value of the spin, since we already know that this must be an integer. If we use the original data (Cartwright *et al.*, 1953) for the reaction $pp\to\pi^+d$ from experiments carried out at Berkeley at an incident proton energy of 340 MeV, we have a cross-section of 0.18 ± 0.06 mb. This energy for the incident protons corresponds to a meson energy of 22.3 MeV in the cms. For the reaction $\pi^+d\to pp$ the cross-section for π^+-mesons incident on deuterium, and having an energy of 29 MeV, is 3.1 ± 0.3 mb (Durbin *et al.*, 1951). This incident energy corresponds to a meson energy of 25 MeV in cms, sufficiently close to that in the proton–proton work mentioned above. To obtain the value of the ratio p_π/p_p we may use the non-relativistic relationships, since the energies are quite low. For a kinetic energy of 23 MeV for the pion, and a corresponding energy of 85 MeV for each proton in the cms, we obtain a ratio $p_\pi/p_p = 0.20$. Substituting in the equation we then have

$$(0.18\pm0.06) = (3.1\pm0.3)\times 1.5\times 0.04(2s_\pi+1).$$

Thus $(2s_\pi+1) = 1$ and the pion spin is 0. Data on the same reactions at other energies have confirmed this conclusion. A further indication of zero spin comes from the observation that the decay of charged mesons is found to be closely isotropic in the cms. We shall assume that the spin of the negative pion is the same as for the positive pion. This assumption is found to lead to consistent interpretations of a large variety of processes involving such particles. Discussion of the spin of the neutral pion is delayed until we have dealt with the subject of parity. We should, of course, expect that its spin also would be the same as for the charged pions.

We may note that our analysis depended on having an incident pion beam for the π^+p-absorption in which all spin states were equally probable. If, in fact, the pion spin was not zero, and if the pions in the beam were polarised, then this assumption would be invalid. It is, however, highly unlikely that the beam would be substantially polarised, and in such a way as to simulate the result for a beam of spin-zero particles.

2.8   **Resumé concerning parity and the parity of the charged π-meson**
    We shall deal with the significance of conservation laws in general in the following chapters. In order to complete the discussion of the pion properties, however, it is convenient to discuss the pion parity here. Such a discussion should also lend some degree of reality to the subsequent more general treatment.

The parity operator reverses each of the coordinate axes, i.e. it produces the transformation

$$x \to -x$$
$$y \to -y$$
$$z \to -z.$$

If a wave function subjected to this transformation is unchanged, then it is said to have positive parity, i.e.

$$P\psi(x, y, z) = \psi(-x, -y, -z) = +1\psi(x, y, z),$$

where P is the parity operator producing the above transformation. Alternatively we may have

$$P\psi(x, y, z) = \psi(-x, -y, -z) = -1\psi(x, y, z)$$

In the first case we say that the parity of the wave function is even, or sometimes we say it is $+1$, meaning that the eigenvalue of the wave function under P is $+1$, while in the second case we say that the parity is odd or -1. All physically-meaningful wave functions must fall into one or other category since $|\psi|^2$ must be invariant under the parity transformation. This property is clearly related to the symmetry of the wave function as is illustrated in fig. 2.11. *Invariance* of a system under the parity transformation implies that the eigenvalues of the wave function under this transformation are conserved quantities.

We may look on the parity transformation as consisting of two steps:
(a) rotation by 180° around the *x*-axis giving

$$y \to -y$$
$$z \to -z;$$

(b) reversal of the *x*-axis

$$x \to -x,$$

i.e. a 180° rotation followed by a 'mirror-image' transformation. As we shall see, conservation of angular momentum implies invariance under rotations so that the additional invariance implied by the parity transformation is invariance under mirror imaging.

In practice, of course, we are always concerned with systems consisting of at least two particles. In such a system the parity of the whole wave function can be considered as the product of the 'intrinsic' parities of the individual particles and the parity of the orbital angular momentum part of the wave function. In this we simply state that we can represent, for instance, the spatial wave function of a pion and a proton in the form

$$\psi(\mathrm{p})\psi(\pi)\psi(\text{orbital angular momentum } L \text{ of p and } \pi)$$

and then

$$P(\pi\mathrm{p}) = P(\pi) \cdot P(\mathrm{p})$$
$$\cdot P(\text{orbital angular-momentum wave function}).$$

Considering the orbital angular momentum part, we need only substitute $(\pi - \theta)$ for θ and $(\phi + \pi)$ for ϕ in the spherical harmonics required to describe the appropriate wave functions for s, p, d, ... states to see that the

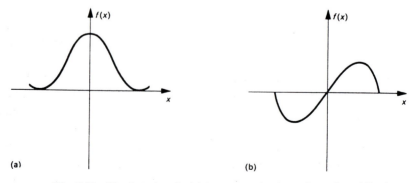

Fig. 2.11. The function in (a) is even under inversion of *x* while that in (b) is odd.

parity is given in terms of the orbital angular momentum quantum number L by $P(L) = (-1)^L$. For instance,

$$\text{for the} \begin{Bmatrix} s\text{-state} \\ p\text{-state} \\ d\text{-state} \end{Bmatrix} \psi(\text{orbital}) \begin{Bmatrix} \propto \text{constant} & L = 0 & \text{parity is } +1 \\ \propto \cos\theta & L = 1 & \text{parity is } -1 \\ \propto (3\cos^2\theta - 1) & L = 2 & \text{parity is } +1. \end{Bmatrix}$$

The intrinsic parities of particles, as we shall use this quantity, have meaning only relative to each other. We choose the parity of the proton to be positive. If the parity of the neutron is also positive then the parity of the deuteron, according to our rule described above, will be positive. Indeed, a knowledge of the deuteron wave function may be used to establish the neutron parity as positive.

We may now return to the question of the parity for the charged mesons. Consider the following processes where the meson is absorbed at rest by a deuteron:

$$\pi^- d \rightarrow nn \qquad \qquad \textbf{2.6}$$
$$\pi^- d \rightarrow nn\gamma \qquad \qquad \textbf{2.7}$$
$$\pi^- d \rightarrow nn\pi^0. \qquad \qquad \textbf{2.8}$$

We must consider first the angular momentum state of the π^--d system when the meson is absorbed at rest. In fact, it has been shown that the time for a π^- to reach the K-orbit in a π^--d atom is only $\sim 10^{-10}$ s. Also, direct nuclear capture of the π^-, even from the 2p-level, is negligible compared with the 2p to 1s transition. Thus all πs will be captured from the s-state in the π^-d atom before they can decay. This means that the parity of the initial state is simply the parity of the pion, while the total angular momentum of the initial state is 1.

For the reaction **2.6** above, the two-neutron state must be a p-state since the Pauli principle forbids a $^3S_1$ state. Thus the final state has odd parity and, assuming that the parity is conserved in the reaction (we shall see later that there is very good evidence that parity is conserved in all 'strong' reactions), the process **2.6** can take place only if the pion parity is odd.

The reaction **2.6** has been clearly demonstrated to take place by direct observation of the neutrons, which have a unique energy for this process. The ratio of **2.6** to **2.7** equals 2.35 ± 0.35. The reaction **2.8** is not observed, as is established by the absence of decay γ-rays from the neutral pion which would have an energy of about 70 MeV. The fact that this reaction does not take place is not surprising since only 2.3 MeV is available as particle kinetic energy in the final state and, since an $L = 1$ combination is required to give a total angular momentum of 1, the process is very unlikely. The reaction also turns out to be forbidden by parity conservation, as we shall see later.

The pion thus has spin parity 0^- and is called a pseudo-scalar particle.

2.9   Spin and parity of the neutral π-meson

For two-photon decay the photons must have equal and opposite momenta in the cms. No generality is lost if we consider both photons as circularly polarised and it will be convenient to do this in the first instance. Thus the final two-photon system can take the four forms shown in fig. 2.12 we use the letters l and r to signify left- and right-circular polarisation for the photon travelling in the $-z$ direction and L and R as the equivalent symbols for the $+z$ direction, where we have defined the z-axis as the direction of propagation. With this definition the photons will always have $J_z = \pm 1$ regardless of the total angular momentum J of the decaying particle. Thus, for the two photons the total z-component of the angular momentum is

$$J_z = \pm 2 \quad \text{for } rL \text{ and } lR: \text{parallel spins}$$
$$J_z = 0 \quad \text{for } lL \text{ and } rR: \text{opposed spins}$$

Since angular momentum is conserved, rL and lR may occur only for $J \geqslant 2$.

We now consider the following functions: rR, lL, rL, lR. We will examine their behaviour under transformations which should leave the system invariant. First we take a rotation of $180°$ about the x-axis so that the $+z$ and $-z$ directions are interchanged. This represents simply an interchange of the two γ-rays, which must leave the system as a whole unchanged. Then all angles θ measured from $+z$ become $(\pi - \theta)$, $l \leftrightarrow L$ and $r \leftrightarrow R$. This transformation conserves J and J_z. We may write the angular dependence of the initial and final states in terms of spherical harmonics $Y_J^{J_z}(\theta)$. Spherical harmonics, moreover, have the general property

$$Y_J^{J_z}(\pi - \theta) = (-1)^J Y_J^{J_z}(\theta).$$

It is then clear that J cannot be equal to 1 since in this case the system would not be invariant under the transformation in question. Thus two-photon decay is forbidden for vector or pseudo-vector particles.

Now let us consider the effect of the parity transformation. We have already seen that this corresponds to a rotation of $180°$ about the z-axis plus inversion of the z-axis. The result is therefore $R \leftrightarrow l, r \leftrightarrow L$. This means that the two photon wave functions transform as follows:

$$Rr \leftrightarrow lL, \quad Rl \leftrightarrow lR, \quad Lr \leftrightarrow rL, \quad Ll \leftrightarrow rR.$$

Thus, for the parity transformation the two-quantum wave functions Rr and Ll are *not* eigenfunctions. However, the combinations $Rr + Ll$ and $Rr - Ll$ are eigenfunctions. We will consider these functions for which zero is the lowest possible spin value, and also the corresponding functions $Rl + Lr$ and $Rl - Lr$ where 2 is the lowest possible value for the spin. Applying the parity transformation to these functions we see that

$$0_+ \rightarrow 0_+$$
$$0_- \rightarrow -0_-$$
$$2_+ \rightarrow 2_+$$
$$2_- \rightarrow 2_-,$$

where we have written 0_+ and 0_- for the symmetrical and antisymmetrical wave functions corresponding to lowest spin 0, and 2_+ and 2_- for the corresponding wave functions for lowest spin 2. Therefore, for systems of even parity only 0_+, 2_+ and 2_- photon wave functions can occur, while for odd parity only the 0_- function can occur.

If we reject the higher spins 2, 4, 6, ..., which lead to trouble in the field theory, and take the lowest value consistent with the data, we thus get the same spin for the neutral as for the charged pions, as we might expect. The parity is then seen to be allowed to be either $+$ or $-$ according to the above argument. A distinction between these two possibilities can be made by measurement of the polarisation of the decay γ-rays.

In order to see what is to be expected in the two possible cases, it is convenient to rewrite the wave functions in terms of plane-polarised photons. We may represent the circularly polarised photon by a sum of two wave functions with appropriate relative phase, one of which represents a photon plane polarised in the x-direction and the other a photon plane polarised in the y-direction. Thus we write

$$L = P_x + iP_y = R^* \quad \text{and} \quad l = p_x - ip_y = r^*,$$

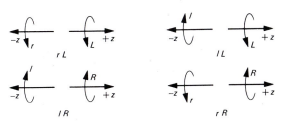

Fig. 2.12. The four possible polarisation states for a two-photon system.

where P and p are real. The various cases are symbolised in fig. 2.13. Then, in terms of the previous photon wave functions, we can write

$$
\begin{aligned}
0_+ &= lL + rR \\
&= (p_x - ip_y)(P_x + iP_y) + \text{complex conjugate} \\
&= 2\operatorname{Re}(lL) \\
&= 2(p_x P_x + p_y P_y).
\end{aligned}
$$

Similarly

$$
\begin{aligned}
2_+ &= 2(p_x P_x - p_y P_y) \\
0_- &= -2i(p_x P_y - p_y P_x) \\
2_- &= -2i(p_x P_y + p_y P_x).
\end{aligned}
$$

If the pion spin is zero (threshold behaviour of π^0-production is certainly consistent with this hypothesis although it cannot prove absolutely that it is true), then this last result affords a means of determining whether the π^0 is a scalar or pseudo-scalar particle. For even-parity the planes of polarisation of the decay photons are parallel, while for odd-parity they are perpendicular.

The polarisation of high-energy photons such as those from the π^0-decay is difficult to measure. However, the π^0-decay into two electron pairs affords the possibility of measuring the polarisation as a correlation between the planes of pairs. Such a correlation has been studied by Plano *et al.* (1959). The double-pair decay is very rare, occurring in only 1 in 30000 decays. In the experiment in question, π^0-mesons were produced by the capture reaction

$$
\pi^- p \rightarrow \pi^0 n.
$$

About one and a half million stopped π^- yielded about two hundred double-pair events. The theoretical correlation then has the form

$$
\begin{aligned}
I_0^+(\theta) &= 1 + A\cos 2\theta \quad \text{scalar } \pi^0 \\
I_0^-(\theta) &= 1 - A\cos 2\theta \quad \text{pseudo-scalar } \pi^0,
\end{aligned}
$$

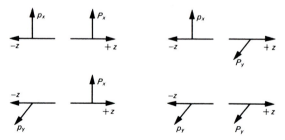

Fig. 2.13. Plane polarisation representations for a two-photon state.

where θ is the angle between the planes of the pairs and the coefficient A is a function of the angles and energy division in the pairs. Even with this relatively small number of events the results indicated quite unambiguously that the parity is odd.

A further evidence for odd parity comes from the considerable data in support of charge independence which requires that the pions form an isotopic spin triplet, the members of which must then have the same spin and parity.

We recall that in the discussion of the π^-d absorption process it was mentioned that the process

$$\pi^-d \rightarrow nn\pi^0$$

turned out to be forbidden by parity conservation. It was shown there that the initial state has the parity of the π^-, i.e. odd, while the final state involved orbital angular momentum $L = 1$. Thus the π^0 parity P_{π^0} would be given by

$$P_{\pi^0} \cdot +1 \cdot -1 = -1, \quad \text{i.e. } P_{\pi^0} = +1$$

if the reaction was observed to take place. This involves the assumption that the π^0 is in an s-state with respect to the n–n pair.

2.10 Isotopic spin for the π-meson

As with parity, we shall discuss this question from a more general point of view in chapter 3. However, it will be useful to have available the example of the π-nucleon system, so that we will deal briefly with the idea and application of isotopic spin, I-spin, at this point. We shall not develop here the complete I-spin formalism.

We first recall that in all properties except those associated with electric charge the neutron and proton are practically identical. It is therefore reasonable to denote these particles as two orientations of a vector in a space which we call charge space or isotopic spin space. The charge of the particle depends only on the *orientation* of the vector. The reason for the I-spin nomenclature is thus clear in that there is a precise analogy between this idea and the concept of ordinary spin. A particle having ordinary spin $\frac{1}{2}$ has two possible orientations, $+\frac{1}{2}$ and $-\frac{1}{2}$, for its angular momentum vector. Similarly the nucleon has two possible states of orientation of the isotopic spin vector, which are manifest as the proton and the neutron.

In angular momentum theory it is shown that the number of states for a system with angular momentum J is $2J+1$. The development of the theory for I-spin is identical, so that for the nucleon the number of states, which we know to be two, shows that the I-spin must be $\frac{1}{2}$. It is natural then to assign the neutron as the state with $I_3 = -\frac{1}{2}$ and the proton to $I_3 = +\frac{1}{2}$, where I_3 is the 'third' component of the isotopic spin. In this case the charge Q is seen to be related to I_3 by

$$Q = I_3 + \tfrac{1}{2}.$$

At this stage, of course, we have learned nothing new but have merely introduced an alternative notation for nucleons. This notation facilitates some interesting conclusions in nuclear physics, but here we proceed to deal with the I-spin of π-mesons before attempting to illustrate the usefulness of the concept in situations with two or more particles.

We know that there exist three π-mesons which are almost identical except for their charges. Thus we write $2I_\pi + 1 = 3$, so that $I_\pi = 1$ and we assign

$$(I_{\pi^-})_3 = -1, \quad (I_{\pi^0})_3 = 0, \quad (I_{\pi^+})_3 = +1.$$

Note that the charge–I-spin relationship is now apparently different from that for nucleons in that $Q = I_3$. This is one of the first symptoms that mesons are fundamentally different from nucleons. The situation is rationalised in an *ad hoc* manner by assigning a 'baryon number' (i.e. heavy particle number) B of $+1$ to the nucleons and 0 to the pions. Then we can write for nucleons *or* mesons

$$Q = I_3 + \tfrac{1}{2}B.$$

Now we make the important hypotheses:

(*a*) Strong interactions (those involving nucleons and mesons only) are independent of I_3, and thus of Q, and depend only on the total I-spin, I;

(*b*) the total I-spin is conserved in strong interactions; i.e. I is a 'good quantum number' in such processes.

We shall justify these hypotheses by comparison of the results obtained by their use with experiments. We note, of course, that I_3 must also be conserved, since Q and B are both conserved quantities.

2.11   Pion–proton scattering

The study of pion–proton scattering provides a first and most striking illustration of the usefulness of the I-spin concept.

We first consider all the possible I-spin states of the pion–nucleon system. The procedure is the same as for ordinary spin so that the total I-spin can either be $I = \frac{3}{2}$ or $I = \frac{1}{2}$. Thus the range of states for the system is:

$$(I, I_3) = (\tfrac{3}{2}, \tfrac{3}{2}); (\tfrac{3}{2}, \tfrac{1}{2}); (\tfrac{3}{2}, -\tfrac{1}{2}); (\tfrac{3}{2}, -\tfrac{3}{2}); (\tfrac{1}{2}, \tfrac{1}{2}); (\tfrac{1}{2}, -\tfrac{1}{2}). \qquad \textbf{2.9}$$

The first hypothesis then asserts that the first four states of **2.9** behave identically to each other as far as strong interactions are concerned, and that the latter two states also behave identically for strong interactions.

Now we must write these states in terms of the *I*-spin wave function of the pions and nucleons. The notation will be that the *I*-spin wave functions will be written as

$$(I_{\text{Pion}}[I_{\text{Pion}}]_3)(I_{\text{Nucleon}}[I_{\text{Nucleon}}]_3)$$

while the left-hand side of the equation will give the wave function for the total state in the corresponding notation. For the $(\frac{3}{2}, \frac{3}{2})$ state we have only one possible pion–nucleon combination: $(1, 1)(\frac{1}{2}, \frac{1}{2})$. For the state $(\frac{3}{2}, -\frac{3}{2})$ the only possible combination is $(1, -1)(\frac{1}{2}, -\frac{1}{2})$. For the other four states two possible pion–nucleon combinations can yield the required state which is thus expressed as the sum of the combinations, weighted with the appropriate probabilities which are given by the relevant Clebsch–Gordan coefficients (see appendix B). Thus we have

$$(\tfrac{3}{2}, \tfrac{1}{2}) = \sqrt{\tfrac{1}{3}}(1, 1)(\tfrac{1}{2}, -\tfrac{1}{2}) + \sqrt{\tfrac{2}{3}}(1, 0)(\tfrac{1}{2}, \tfrac{1}{2})$$
$$(\tfrac{3}{2}, -\tfrac{1}{2}) = \sqrt{\tfrac{2}{3}}(1, 0)(\tfrac{1}{2}, -\tfrac{1}{2}) + \sqrt{\tfrac{1}{3}}(1, -1)(\tfrac{1}{2}, \tfrac{1}{2})$$
$$(\tfrac{1}{2}, \tfrac{1}{2}) = \sqrt{\tfrac{2}{3}}(1, 1)(\tfrac{1}{2}, -\tfrac{1}{2}) - \sqrt{\tfrac{1}{3}}(1, 0)(\tfrac{1}{2}, \tfrac{1}{2})$$
$$(\tfrac{1}{2}, -\tfrac{1}{2}) = \sqrt{\tfrac{1}{3}}(1, 0)(\tfrac{1}{2}, -\tfrac{1}{2}) - \sqrt{\tfrac{2}{3}}(1, -1)(\tfrac{1}{2}, \tfrac{1}{2}).$$

We can now replace the brackets on the right by the names of the particles themselves, in order to exhibit more clearly the nature of the wave functions, although it should still be remembered that in this context the names represent the wave functions. Rewriting in this way we have

$$(\tfrac{3}{2}, +\tfrac{3}{2}) = \pi^+ p$$
$$(\tfrac{3}{2}, +\tfrac{1}{2}) = \sqrt{\tfrac{1}{3}}\pi^+ n + \sqrt{\tfrac{2}{3}}\pi^0 p$$
$$(\tfrac{3}{2}, -\tfrac{1}{2}) = \sqrt{\tfrac{2}{3}}\pi^0 n + \sqrt{\tfrac{1}{3}}\pi^- p$$
$$(\tfrac{3}{2}, -\tfrac{3}{2}) = \pi^- n$$

and

$$(\tfrac{1}{2}, +\tfrac{1}{2}) = \sqrt{\tfrac{2}{3}}\pi^+ n - \sqrt{\tfrac{1}{3}}\pi^0 p$$
$$(\tfrac{1}{2}, -\tfrac{1}{2}) = \sqrt{\tfrac{1}{3}}\pi^0 n - \sqrt{\tfrac{2}{3}}\pi^- p.$$

A little algebra then gives the *I*-spin wave functions for different pion–nucleon combinations.

$$\pi^+ p = (\tfrac{3}{2}, \tfrac{3}{2})$$
$$\pi^- p = \sqrt{\tfrac{1}{3}}(\tfrac{3}{2}, -\tfrac{1}{2}) - \sqrt{\tfrac{2}{3}}(\tfrac{1}{2}, -\tfrac{1}{2})$$
$$\pi^0 p = \sqrt{\tfrac{2}{3}}(\tfrac{3}{2}, \tfrac{1}{2}) - \sqrt{\tfrac{1}{3}}(\tfrac{1}{2}, \tfrac{1}{2})$$
$$\pi^+ n = \sqrt{\tfrac{1}{3}}(\tfrac{3}{2}, \tfrac{1}{2}) + \sqrt{\tfrac{2}{3}}(\tfrac{1}{2}, \tfrac{1}{2})$$
$$\pi^0 n = \sqrt{\tfrac{2}{3}}(\tfrac{3}{2}, -\tfrac{1}{2}) + \sqrt{\tfrac{1}{3}}(\tfrac{1}{2}, -\tfrac{1}{2})$$
$$\pi^- n = (\tfrac{3}{2}, -\tfrac{3}{2}).$$

We note that the π^+p and π^-n states are states of pure I-spin while the other pion–nucleon states consist of mixtures of I-spin $\frac{3}{2}$ and I-spin $\frac{1}{2}$ states.

On the basis of our earlier assumption we can now describe all the meson–nucleon scattering processes in terms of only two scattering amplitudes $A(\frac{3}{2})$ and $A(\frac{1}{2})$, compared with eight amplitudes in the case of no relationship between the different processes. We may now write the amplitudes for the processes

$$\pi^+p \to \pi^+p$$
$$\pi^-p \to \pi^-p$$
$$\pi^-p \to \pi^0n.$$

We use the result that the cross-section for any process is proportional to the square of the matrix element between initial and final states:

$$\sigma \propto |\psi_f|H|\psi_i|^2$$

where H is the appropriate matrix element.

Then using the I-spin wave functions as above we obtain

$$\sigma(\pi^+p \to \pi^+p) \propto |(\tfrac{3}{2},\tfrac{3}{2})H(\tfrac{3}{2})(\tfrac{3}{2},\tfrac{3}{2})|^2$$

for π^+–proton elastic scattering,

$$\sigma(\pi^-p \to \pi^-p) \propto |(\sqrt{\tfrac{1}{3}}(\tfrac{3}{2},-\tfrac{1}{2})H(\tfrac{3}{2})\sqrt{\tfrac{1}{3}}(\tfrac{3}{2},-\tfrac{1}{2}))$$
$$+ (\sqrt{\tfrac{2}{3}}(\tfrac{1}{2},-\tfrac{1}{2})H(\tfrac{1}{2})\sqrt{\tfrac{2}{3}}(\tfrac{1}{2},-\tfrac{1}{2}))|^2$$

where we have used the conservation of total I-spin, for π^-–proton elastic scattering, and

$$\sigma(\pi^-p \to \pi^0n) \propto |(\sqrt{\tfrac{2}{3}}(\tfrac{3}{2},-\tfrac{1}{2})H(\tfrac{3}{2})\sqrt{\tfrac{1}{3}}(\tfrac{3}{2},-\tfrac{1}{2}))$$
$$- (\sqrt{\tfrac{1}{3}}(\tfrac{1}{2},-\tfrac{1}{2})H(\tfrac{1}{2})\sqrt{\tfrac{2}{3}}(\tfrac{1}{2},-\tfrac{1}{2}))|^2.$$

for charge-exchange scattering. If we then write

$$A(\tfrac{3}{2}) = (\tfrac{3}{2},\ldots)H(\tfrac{3}{2})(\tfrac{3}{2},\ldots)$$
$$A(\tfrac{1}{2}) = (\tfrac{1}{2},\ldots)H(\tfrac{1}{2})(\tfrac{1}{2},\ldots)$$

we obtain

$$\sigma(\pi^+p \to \pi^+p) \propto |A(\tfrac{3}{2})|^2$$
$$\sigma(\pi^-p \to \pi^-p) \propto |\tfrac{1}{3}A(\tfrac{3}{2}) + \tfrac{2}{3}A(\tfrac{1}{2})|^2$$
$$\sigma(\pi^-p \to \pi^0n) \propto |\sqrt{\tfrac{2}{9}}A(\tfrac{3}{2}) - \sqrt{\tfrac{2}{9}}A(\tfrac{1}{2})|^2.$$

This result enables us to compare cross-sections for the three processes at the same angles and energies. Thus

$$\sigma(\pi^+p \to \pi^+p) : \sigma(\pi^-p \to \pi^-p) : \sigma(\pi^-p \to \pi^0n)$$
$$= |A(\tfrac{3}{2})|^2 : |\tfrac{1}{3}A(\tfrac{3}{2}) + \tfrac{2}{3}A(\tfrac{1}{2})|^2 : |\sqrt{\tfrac{2}{9}}A(\tfrac{3}{2}) - \sqrt{\tfrac{2}{9}}A(\tfrac{1}{2})|^2.$$

Further progress in predicting the ratios of the cross-sections depends on a knowledge of $A(\frac{3}{2})$ and $A(\frac{1}{2})$. We examine some special cases:

(a) $A(\tfrac{3}{2}) \gg A(\tfrac{1}{2})$ then

$$\sigma(\pi^+p \to \pi^+p) \; : \; \sigma(\pi^-p \; \to \pi^-p) \; : \; \sigma(\pi^-p \to \pi^0n)$$
$$= \qquad 1 \qquad : \qquad \tfrac{1}{9} \qquad : \qquad \tfrac{2}{9}$$

(b) $A(\tfrac{1}{2}) \gg A(\tfrac{3}{2})$ then

$$\sigma(\pi^+p \to \pi^+p) \; : \; \sigma(\pi^-p \to \pi^-p) \quad : \; \sigma(\pi^-p \to \pi^0n)$$
$$= \qquad 0 \qquad : \qquad \tfrac{4}{9} \qquad : \qquad \tfrac{2}{9}$$

(c) $A(\tfrac{3}{2}) = A(\tfrac{1}{2})$ then

$$\sigma(\pi^+p \to \pi^+p) \; : \; \sigma(\pi^-p \to \pi^-p) \quad : \; \sigma(\pi^-p \to \pi^0n)$$
$$= \qquad 1 \qquad : \qquad 1 \qquad : \qquad 0.$$

Now let us turn to the measured values. The experiments involve the production of a beam of positive or negative mesons, normally by means of a cyclotron. The π-meson beam is focused and analysed in momentum, by passing through a bending magnet followed by a suitable collimator, before being allowed to pass into a target of liquid hydrogen. In the elastic-scattering processes the charged meson and the proton are then detected by counter telescopes in coincidence, in which may be measured both the energy and the rate of loss of energy of the particles, so that they may be identified. In such a two-particle process, with a mono-energetic incident beam, the conservation of energy and momentum impose tight conditions on the energies and angles of the outgoing particles. The elastic-scattering events may thus be easily and unambiguously identified,

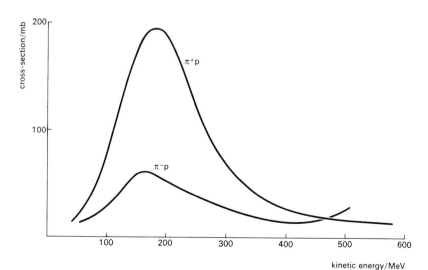

Fig. 2.14. Cross-sections for the scattering of π^+ and π^--mesons on protons showing the $\Delta(1232)$ resonance. The π^-p cross-section is the sum of the elastic and charge-exchange data.

especially in the region where the energy is not too high. In the charge-exchange scattering the π^0 is normally detected by observations on the decay γ-rays. The results for the total cross-section for π^+p and π^-p scattering as a function of the kinetic energy of the incident pion, are shown in fig. 2.14. The most striking feature of the data is the strong peak in the cross-section for a pion kinetic energy of about 180 MeV. In the region of this peak, or 'resonance', the cross-section values are found to be 195, 23 and 45 mb, in close agreement with the 9:1:2 ratio, indicating that this resonance arises from an interaction which takes place very predominantly in the $I = \frac{3}{2}$ state. The success of this analysis is a compelling proof of the correctness of the earlier hypotheses. Note that this result is certainly not apparent on the basis of our ordinary ideas concerning charge conservation and charge independence.

We shall return to a fuller discussion of the properties of this $I = \frac{3}{2}$ resonance in chapter 9.

3

Conservation laws

3.1   General features of conservation laws

In general terms we may say that a theory gives us the equation of motion for a system. Examples are the Schrödinger equation, the Dirac equation, Maxwell's equations, and the Lagrange equations. To solve a given problem it is necessary to integrate such differential equations of motion, and occasionally this can be done analytically. The equations are generally of second order in time, so that two integrations are necessary for their solution. In many cases the first integral gives a conservation law.

The most familiar of such equations are probably the Lagrange equations. For a system with Lagrangian $L = T - V$ the generalised momenta are given by

$$p_i = \frac{\partial L}{\partial \dot{q}_i}$$

and the Lagrange equations for the system are

$$\frac{\mathrm{d}}{\mathrm{d}t}\frac{\partial L}{\partial \dot{q}_i} - \frac{\partial L}{\partial q_i} = 0$$

where q_i are the generalised coordinates and T and V the kinetic and potential energy respectively. Now, suppose that a particular coordinate is missing from the Lagrangian of a system. In such a case the Lagrangian is independent of the quantity in question, or symmetrical with respect to transformations of this coordinate. For instance, the Lagrangian may be independent of θ, the angle of rotation about the z-axis, for fields symmetrical about this axis. Then

$$\frac{\partial L}{\partial q_i} = 0$$

and we have

$$\frac{\mathrm{d}}{\mathrm{d}t}\frac{\partial L}{\partial \dot{q}_i} = 0$$

or, since

$$\frac{\partial L}{\partial \dot{q}_i} = p_i, \quad \text{then} \quad \frac{\mathrm{d}p_i}{\mathrm{d}t} = 0$$

so that p_i is a constant. This means that the generalised momentum conjugate to a coordinate, with respect to which the system is symmetrical, is conserved. In the example already mentioned, if we write $q_i = \theta$ and if we take V to be independent of velocity, then for a particle of mass m

$$T = \tfrac{1}{2}m\dot{\theta}^2 r^2$$

so that

$$p_i = m\dot{\theta}r^2 = \text{angular momentum.}$$

Thus the statement that the angular momentum is conserved is equivalent to the statement that the system is invariant with respect to rotations about the relevant axis, or is symmetrical with respect to such an axis. Another way of describing such a symmetry is to say that it implies that there is some aspect of the physical system which is irrelevant to its behaviour.

Similarly for the conservation of linear momentum we may write the Lagrangian for motion of a particle of mass m along the x-axis in a potential V which is independent of x as (non-relativistic case)

$$L = \tfrac{1}{2}m\dot{x}^2 - V(y, z)$$

and the momentum $p_x = m\dot{x}$ so that the Lagrange equation gives p_x as constant. Thus we see that if the system is invariant under translation in the x-direction then the component of the momentum in the x-direction is conserved, while the converse is also true.

3.2   Conservation laws in quantum theory

It is useful to note some formal properties of the operators and wave functions of quantum mechanics under transformation.

If $\psi(\mathbf{r}_i)$ is a wave function in system 1, then in system 2 we can write a wave function $\psi'(\mathbf{r}_i)$ which is related to $\psi(\mathbf{r}_i)$ by an operator U such that

$$\psi'(\mathbf{r}_i) = U\psi(\mathbf{r}_i).$$

It can readily be shown that if ψ and ψ' are independently normalised to unity then U is a unitary operator, i.e.

$$U^+U = 1,$$

where U^+ is the Hermitian conjugate of U and

$$U^{-1} = U^+.$$

Now, associated with every observable we have an operator Q; for instance, with linear momentum along the x-axis we associate $-i\hbar(\partial/\partial x)$. Then the expectation value $\langle Q \rangle$ for the observable is

$$\langle Q \rangle = [\psi(\mathbf{r}_i)|Q|\psi(\mathbf{r}_i)] = \int \psi^* Q \psi \, d\tau.$$

In the transformed coordinate system we can get the expectation value $\langle Q \rangle'$ from the expression

$$\langle Q \rangle' = [\psi'(\mathbf{r}_i)|Q|\psi'(\mathbf{r}_i)]$$

and using the relationship between $\psi(\mathbf{r}_i)$ *and* $\psi'(\mathbf{r}_i)$

$$\langle Q \rangle' = [U\psi(\mathbf{r}_i)|Q|U\psi(\mathbf{r}_i)]$$
$$= [\psi(\mathbf{r}_i)|U^{-1}QU|\psi(\mathbf{r}_i)]$$

where we have used the definitions of Hermitian conjugate and unitary operators. Now if $\langle Q \rangle$ is invariant under the transformation induced by U, then $\langle Q \rangle = \langle Q \rangle'$ so that

$$U^{-1}QU = Q,$$

but

$$U^{-1}U = 1,$$

therefore

$$U^{-1}QU = QU^{-1}U,$$

i.e. Q commutes with U,

$$[Q, U] = 0.$$

In other words, if an operator is invariant under a coordinate transformation induced by the unitary operator U, then these two operators commute.

3.3   Parity conservation

We have already made some study of the parity transformation. Here, we may look at it from a somewhat different point of view, since it illustrates well the theory just described. If we write the Schrödinger equation as

$$H(\mathbf{r}_i)\psi(\mathbf{r}_i) = E\psi(\mathbf{r}_i)$$

where H is the Hamiltonian operator and E the energy eigenvalue, then it is clear that, for isolated systems, H and E do not change if we invert all the coordinates simultaneously through the origin, i.e. if we apply the parity transformation. Thus

$$H(\mathbf{r}_i) = H(-\mathbf{r}_i),$$

and replacing $\mathbf{r}_i$ by $-\mathbf{r}_i$ in the Schrödinger equation we get

$$H(\mathbf{r}_i)\psi(-\mathbf{r}_i) = E\psi(-\mathbf{r}_i)$$

so that $\psi(\mathbf{r}_i)$ and $\psi(-\mathbf{r}_i)$ satisfy the same equation. If there is no degeneracy (degeneracy may be removed formally by a small perturbation), then $\psi(\mathbf{r}_i)$ must be proportional to $\psi(-\mathbf{r}_i)$,

$$\psi(\mathbf{r}_i) = k\psi(-\mathbf{r}_i)$$

If we exchange $\mathbf{r}_i$ and $-\mathbf{r}_i$ we get

$$\psi(-\mathbf{r}_i) = k\psi(\mathbf{r}_i)$$

so that $k = \pm 1$.

We now see that the conservation of the parity of a system is a consequence of the equation of motion, and that it is true for systems obeying the Schrödinger equation. If we call the parity operator P and take a potential V which is symmetrical about the origin, so that the Hamiltonian H is invariant under the parity transformation, then P and H will commute. They thus have simultaneous eigenfunctions. The eigenvalues of P, ± 1, are also conserved, since we can write

$$ih\frac{\mathrm{d}P}{\mathrm{d}t} = [P, H]$$

and if $[P, H] = 0$ then $\mathrm{d}p/\mathrm{d}t = 0$ and the eigenvalues of P are constants of the motion. Here we have the analogy to our earlier classical treatment where we wrote

$$\frac{\mathrm{d}}{\mathrm{d}t}\left(\frac{\partial L}{\partial \dot{q}_i}\right) = \frac{\partial L}{\partial q_i}.$$

For the classical case, the requirement for conservation of a quantity was that the Lagrangian be independent of that quantity. The parallel requirement in the quantum case is that the operator associated with a conserved observable must commute with the unitary operator which induces the transformation.

3.4   Operators and transformations in quantum theory

Often we are interested in transformations associated with the actual operators of quantum mechanics. First, we require the connection between such operators (which are in fact Hermitian operators) and the unitary transformation operators, i.e. what unitary transformation is generated by a given Hermitian operator Q? These are related by the expression

$$U = e^{iQ} = 1 + iQ + \frac{(iQ)^2}{2!} + \frac{(iQ)^3}{3!} + \cdots.$$

Now consider the example of a transformation between two frames with only relative linear motion of constant velocity along the x-axis, so that the y- and z-coordinates are identical in both frames. The relativistic transformation equations (see appendix A) then give for the coordinates in the two frames

$$x_1^2 - t_1^2 = x_2^2 - t_2^2 \quad \text{(writing } c = 1 \text{ as usual)}$$
$$y_1 = y_2$$
$$z_1 = z_2.$$

We need a U such that $\psi' = U\psi$ or a Q such that

$$\psi' = \psi + iQ\psi + \frac{(iQ)^2}{2!}\psi + \cdots.$$

Consider frame 2 moved a small distance a relative to frame 1 from the moment when the origins are coincident, so that initially $x_1 = x_2$ and $t_1 = t_2 = 0$. Then, using Taylor's theorem, we have

$$\psi'(x, t) = U\psi(x, t) = \psi(x', t')$$

and

$$\psi(x', t') = \psi(x, t) + a\frac{\partial}{\partial x}\psi(x, t) + t\frac{\partial}{\partial t}\psi(x, t)$$

$$+ \frac{a^2}{2!}\frac{\partial^2}{\partial x^2}\psi(x, t) + \cdots.$$

Suppose now that ψ is constant in time so that $\partial\psi/\partial t = 0$. Then

$$\psi(x', t') = \psi(x, t) + a\frac{\partial\psi}{\partial x} + \frac{a^2}{2!}\frac{\partial^2\psi}{\partial x^2} + \cdots$$

and we obtain

$$U = 1 + a\frac{\partial}{\partial x} + \frac{a^2}{2!}\frac{\partial^2}{\partial x^2} + \cdots.$$

Now let a be a small unit distance along the x-axis so that we can write

$$U = 1 + \frac{\partial}{\partial x} + \frac{1}{2!}\frac{\partial^2}{\partial x^2} + \cdots$$

$$= 1 + i\left(\frac{1}{i}\frac{\partial}{\partial x}\right) + \frac{i^2}{2!}\left(\frac{1}{i^2}\frac{\partial^2}{\partial x^2}\right) + \cdots.$$

Thus we have

$$Q = \frac{1}{i}\frac{\partial}{\partial x}$$

corresponding, apart from the constant $\hbar$, to the linear momentum operator of quantum theory. Although we have considered only an infinitesimal transformation, we can perform a series of such transformations to achieve the finite transformation.

Consider now a displacement only along the t-axis. The derivative in question corresponding to Q is $i\hbar(\partial/\partial t)$, which we see to be proportional to the energy operator. We may recapitulate the steps in the argument linking symmetry under transformation and conservation laws as follows:

If we can show that $[Q, H] = 0$, then we can say that the eigenvalues of Q are constants, and we can see that the invariance of the eigenvalues of Q corresponds to symmetry under the transformation induced by Q. Thus, since the momentum operator commutes with H, the corresponding momentum eigenvalue is a constant, and the system is invariant under translation along the appropriate axis. The conservation of the energy eigenvalues corresponds to invariance under displacement in time.

3.5   Angular momentum conservation in quantum theory

We have already discussed the conservation of angular momentum in the classical case. We may apply the principle of the previous section to the quantum theory case. If we consider a rotation of the coordinate system by a small amount $\delta\phi$ about the z-axis then the wave function (x, y, z) becomes

$$\psi'(x, y, z) = \psi(x + \delta x, y + \delta y, z)$$

where

$$\delta x = y\,\delta\phi \quad \text{and} \quad \delta y = -x\,\delta\phi \quad \text{and} \quad \delta z = 0.$$

Using Taylor's theorem we can write

$$\psi'(x, y, z) = \psi(x, y, z) + y\,\delta\phi\frac{\partial\psi}{\partial x} - x\,\delta\phi\frac{\partial\psi}{\partial y} + \cdots$$

$$= \psi + \left(y\frac{\partial}{\partial x} - x\frac{\partial}{\partial y}\right)\psi\,\delta\phi + \cdots.$$

However, we have said that the angular momentum operator L_z is given by

$$-i\hbar\left(x\frac{\partial}{\partial y} - y\frac{\partial}{\partial x}\right)$$

and for an infinitesimal rotation we can neglect the higher-order terms and write

$$\psi'(x, y, z) \simeq \left(1 - \frac{i}{\hbar}L_z\delta\phi\right)\psi(x, y, z)$$

or, including the higher-order terms,

$$\psi'(x, y, z) = \psi(x, y, z) - \frac{i}{\hbar}L_z\delta\phi\psi + \frac{1}{2!}\left(-iL_z\frac{\delta\phi}{\hbar}\right)^2\psi\cdots$$

$$= e^{-iL_z\delta\phi/\hbar}\psi(x, y, z).$$

Compounding many infinitesimal rotations to obtain the rotation we then obtain

$$\psi'(x, y, z) = e^{-iL_z\phi/\hbar}\psi(x, y, z).$$

The general expression for a rotation θ about a unit vector n is then

$$\psi'(x, y, z) = e^{-i\theta\mathbf{L}\cdot\mathbf{n}/\hbar}\psi(x, y, z).$$

Thus $e^{-i\theta\mathbf{L}\cdot\mathbf{n}/\hbar}$ corresponds to the unitary transformation operator and the operator $-\theta\mathbf{L}\cdot\mathbf{n}/\hbar$ commutes with operators associated with invariants. The energy is invariant under rotation in a system symmetrical about the axis of rotation so that

$$\left[\frac{-\theta\mathbf{L}\cdot\mathbf{n}}{\hbar}, H\right] = 0$$

or, removing the numerical factor,

$$[L, H] = 0,$$

and the eigenvalues of L, i.e. the angular momenta, are conserved. Here we have neglected the effects of spin and also relativistic effects.

3.6 Conservation of isotopic spin

We shall deal here slightly more formally with this quantity, which we have already introduced in chapter 2.

The idea of isotopic spin was suggested for nucleons by Heisenberg shortly after the discovery of the neutron. Thus a nucleon is represented as a function of a dichotomic variable which can take one of two values,

$$f = \begin{bmatrix} a \\ b \end{bmatrix}.$$

The nature of such dichotomic functions is well known from the properties of particles of ordinary spin $\frac{1}{2}$. The standard linear operators which operate on such variables are the Pauli spin matrices and the unity operator. As usual, it is convenient to use the half-values of the operators. These operators and their combinations are the only ones which can act upon a dichotomic function. Thus we have

$$2\tau_1 = \begin{bmatrix} 0 & 1 \\ 1 & 0 \end{bmatrix} \quad 2\tau_2 = \begin{bmatrix} 0 & -i \\ i & 0 \end{bmatrix} \quad 2\tau_3 = \begin{bmatrix} 1 & 0 \\ 0 & -1 \end{bmatrix} \quad 2\tau_4 = \begin{bmatrix} 1 & 0 \\ 0 & 1 \end{bmatrix}.$$

We represent the proton as $\begin{bmatrix} 1 \\ 0 \end{bmatrix}$ and the neutron as $\begin{bmatrix} 0 \\ 1 \end{bmatrix}$. Then

$$2\tau_1 p = \begin{bmatrix} 0 & 1 \\ 1 & 0 \end{bmatrix}\begin{bmatrix} 1 \\ 0 \end{bmatrix} = \begin{bmatrix} 0 \\ 1 \end{bmatrix} = n$$

and similarly $2\tau_1 n = p$, so that τ_1 transforms $n \leftrightarrow p$, while the effects of τ_2 and τ_3 may be worked out as an example. We may also examine the operators $(\tau_4 + \tau_3)$ and $(\tau_4 - \tau_3)$:

$$(\tau_4 + \tau_3) = \begin{bmatrix} \frac{1}{2} & 0 \\ 0 & \frac{1}{2} \end{bmatrix} + \begin{bmatrix} \frac{1}{2} & 0 \\ 0 & -\frac{1}{2} \end{bmatrix} = \begin{bmatrix} 1 & 0 \\ 0 & 0 \end{bmatrix}$$

$$(\tau_4 + \tau_3)p = p \quad \text{and} \quad (\tau_4 + \tau_3)n = 0$$

and similarly

$$(\tau_4 - \tau_3)p = 0 \quad \text{and} \quad (\tau_4 - \tau_3)n = n.$$

Thus these operators will project out the proton or neutron parts respectively of a mixture of states. The step operators $(\tau_1 \pm i\tau_2)$ are also sometimes useful, as in the angular momentum case:

$$(\tau_1 \pm i\tau_2) = \begin{bmatrix} 0 & 1 \\ 0 & 0 \end{bmatrix} \quad \text{and} \quad \begin{bmatrix} 0 & 0 \\ 1 & 0 \end{bmatrix}.$$

They have the properties

$$(\tau_1 + i\tau_2)n = \tau_+ n = \begin{bmatrix} 0 & 1 \\ 0 & 0 \end{bmatrix}\begin{bmatrix} 0 \\ 1 \end{bmatrix} = \begin{bmatrix} 1 \\ 0 \end{bmatrix} = p$$

while

$$(\tau_1 - i\tau_2)p = \tau_- p = n.$$

The τs, of course, obey the usual commutation rules for spin operators.

If we have more than one particle we use the method of product spaces so that, for instance, two particles are described in a four-dimensional space and n particles in a 2^n-dimensional space. In the two-particle case we may write vectors

$$\mathrm{pp} = \begin{bmatrix} 1 \\ 0 \\ 0 \\ 0 \end{bmatrix} \quad \mathrm{pn} = \begin{bmatrix} 0 \\ 0 \\ 1 \\ 0 \end{bmatrix} \quad \mathrm{np} = \begin{bmatrix} 0 \\ 1 \\ 0 \\ 0 \end{bmatrix} \quad \mathrm{nn} = \begin{bmatrix} 0 \\ 0 \\ 0 \\ 1 \end{bmatrix}.$$

Since the I-spin for a nucleon is compounded of I_1, I_2 and I_3, which are proportional to the Pauli matrices, we can carry over all the normal rules of angular momentum to the I-spin 'space'. A properly-constructed Hamiltonian involving forces which are charge independent is invariant under rotations in the I-spin space. Such a Hamiltonian then commutes with the corresponding rotation operators, and the I-spin is a constant of the motion. If the Coulomb force for, say, a pair of protons, is included in the Hamiltonian, then H no longer commutes with the total I-spin. Here we have a clear example of an invariance principle which is true for the 'strong' interactions but not for the electromagnetic interactions. We have already examined the relationship between I-spin and charge. We can also treat this in terms of a charge operator, the eigenvalues of which for the nucleon system must be $+1$ and 0. We write

$$Q = (\tfrac{1}{2} + I_3)$$

so that $Q\mathrm{p} = +1\mathrm{p}$ and $Q\mathrm{n} = 0\mathrm{n}$.

For two nucleons we have $Q = Q_1 + Q_2 = 1 + (I_3^1 + I_3^2)$. For N nucleons

$$Q = \tfrac{1}{2}N + \sum_N I_3$$
$$= \tfrac{1}{2}N + T_3$$

where we have written T_3 for the sum of the I-spin third components. Since Q and N have conserved eigenvalues (N here is simply a 'number operator' for nucleons) so must T_3. The eigenvalues of T_3 are then $-T$, $-T+1, \ldots, +T$, where T is the eigenvalue of the total I-spin operator I^2, i.e.

$$I^2\psi = T(T+1)\psi.$$

Thus the operator Q has eigenvalue q where

$$q = -T + \tfrac{1}{2}N, -T+1+\tfrac{1}{2}N, \ldots, +T+\tfrac{1}{2}N.$$

The centre of charge of this isotopic-spin multiplet is $\frac{1}{2}N$. In fact, we shall see that it is always true for 'ordinary' (non-strange, non-charmed ... see later sections) particles that

centre of charge $= \frac{1}{2}$ (baryon number),

where the baryon number in the present case is simply the total number of nucleons, and will be discussed further in section 3.9.

For $I = 1$ we have an I-spin triplet with $I = -1, 0, +1$. In nuclei this triplet corresponds to the three neighbouring isobars having charges $\frac{1}{2}N-1$, $\frac{1}{2}N$, $\frac{1}{2}N+1$. For $I = \frac{1}{2}$ we have isotopic nuclear doublets with charges $\frac{1}{2}N-\frac{1}{2}$ and $\frac{1}{2}N+\frac{1}{2}$. Such an I-spin doublet is the pair $_3^7\text{Li}, _4^7\text{Be}$,

$$_3^7\text{Li} \quad 3 \text{ protons } q = 3, \; T = \tfrac{1}{2}, \; T_3 = -\tfrac{1}{2},$$
$$_4^7\text{Be} \quad 4 \text{ protons } q = 4, \; T = \tfrac{1}{2}, \; T_3 = +\tfrac{1}{2}.$$

In accord with charge independence, the ground and first excited states of the members of this doublet are indeed found to be closely similar when allowance is made for the Coulomb effects.

We have already seen that the existence of three charged states for the pion implies an I-spin of 1 for this particle. This is in accord with the requirement from the basic Yukawa process

$$N \rightarrow N + \pi \quad (\text{nucleon} \rightarrow \text{nucleon} + \text{pion}),$$

in which I-spin conservation would limit the pion I-spin to 0 or 1. Note that here also the relationship between charge I and baryon number is

$$Q = I_3 + \tfrac{1}{2}N,$$

and that the centre of charge is again given by $\frac{1}{2}N$, since for pions $N = 0$.

The I-spin operators for pions are then the same as those for a spin 1 system

$$\rho_1 = \frac{1}{\sqrt{2}} \begin{vmatrix} 0 & 1 & 0 \\ 1 & 0 & 1 \\ 0 & 1 & 0 \end{vmatrix} \quad \rho_2 = \frac{1}{\sqrt{2}} \begin{vmatrix} 0 & -i & 0 \\ i & 0 & -i \\ 0 & i & 0 \end{vmatrix} \quad \rho_3 = \begin{vmatrix} 1 & 0 & 0 \\ 0 & 0 & 0 \\ 0 & 0 & -1 \end{vmatrix}$$

while the I-spin wave functions are

$$\pi^+ = \begin{vmatrix} 1 \\ 0 \\ 0 \end{vmatrix} \quad \pi^0 = \begin{vmatrix} 0 \\ 1 \\ 0 \end{vmatrix} \quad \pi^- = \begin{vmatrix} 0 \\ 0 \\ 1 \end{vmatrix}.$$

It is easy to check the effects of these operators, e.g.

$$\rho_1 \pi^+ = \frac{1}{\sqrt{2}} \pi^0 \quad \rho_1 \pi^0 = \frac{1}{\sqrt{2}} (\pi^+ + \pi^-) \quad \rho_1 \pi^- = \frac{1}{\sqrt{2}} \pi^0.$$

We have already considered the problem of dealing with systems consisting of pions and nucleons which is exactly analogous to the combination of angular momenta $\frac{1}{2}$ and 1; we shall not develop this formalism further here.

3.7   The generalised Pauli principle

We have now studied the ideas which are necessary to make a generalisation of the Pauli exclusion principle. We recall that Pauli proposed that the electron structure in atoms was consistent with the hypothesis that no quantum state could be occupied by more than one electron. This result follows from certain anticommutation relations among the operators for the Dirac field appropriate for spin $\frac{1}{2}$ particles, and it is beyond the scope of the present discussion to deal with this aspect more deeply. However, we can show that the Pauli principle as stated above implies that only antisymmetric overall wave functions are possible for the electron or other fermion systems.

Formally this result can be seen to arise in the following way. We write a wave function for two identical particles in terms of quantum numbers for each particle, which we group as a vector $\mathbf{x}$. For particles 1 and 2 we write

$$\psi(\mathbf{x}_1, \mathbf{x}_2).$$

Since the physical situation described by the wave function must be the same when the identical particles are interchanged we have

$$|\psi(\mathbf{x}_1, \mathbf{x}_2)|^2 = |\psi(\mathbf{x}_2, \mathbf{x}_1)|^2$$

so that

$$\psi(\mathbf{x}_1, \mathbf{x}_2) = \pm e^{i\phi}\psi(\mathbf{x}_2, \mathbf{x}_1) \qquad \textbf{3.1}$$

and we write the unobservable phase factor equal to one, since if we repeat the interchange of particles we retrieve the original state and the phase factor becomes

$$e^{2i\phi} = 1 \quad \text{so that} \quad e^{i\phi} = \pm 1.$$

We now write $\psi(\mathbf{x}_1, \mathbf{x}_2)$ in terms of a linear superposition of product wave functions $\psi_1(\mathbf{x}_1)\psi_2(\mathbf{x}_2)$. Including a normalisation factor 2 we can write

$$\psi(\mathbf{x}_1, \mathbf{x}_2) = \frac{\psi_1\psi_2 + \psi_2\psi_1}{\sqrt{2}} \qquad \textbf{3.2}$$

and

$$\psi(\mathbf{x}_1, \mathbf{x}_2) = \frac{\psi_1\psi_2 - \psi_2\psi_1}{\sqrt{2}} \qquad \textbf{3.3}$$

both of which satisfy the relation **3.1** but the first of which is symmetric, since it does not change sign on interchange of particles 1 and 2, and the second is antisymmetric. However, if $\psi_1 \equiv \psi_2$ the expression **3.3** vanishes while **3.2** does not, so that satisfaction of the Pauli principle is seen to require a wave function which is antisymmetric. The argument may be extended to show that, for any number of identical particles, the wave function must be antisymmetric under interchange of any two.

For bosons, on the other hand, the total wave function is symmetric and the exclusion principle as such no longer holds.

These results concerning symmetry have some interesting consequences for the possible states of systems consisting of bosons or fermions. We consider a system consisting of two identical fermions of spin $\frac{1}{2}$. The wave function for such a system may be written as the product of a space-dependent part $\xi(\mathbf{r}_1, \mathbf{r}_2)$ and a spin-dependent part $\chi(S, m_1, m_2)$ where S is the total spin and m_1 and m_2 are the individual z-components, $\pm\frac{1}{2}$. Clearly, parallel spin states $(S = 1)$ are symmetric, while antiparallel spin states $(S = 0)$ are antisymmetric, as can be seen by writing **3.2** and **3.3** for the spin wave function:

$$\chi = \chi_1 \chi_2 = \frac{\chi_1 \chi_2 \pm \chi_2 \chi_1}{\sqrt{2}}.$$

The orbital angular momentum wave function we have already seen to be symmetric or antisymmetric according as the orbital angular momentum quantum number l is even or odd (section 2.8). Thus we may summarise the behaviour of the overall wave function under exchange of the two particles as

$$(-1)^l (-1)^{S+1}$$

and according to the Pauli principle we must make this antisymmetric, i.e. equal to -1, so that $(l + S)$ must be even. Thus parallel spin can exist only for odd orbital angular momentum, and vice versa. In the usual spectroscopic notation this means that permissible states for two spin $\frac{1}{2}$ particles are

$$^1S_0, {}^3P_{0,1,2} \, {}^1D_2 \cdots.$$

For spinless bosons, on the other hand, l must be even.

We may readily extend the wave function and the Pauli principle to include the isotopic spin. As with the ordinary spin, we can write an I-spin wave function $\phi(I, I_{1z}, I_{2z})$. The Pauli principle then yields

$$(-1)^l (-1)^{S+1} (-1)^{I+1} = -1,$$

so that $(l + S + I)$ must be odd.

One consequence of this rule is that a proton–neutron pair in an $I = 1$ state must have even $(l + S)$ and in an $I = 0$ state odd $(l + S)$.

3.8   Additive conservation laws

A somewhat different kind of conservation law from those discussed in the preceding sections of this chapter is the so-called 'additive' conservation law which requires that the algebraic sum of certain quantum numbers remains constant throughout all possible processes, or throughout interactions of a particular kind. The most familiar such quantum number, which is conserved in all known interactions, is electric charge. This is sometimes known as an 'internal' quantum number since it cannot be visualised as having any relation to an external coordinate system. As we shall see, there are other such internal quantum numbers.

In view of the relationship between symmetry under transformation and conservation laws which we have discussed already, it is natural to ask what transformations are involved for the additive conservation laws. The invariance is under the so-called gauge transformations of the first kind.

For such a transformation the wave functions of all the particles in the state in question are multiplied by a phase factor:

For the ith particle

$$\psi_i(x, y, z, t) \rightarrow e^{i\alpha}\psi_i(x, y, z, t)$$

where α is a real constant.

For such a transformation of the wave functions we require corresponding transformations of the field amplitudes (recall the Yukawa equation of chapter 2 involving the field amplitude ϕ). The transformations are

$$\phi_i(x, y, z, t) = e^{i\alpha A}\phi_i(x, y, z, t)$$

where A is a real number. It may be shown that invariance of the Lagrangian under such transformations leads to a continuity relationship for the particle and current densities. The continuity equation implies conservation of the algebraic sum of the appropriate quantum numbers. We return to a more-detailed study of gauge transformations in relation to the weak interaction in chapter 11.

3.9   Antiprotons and the conservation of baryons

An antiparticle to the electron (and thus also for other fermions such as the proton and neutron) was predicted by Dirac in 1930, as a consequence of his relativistic wave equation. Particle and antiparticle were predicted to annihilate into quanta of the interaction field. The antiparticle of the electron was observed by Anderson in a cloud chamber in 1932, during studies of the cosmic radiation, as a positively charged electron or positron. Subsequent studies of positrons have demonstrated that they annihilate with electrons into γ-rays.

The antiproton, according to the argument to be presented below, can only be produced in proton–nucleon collisions at energies greater than 5.6 GeV, while in a nuclear target the threshold is ~ 4.3 GeV due to the Fermi motion. Thus only when the Bevatron accelerator with an accelerated proton energy of 6 GeV came into operation could antiproton production in the laboratory be expected. The antiprotons were sought and found in an experiment carried out in 1955 by Chamberlain, Segré, Wiegand and Ypsilantis. The fundamental problem was to sort out the rather small number of antiprotons produced at this energy from the very much greater number of negative mesons produced in the bevatron target, a background which exceeds the number of antiprotons by a factor of about 10^5.

The experimental arrangement is shown in fig. 3.1. Only negative particles originating at the Bevatron target and having a momentum such that the magnets M_1 and M_2 guide them through the system, were counted. The selected momentum was 1.19 GeV/c. At this momentum pions have a velocity of $0.99c$, kaons of $0.93c$, and protons of $0.78c$. The particles were then distinguished by a measurement of their velocity by means of the scintillation counters S_1 and S_2 placed at beam foci produced by the quadrupoles Q_1 and Q_2. For protons, the time to traverse the 12 m between S_1 and S_2 was 51 ns (1 ns = 10^{-9} s), while for kaons it was 43 ns. Thus the S_1 signal was delayed by 51 ns, and coincidences sought between the delayed S_1 signal and the pulse from S_2. With a background of magnitude such as in this experiment additional means of discrimination are desirable, and were provided by the Cerenkov counters C_1 and C_2; C_1 is sensitive to all charged particles with velocity greater than $0.79c$. By the time an antiproton had passed through S_2, C_1 and C_2 its velocity was reduced to $0.765c$ and C_2 was a counter designed to count particles with velocity between $0.75c$ and $0.78c$. The signal for an antiproton is thus counts in S_1, S_2, S_3 and C_2 with no count in C_1 (conventionally written as $S_1 S_2 S_3 C_2 \bar{C_1}$) and an $S_1 - S_2$ delay of 51 ns. This system eliminates practically all forms of background. The result obtained is illustrated in fig. 3.2, where the solid curve gives the mass resolution as obtained by running protons through the system, while the triangles indicate the antiproton measurements.

In the years since this initial observation, the advent of higher-energy accelerators and the development of particle-beam technology have rendered the production of secondary beams of antiprotons standard practice in high-energy laboratories while p̄p collisions in storage rings have been used in experiments at CERN (see sections 1.2.2, 1.2.5 and 11.6.2).

We have already met the baryon number in our discussion of isotopic spin, where we saw that the charge was related to the third component of the *I*-spin by $Q = I_3 + \frac{1}{2}N$, where for nucleons $N = 1$ and for mesons $N = 0$. Before the discovery of some very heavy meson resonances, it was possible to define a baryon as a strongly interacting particle of mass equal to, or greater than, that of the nucleons. However, it is now known that there exist mesons and a lepton with masses greater than the nucleon mass, and this definition is no longer adequate. For 'ordinary' particles

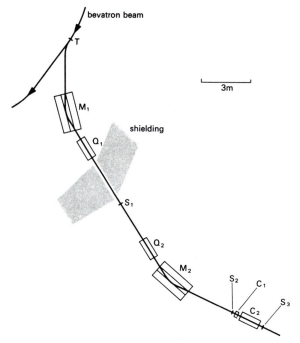

Fig. 3.1. The experimental arrangement used by Chamberlain, Segré, Wiegand and Ypsilantis (1955) in the first observation of antiprotons. M_1 and M_2 are bending magnets, Q_1 and Q_2 quadrupole focusing magnets, S_1, S_2 and S_3 plastic scintillation counters, C_1 a Cerenkov counter containing $C_8F_{16}O$ ($\mu = 1.276$) and C_2 a Cerenkov counter of fused quartz ($\mu = 1.458$).

(i.e. with zero strangeness, charm ... see chapters 5 and 12) if we know the *I*-spin for a multiplet of particles we can obtain the baryon number from the above relationship, in general however we depend on the conservation law for baryon number, which is experimentally shown to be true for nucleons and mesons and which is confirmed by, or found to be consistent with, all reactions involving the known particles. The conservation law operates as follows: baryon number +1 is assigned to the nucleons and −1 to the antinucleons. We then see that the conservation of the baryon number will allow some processes but forbid others. For instance, consider the production of antiprotons in proton–proton collisions. If

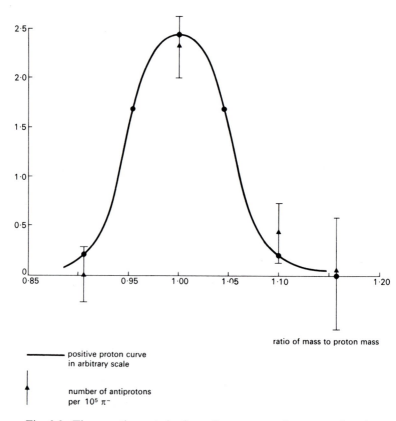

ratio of mass to proton mass

—— positive proton curve
in arbitrary scale

number of antiprotons
per 10^5 π^-

Fig. 3.2. The counting rate in the antiproton experiment as a function of particle mass, as determined by changing the momentum selection and keeping the velocity selection unchanged. The dot points were determined using protons. The antiproton points are seen to be consistent with the same curve.

baryon number need not be conserved, then antiprotons could be produced in the reaction

$$p + p \rightarrow p + \bar{p} + \pi^+ + \pi^+$$

the threshold for which is only 600 MeV. However, when baryon number conservation is required, the energetically-cheapest way of producing antiprotons is by means of the process

$$p + p \rightarrow p + p + p + \bar{p}$$

for which the threshold is 5.6 GeV kinetic energy for the incident proton. The fact that antiprotons are experimentally observed to be produced in the second reaction, but not in the first, is a confirmation of the baryon conservation law. Other examples may be quoted, such as in antiproton annihilation where

$$\bar{p} + p \rightarrow \bar{n} + n \quad \text{is allowed}$$
$$\bar{p} + p \rightarrow n + n \quad \text{is forbidden}$$
$$\bar{p} + p \rightarrow \pi^+ + \pi^- \quad \text{is allowed}$$
$$\bar{p} + p \rightarrow p + \pi^- \quad \text{is forbidden}.$$

As we shall see later, the law as applied to the nucleons, mesons and other particles is a consequence of the conservation of the quarks of which the particles are built (chapter 10).

We see that the assignment of baryon number -1 to the antinucleons is consistent with the charge–I_3 relationship. If we investigate the corresponding invariance to this conservation law, we find that it implies invariance under a gauge transformation of the field variables

$$\phi_i = e^{i \alpha B_i} \phi_i$$

where B_i is the baryon number for the ith particle, and B has been used rather than N to emphasise that the hyperons (see chapter 5) are also baryons.

3.10 Lepton conservation

The leptons may be defined as those particles specifically associated with 'weak' interactions (see chapter 4). Before the discovery of the τ-lepton the known leptons all had masses less than the π-meson mass, i.e. the electron, the muon, the neutrino and their corresponding antiparticles – hence the name, meaning light particle. Developments since the early days of the subject have demonstrated that the classification by mass – lepton, meson, baryon – does not have fundamental significance but the nomenclature has remained. As with the baryons, if we assign a 'lepton number' which is $+1$ for particles and -1 for the antiparticles (it does not matter which convention one adopts), then we find that this number is conserved for any system. We may state the law alternatively by saying that the difference between the number of leptons and the number of antileptons in any weak interaction is a constant. The corresponding gauge transformation is then

$$\phi_i = e^{i\alpha l_i}\phi_i$$

where l_i is the lepton number for the ith particle.

We may assign lepton numbers, in the first instance, by examining some of the observed processes, and we may then examine whether these assignments are consistent with all the processes observed and apparently forbidden. If we take the usual β-decay process

$$n \rightarrow p + e^- + \bar{v}$$

then the electron and the antineutrino must have opposite lepton numbers. Similarly for the decay

$$\pi^- \rightarrow \mu^- + \bar{v}$$

we see that the μ^- and the antineutrino have opposite lepton numbers. Let us assign lepton number $+1$ to the e^-, μ^- and v, and lepton number -1 to their antiparticles e^+, μ^+ and $\bar{v}$. Then we see that in the muon decay

$$\mu^+ \rightarrow e^+ + v + \bar{v}$$

the two neutrinos must be one antineutrino and one normal neutrino. We also note that the muon cannot decay into two leptons with conservation of lepton number.

It is clear that in fact the baryon and lepton numbers behave in exactly the same way as the electric charge of a particle. They are intrinsic properties of the particle and the total baryonic, leptonic and electric charges of any system are conserved to a high degree of precision at accessible energies.

3.11   Muon and electron conservation and the two neutrinos

As we shall see below, it has been conclusively shown that there exist two distinct varieties of neutrino, one of which is always associated with electrons, and the other always with muons. We may thus make the conservation law for leptons even more specific, and state that the total 'electron' number and the total 'muon' number are separately conserved. By electron, in this context, we mean electron, positron, electron neutrino and electron antineutrino, and by muon we mean positive and negative μ-meson with the corresponding muon neutrino and antineutrino. The difference between these two neutrinos, and the corresponding separate muonic and electron number conservation laws, can only be established by means of a study of neutrino absorption, and not from decay processes, since in decay processes we have no means of checking the nature of the neutrino produced.

An experiment which conclusively established the existence of two varieties of the neutrino was performed at the Brookhaven Laboratory proton synchrotron in 1962. The principle of this experiment was to establish whether neutrinos arising from the decays

$$\pi^+ \to \mu^+ + \nu, \quad \pi^- \to \mu^- + \bar{\nu}$$

would, when absorbed by protons or neutrons, give rise only to muons or to muons and electrons according to the reactions

$$\nu + n \to p + \mu^- \quad \nu + n \to p + e^-$$
$$\bar{\nu} + p \to n + \mu^+ \quad \bar{\nu} + p \to n + e^+.$$

The cross-section for the absorption reactions was expected to be very small, of the order of 10^{-38} cm^2, on the basis of weak interaction calculations. This cross-section is for neutrinos having an energy of about one to two thousand million electron volts. At lower energies the cross-sections are even smaller. Thus in order to study these processes one requires a very high flux of neutrinos having a high energy. The advent of the very high energy accelerators at Brookhaven and at the European Laboratory CERN, at Geneva, made it possible to produce beams of neutrinos satisfying these conditions. The arrangement used in the original experiment at Brookhaven was one in which the circulating proton beam of intensity about 2×10^{11} protons per second, was allowed to strike a target of beryllium in a 3 m straight section. An intense flux of pions (as well as other particles) was emitted from the target, predominantly in the forward direction. The pions were allowed to decay along a 21 m flight path, and the muons and neutrinos from this decay were also collimated in the forward direction due to the centre-of-mass motion. About 10% of the pions decayed in the flight path. At the end of the flight path all particles except the neutrinos were stopped by a wall consisting of 13 m of solid iron. The neutrinos, being so weakly interacting, were largely unaffected by this material and passed into an arrangement of large spark chambers, with their lines of flight undeviated from the original direction. The spark chambers used in the Brookhaven experiment had dimensions 1.3 m × 1.3 m × 0.3 m in thickness, and were each composed of nine 25 mm-thick aluminium plates. The arrangement is shown schematically in fig. 3.3. The total mass of material in the chambers was about 10^4 kg. Since the cross-section for neutrino interactions is so small, it is essential to avoid even very weak backgrounds arising from cosmic rays or other sources. This was achieved by two methods. First, the spark chambers were surrounded by scintillation counters in anticoincidence with the triggering of the scintillation counters placed between the spark chambers. This arrangement eliminated to a large extent the cosmic-ray background and also any very energetic muon penetrating the shield. The requirement for triggering was a count in any two of the internal counters, with no count in the surrounding anticoincidence counters. Second, the chambers were switched on only for the very brief period when particles were actually available from the target of the synchrotron. In fact, the beam pulse lasted only 25 µs, and within

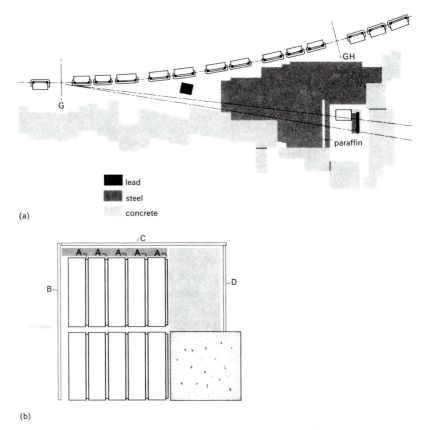

(a)

(b)

Fig. 3.3. (a) Plan view of the arrangement for the Brookhaven experiment showing part of the accelerator, the shielding and position of the spark chamber arrangement. (b) The spark chamber and counter arrangement in the experiment. *A* are the trigger counters, *B*, *C* and *D* are anticoincidence counters (Danby *et al.*, 1962).

this there was a structure consisting of 12 bunches, each of which was about 20 ns long, separated by 220 ns. The chambers were then gated by means of a signal from a Cerenkov counter looking at the synchrotron target, and were opened for only 3.5 µs per pulse. This resulted in a most remarkable feature of the experiment in which, although 1.7×10^6 pulses were accepted during the experiment, the total sensitive time of the chamber was only 5.5 s, thus effecting a most useful limitation of cosmic-ray background. Such 'gating' of the detection equipment is now standard in experiments.

The results of the experiment were the observation of over a hundred events in the spark chamber. The problem was then to determine (a) whether these events are in fact due to genuine neutrino absorptions and, if so, (b) whether the absorptions resulted in the production of muons and of electrons, or of muons only. We first consider possible production of events by cosmic rays. The apparatus was operated for a period while the accelerator was not running, in order to give a measure of the expected number of cosmic-ray background events. This indicated that about five events were to be expected in which a long track, having a momentum greater than 300 MeV/c if it were a muon, was produced. In fact, 34 such events were observed in the experiment. Secondly, it was necessary to test the possibility that the events might be due to neutrons which had somehow penetrated the shield. In fact, in the early part of the run there was some indication of such events, originating in a certain section of the chamber, due to neutrons which had leaked past the shield; they were eliminated in the second section of the run by the addition of extra shielding. These gave rise to short-track events. For the long tracks, and an additional 22 events showing a vertex, the evidence against production by neutrons is (a) there was no attenuation in the production of events along the length of the chamber, and (b) the removal of 1.2 m of iron from the main shielding did not produce any observable increase in the event rate, where a substantial increase would have been expected for neutrons in the beam direction. That the events were produced by pion- (or kaon-, see later) decay products was established by removing 1.2 m of iron from the shielding wall and replacing it by lead at a distance of only 1 m from the target, thus maintaining the same total interaction path but reducing the flight path of the pions so that 90% of them were eliminated by interaction before they had time to decay. The event rate was found to be reduced to 0.3 ± 0.2 events per 10^{16} protons, compared with the previous number of 1.46 ± 0.2 events per 10^{16} protons, quite consistent with the expected numbers if the events were due to the neutrinos from pion decay.

Having established that the events were indeed due to neutrino absorptions, the problem was to identify the secondary tracks. If the secondary tracks were muons they should not interact in the spark chamber plates. In the second half of the run, where the background had been minimised, 8.2 m of aluminium was traversed by single tracks, no case of nuclear interaction being observed. Had these tracks been π-mesons a total of eight nuclear interactions would have been expected on the basis of calibration experiments. In addition, had the charged tracks been π-mesons one would have expected also about 15 neutral pions. No such neutral pions were observed. The chamber sensitivity to electron showers was calibrated by putting electrons of various energies into the chamber. These produced fairly characteristic short tracks with discontinuities. In the clean part of the run no such showers were observed.

Thus it seems clear that the absorption of these muon-associated neutrinos gives rise to production of muons, while, if there is only one variety of neutrino, we might have expected equal numbers of muons and electrons.

Both the theoretical interpretation and the quantity and quality of the experimental data have been vastly improved since the first experiment. Detailed consideration of the possible interactions for the absorption mechanism has produced no theory which is consistent with the experimental data and would allow an explanation in terms of only one type of neutrino. Following the Brookhaven experiment a more elaborate experiment was performed at CERN in which the flux of neutrinos achieved was very much improved, due particularly to the use of an extracted proton beam from the accelerator. This beam was allowed to produce pions in a target placed within a special magnetic focusing device (the magnetic 'horn') which concentrated into the forward direction pions having a wide range of momenta and angles of emission. In addition, the detectors used in this work were more sophisticated, consisting of a spark-chamber arrangement in which the tracks were bent in a magnetic field, facilitating momentum measurement, and in addition a large heavy-liquid bubble chamber. Although a larger number of events was observed in the spark chamber, those observed in the bubble chamber allowed a very complete analysis. The data obtained confirmed the original two-neutrino result and allowed a more detailed study of the neutrino absorption process.

3.12   The τ-lepton

A new lepton, named the τ, was discovered in the study of $e^+ - e^-$ interactions in the SPEAR storage ring at the Stanford Linear Accelerator

Centre (SLAC) in 1975 (Perl *et al.*, 1975). The initial evidence for the τ was the observation of 24 events, each of which included both a high-energy electron and a high-energy muon.

After the analysis and exclusion of a number of possible backgrounds these events were interpreted as due to the production and decay of pairs of a new lepton

$$e^+ + e^- \rightarrow \tau^+ + \tau^-$$
$$\quad\quad\quad\quad\quad\mathrel{\mathop{\vert}}\mathrel{\mathop{\vert}}$$
$$\quad\quad\quad\quad\quad\quad\quad \hookrightarrow \mu^- \bar{\nu}_\mu \nu_\tau \quad\quad\quad\quad \textbf{3.4}$$
$$\quad\quad\quad\quad\quad \hookrightarrow e^+ \nu_e \bar{\nu}_\tau.$$

The threshold for such events was found to be 3.56 GeV, leading to a mass of 1.78 GeV/c^2 for the τ. These purely leptonic decays establish the τ as itself a lepton. The spectrum of energy observed for the electrons and muons requires the presence of more than one neutral particle in the τ-decay and our knowledge of lepton conservation suggests the neutrino suffixes of equation **3.4**. Of course at this stage it is not clear that there should exist a neutrino specifically associated with the τ. Measurement of the τ-lifetime (see below) is not consistent with coupling to ν_e and ν_μ. A direct verification using interactions of neutrinos from τ-decay presents formidable difficulties.

Since its first detection via μ- and e-decays the τ-lepton has been observed in a variety of other decay modes. In particular, significant decay channels are

$$\tau^- \rightarrow \pi^- \nu_\tau$$
$$\pi^- \pi^0 \nu_\tau$$
$$\pi^- \pi^- \pi^+ \nu_\tau$$
$$\pi^- \pi^- \pi^+ (+\pi^0 \text{s}) \nu_\tau.$$

The *lifetime* of the τ has been measured using a large cylindrical spectrometer (Mark II) at the PEP colliding-beam machine at SLAC. The detector allows the reconstruction of particle tracks and the lifetime was measured by reconstructing the vertex of three-pronged decays such as $\tau^- \rightarrow \pi^- \pi^- \pi^+ \nu$ and calculating the flight distance between the interaction point and the vertex. The mean flight distance is, in fact, less than the resolution, so that statistical averaging and a good understanding of the systematic errors is necessary. The events selected were almost all τ-pair production

$$e^+ + e^- \rightarrow \tau^+ + \tau^-$$

and, since the e^+e^- centre-of-mass energy was 29 GeV, each τ had an energy of 14.5 GeV. Since the τ-velocities are known, the distribution of flight distance can be used to obtain the lifetime. The result was

$$\tau_\tau = (4.6 \pm 1.9) \times 10^{-13} \text{s}.$$

The mean flight distance was ~ 0.7 mm.

The energy spectrum of the decays and the threshold behaviour of the cross-section favour a spin of $\frac{1}{2}$ for the τ-lepton as also for the e and the μ.

4

Strong, weak and electromagnetic interactions

4.1 Types of interactions

In chapter 2 we have seen that the nuclear interaction is very much stronger than the electromagnetic interaction; their relative strengths are characterised by the difference in the 'coupling constants' which are discussed in more detail below. We may recognise at this stage two other forms of interaction, gravitational forces and the so-called 'weak' interaction. One of the most notable milestones of recent years has been the unification of the weak and electromagnetic interactions as different aspects of the same underlying force, in the theory of Glashow, Weinberg and Salam, and the discovery of the W and Z particles, the additional quanta of the unified interaction, with the properties predicted. This topic is treated in more detail in chapter 11.

We may summarise as follows:

Strong interactions are responsible for the interactions between nucleons, nucleons and mesons and a number of other particles. The mesons act as the quanta of the strong interaction on the nuclear scale. These interactions reflect the interaction between quarks due to the exchange of gluons on the sub-nucleon level.

Electromagnetic interactions are responsible for the force between electrically-charged particles and are mediated by the exchange of photons.

Weak interactions are responsible for many particle decays such as radioactive decay (the basic process $n \rightarrow p + e^- + \bar{v}_e$), pion and muon decay and a number of other decay processes.

Gravitational interactions exist between all particles having mass and are believed to be mediated by the so far undetected gravitons. Although the gravitational interactions are familiar and important on the macroscopic scale this force is unimportant for elementary particle phenomena.

In general, the uncertainty principle argument of section 2.1 demonstrates that the shorter the range of the force the heavier will be the associated exchange quantum. For the infinite range, inverse square law electromagnetic and gravitational forces the quantum is massless. The heavy quanta associated with the weak interaction reflect a very short range force.

4.2   Coupling constants: dimensions

In order to compare the nature of observed processes we frequently wish to determine the coupling constants involved. The processes in question are either reactions between bombarding and target particles, for which we can measure the cross-section, or the decay of particles, for which we can measure the lifetime. We shall treat these processes in more detail below, but it is instructive first to examine the dimensions to be expected for the coupling constants. A simple and useful description of processes for this purpose was given by Jauch (1959).

We note first that in considering interactions it is necessary to distinguish between bosons and fermions, since bosons can be absorbed or emitted singly but fermions only in pairs. If we then write the symbol ϕ to indicate boson emission or absorption and the symbols $\psi, \bar{\psi}$ to indicate fermion emission or absorption, we can represent any process in terms of these quantities. Thus, for instance, for a process involving one boson and two fermions, such as the decay

$$\pi \rightarrow \mu v$$

we may draw a diagram as in fig. 4.1. g is the coupling constant measuring the interaction strength, and we can write a Hamiltonian density for the process as $g\phi\psi\bar{\psi}$. The notation is useful either if we know the actual forms of ϕ and ψ or in comparing reactions, even where the actual forms are not known.

Fig. 4.1.

We can obtain the dimensions of g if we write the general term for an interaction as

$$g\phi^B(\psi\bar\psi)^F$$

(where B is the number of bosons and F the number of fermion pairs) and examine the dimensions of ϕ and ψ. Using units such that $\hbar = c = 1$, leaving only the free dimension L then we have for mass the dimensions:

$$\text{(joule seconds)} \, \text{(velocity)}^{-2} \, \text{(time)}^{-1}: (\hbar)(cL)^{-1} = L^{-1}.$$

Similarly, both momentum and energy have dimension L^{-1}. We now need two results from field theory, namely that the energy density in a boson field, due to the self-energy of a particle of mass m, is $(m^2\phi^2)$ and correspondingly in a fermion field it is $(m\psi\bar\psi)$. In our system of units, energy density has the dimensions of L^{-4}, so that ϕ^2 has dimension L^{-2}, and ϕ dimension L^{-1}, while $\psi\bar\psi$ has dimension L^{-3}, and ψ has dimension $L^{-\frac{3}{2}}$. Using the above results we can write

$$g\phi^B(\psi\bar\psi)^F = gL^{-B-3F} = L^{-4}$$

so that g has dimension $L^{-4+B+3F}$. Thus, for the pion-decay interaction, g has dimension L^0. The electromagnetic interaction is also like this; for instance, pair production corresponds to the same diagram, where ϕ represents the photon, ψ and $\bar\psi$ the electron–positron pair, and where g is again dimensionless, being given by $\frac{1}{137}$. The Yukawa process $N \leftrightarrow N + \pi$ is also of this form. The only other dimensionless coupling constant is ϕ^4.

4.3   Coupling constants and decay lifetimes

The problem we shall consider in this section is how to calculate, from the measured value of the decay lifetime, the coupling constant effective in a decay process. We start from the usual expression for a transition rate

$$R = 2\pi|M|^2\frac{\mathrm{d}N}{\mathrm{d}W} \qquad\qquad \textbf{4.1}$$

where we have used units $\hbar = c = 1$, $|M|^2$ is the square of the modulus of the appropriate transition matrix element and where $\mathrm{d}N/\mathrm{d}W$ is the density of final states. For interactions of the type involving one boson and two fermions $(g\phi\psi\bar\psi)$ it can be shown that M, the non-Lorentz invariant matrix element (see appendix A), may be written as

$$\frac{gSV^{-\frac{1}{2}}}{\sqrt{(2e_B)}}$$

where e_B is the boson energy and S is a factor depending upon the spins of the particles involved. This spin factor we shall set equal to unity for the purpose of the present calculation, which is intended to obtain the order of magnitude of g rather than any precise value. V is a normalisation volume and the factor $V^{-\frac{1}{2}}$ arises since we must ensure that

$$\int \psi^* \psi \, d\tau = 1.$$

From equation **4.1** we can see that the factor $1/\sqrt{(2e_B)}$ is necessary, for dimensional reasons, in view of our arguments concerning the dimensions of the coupling constants. It can be justified properly only by deeper analysis than can be given here.

Using the above result we then obtain the probability of, for instance, the decay $\Lambda \rightarrow p + \pi$

$$\frac{g^2 \pi}{V e_\pi} \frac{dN}{dW}$$

or for the lifetime

$$\tau = \frac{e_\pi V}{g^2 \pi} \left(\frac{dN}{dW}\right)^{-1}. \tag{4.2}$$

We require an expression for the phase-space factor, which for the two-particle decay can be written (see appendix A),

$$\frac{dN}{dW} = \frac{V p^2}{2\pi^2 (v_1 + v_2)} \tag{4.3}$$

where p is the cms momentum of the decay products, and v_1 and v_2 are the velocities of the products, also in the cms. If we write M_0 as the mass of the decaying particle, and e_1 and e_2 as the energies of the decay products in the cms, then we obtain for the phase-space factor

$$\frac{dN}{dW} = \frac{V p e_1 e_2}{2\pi^2 M_0}.$$

When substituted into the equation **4.2** this yields an expression for the lifetime

$$\tau = \frac{4\pi}{g^2} \frac{M_0}{2pe_p}$$

where e_p is the decay-proton energy. This is sometimes written in units of $\tau_0 = \hbar/m_e c^2 \simeq 10^{-21}$ s, giving the result

$$\tau = \tau_0 \frac{4\pi m_e M_0}{g^2 2pe_p}.$$

This relationship enables us to calculate the value of the coupling constant from the measured value for the lifetime. If we take the observed value for the lifetime of the lambda decay $\Lambda \rightarrow p\pi$ to be 2.5×10^{-10} s, then we obtain a value for $g^2/4\pi$ of 1.5×10^{-14}. As we shall see later, very similar values are obtained for a number of other decays involving one boson and two fermions.

4.4 Coupling constants and reaction cross-sections

We may also study coupling constants by measuring reaction cross-sections. However, in such situations we have at least two particles in the final state as well as two in the initial state. Thus the form of the interaction for processes such as pion–proton scattering is $\phi^2\psi\bar{\psi}$, since we have two bosons and two fermions. The dimensions of the coupling constants are then seen to be L^{+1}. The matrix element now has the form

$$M = \frac{g}{(2e_{B_1})^{\frac{1}{2}}(2e_{B_2})^{\frac{1}{2}}V}$$

where e_{B_1} and e_{B_2} are the boson energies. Again the energy factors can be rigorously obtained only from field theory, although, as before, we can see that such factors are necessary for dimensional reasons. V represents the normalisation volume. In this case what we can measure is the value of the cross-section, so that we must derive a relationship between the cross-section and the transition rate. This may be done in the following way. Suppose that the target particle is located inside the volume V. The time spent by the projectile in the volume V is L/v where V is taken to be a cube of side L. The 'number of transitions' is then given by

$$N = \frac{RL}{v}$$

where R is the transition rate as in **4.1**, while the cross-section is given by

$$N = \frac{\sigma}{\text{area}} = \frac{\sigma}{L^2}$$

so that

$$\sigma = \frac{L^2 RL}{v} = \frac{VR}{v}.$$

In the cms $v = v_1 + v_2$, and if q is the initial momentum of each particle in the cms, then we can write

$$\frac{1}{v} = \left[q\left(\frac{1}{e_I} + \frac{1}{e_T}\right)\right]^{-1}$$

where e_I and e_T are the incident and target total cms energies. Substituting in the transition rate formula we then obtain (using **4.1** and **4.3**)

$$\sigma = \frac{g^2}{4\pi} \frac{p}{q} \frac{e_{F_1} e_{F_2}}{E^2}$$

where the e_F are the fermion energies, p are the final-state momenta, and we have written $E = e_1 + e_2 = e_I + e_T$ for the total cms energy. At high energies

$$\frac{p}{q} \frac{e_{F_1} e_{F_2}}{E^2} \to 1 \quad \text{so that} \quad \sigma \sim \frac{g^2}{4\pi}.$$

If we use as a unit of cross-section $\sigma_0 = (m_\pi)^{-2} \simeq 2 \times 10^{-26}$ cm^2, then we have that at high energies

$$\sigma \sim \sigma_0 \frac{1}{4\pi} (gm_\pi)^2.$$

If we now insert into this formula the value of, for instance, the cross-section for the process $\pi^- p \to \Lambda K^0$ at high energies, which is approximately one millibarn, we obtain

$$g \sim (m_\pi)^{-1}.$$

We note that, as expected, this coupling constant has dimensions L^{+1}. If we wish to compare with the coupling constant for the decay of a boson into two fermions, which is dimensionless, then we must divide by the appropriate unit of length. We may take this to be the pion Compton wavelength $\hbar/m_\pi c = (m_\pi)^{-1}$ in our units. Thus we arrive at a value for $g^2/4\pi \simeq 0.1$. Such a coupling constant is characteristic of the strong interactions, and in extreme contrast to the value of about 10^{-15} which is characteristic of the weak decays.

5

Strange particles

5.1 V⁰ particles

5.1 V^0 particles

As with the π-mesons, all the early work on what have been called the 'strange' particles was done by means of cosmic-ray studies, using cloud chambers at sea level and at mountain altitudes, and using nuclear emulsions flown in high-altitude balloons. The first example of a particle other than those we have already discussed was reported by Leprince-Ringuet in 1944. A secondary cosmic-ray particle, which crossed the cloud chamber, produced a delta ray, or recoil electron, having substantial energy and emitted at a measurable angle. From the measured curvatures of the tracks in the magnetic field, and the scattering angle, it was possible to determine the mass of the incident particle, which was found to be $500 \pm 50 \text{ MeV}/c^2$. It is now clear that this particle must have been a K-meson, but, at the time, when even the pion had not been identified, the significance of this single event was not clear.

The first clear examples of the new particles were observed in 1947 by Rochester and Butler at the University of Manchester. In these events the decays of the particles were observed, allowing a more convincing conclusion than any that could be obtained from observations on a single track, even when a delta ray was produced. Rochester and Butler operated a cloud chamber in a magnetic field and triggered it when an arrangement of Geiger counters near the chamber detected a penetrating cosmic-ray shower. In the course of about one year of operation at sea level they obtained two photographs out of fifty which showed examples of what were called V-particles. These photographs are shown in fig. 5.1. In the first of these the V appears to have originated just below a 30 mm lead plate across the centre of the chamber. A number of possible explanations were tested for these events:

(*a*) It is possible that the V might represent a very large angle scatter of one of the particles at the vertex. In such a case one might expect to see a short track due to the recoiling nucleus, and such a hypothesis could be quantitatively tested by measurements of the curvatures and drop density of the tracks. Such a test showed that this hypothesis was untenable. In addition, in later work, when an increasing number of such events was collected, it became clear that such possible 'scatters' did not take place any more frequently in the solid metal plates used in the cloud chambers than in the gas itself.

(*b*) Another possibility was that the V of fig. 5.1 might be the decay of a charged particle at the apex point. However, this, too, could be shown to be inconsistent with the measured quantities.

(*c*) The third possibility was that the event represented the decay of a neutral particle formed in the interaction of one of the cosmic-ray particles in the lead plate, the neutral particle decaying at the apex point into a pair of oppositely-charged particles. Such a hypothesis was found to be entirely consistent with the data.

Similar cloud-chamber experiments were continued by the Manchester group at high altitudes and also by a number of other groups, including that of R. W. Thompson in the U.S.A. Thompson achieved a high degree of precision, and, with a growing number of events, the pattern of the data was gradually resolved. It is of interest to look more closely at the nature of the problem of identifying such particles. In the case where the origin of the neutral V-particle is known, it is possible to check whether the neutral particle and the two charged particles are coplanar. If this condition is found to hold, it is very unlikely that any neutral particles are present other than the primary. Secondly, it is possible to check the conservation of transverse momentum in a direction at right angles to that of the primary. Further analysis is more difficult unless it is possible to identify the nature of the charged particles taking part in the decay. This may be done by a combination of curvature and drop-density measurements in the cloud chamber, or multiple scattering and grain-density measurements in a nuclear emulsion. Unfortunately, this was only possible in very few cases and in none of the very first events. However, a technique was introduced by the Manchester group which enabled them to show that two kinds of V^0-particle decay existed, having different decay products and Q-values. For each event they calculated a quantity

$$\alpha = \frac{p_+^2 - p_-^2}{p^2} = \frac{m_+^2 - m_-^2}{M^2} + 2p^*\left(\frac{1}{M^2} + \frac{1}{p^2}\right)^{\frac{1}{2}} \cos\theta^*$$

where p_+ and p_- and m_+ and m_- are the momenta and masses of the decay particles, p is the momentum and M the mass of the incident neutral particle, and θ^* is the centre-of-mass angle between the incident and decay particles, for which p^* is the centre-of-mass momentum. In the cms the average value of $\cos\theta^*$ is zero, so that we can write

$$\langle\alpha\rangle = \frac{m_+^2 - m_-^2}{M^2}.$$

Thus the distribution of α will show a peak for any given decay of a primary particle, of definite mass, into two given decay products. In fact, the Manchester group were able to show that the distribution in α had at least two peaks. The results obtained by this group, and by Thompson and other workers, established unambiguously the existence of the two decays

$$\Lambda \to p + \pi^-$$
$$K^0 \to \pi^+ + \pi^-.$$

Fig. 5.1. The first V^0-particles observed by Rochester and Butler (1947) in a counter-controlled cloud chamber. In the left-hand photograph the V^0 (a, b) originates just below the lead plate across the chamber. In the right-hand photograph the V (c, d) originates near the top-right perimeter.

We may note that once the product particles had been identified, the mass of the neutral primary could be obtained from the relationship

$$M^2 = (E_+ + E_-)^2 - (\mathbf{p}_+ + \mathbf{p}_-)^2$$

where E_+ and E_- are the total energies of the decay particles. This expression for the missing mass, which is a Lorentz-invariant quantity, will also be found useful when we discuss the strongly-decaying resonances.

Following the early cloud-chamber work, much success in this field was achieved in nuclear emulsion studies. This was so for the following reasons:

(a) The emulsions could be flown at high altitudes, collecting data for prolonged periods. The flux of the new particles was much greater at such altitudes than at sea level or at mountain altitudes.

(b) The stopping power of the emulsion is high, but, even so, it is possible to measure very short tracks. For instance, the range of a proton of 14 MeV is 1 mm, but measurements may be made to within a few microns.

(c) It often proved possible by flying stacks of plates or blocks of emulsion to observe both the origin and the decay of the new particles.

As with the study of the π-mesons, the most precise determinations of the masses, lifetimes and decay modes of these particles, as well as the study of their production and the determination of their other quantum numbers, had to wait their production by accelerators. In the more detailed discussion of the new particles in the rest of this chapter we shall use the most significant data available, regardless of date.

5.2   Charged K-mesons

In the previous section we have seen evidence for the existence of what is now called a K^0-meson, decaying into a pair of oppositely charged pions. It was shortly after this discovery that evidence was also obtained for the existence of charged K-mesons. The identification of the charged kaons is easier for the positive than for the negative particle; these particles turn out to have strong interactions with nucleons so that the negative kaon is attracted by nuclei, and interacts before it can decay, while the positive kaon, due to the Coulomb repulsion, is more frequently able to decay beyond the range of the strong forces. We shall therefore first consider the decay modes of the positive kaons.

(*a*) The most striking decay mode is that into three charged pions

$$K^+ \to \pi^+\pi^+\pi^-.$$

The particle decaying in this way was originally referred to as the τ-meson. The first examples of this mode were observed in nuclear emulsions exposed to the cosmic radiation. In the early days of emulsion work, the emulsions were not sensitive to tracks of minimum ionisation. In 1948, so-called electron-sensitive emulsions became available with a resulting series of new discoveries concerning elementary particles. A particularly striking example of this decay is shown in fig. 5.2. In this case the distinctive three-prong star results from a track arising from a disintegration in the emulsion due to a cosmic-ray particle of high energy. The unusual feature of this example is that two of the three secondary prongs from the decay can be identified by their decays, or interactions, in the emulsion. One of them comes to rest in the emulsion and decays, and the secondary decay particle is itself observed to decay into an electron. In fact, this particle, and the one which leaves the emulsion without decaying or interacting, may be identified as π-mesons by means of the grain density, multiple scattering and range. The third particle can be seen to come to rest and interact in the emulsion, producing a track identifiable as a proton, suggesting that it is in fact a π^--meson. The three secondary particles can be shown to be coplanar, indicating the absence of any neutrals in the decay. For such a decay it is possible to obtain fairly precise values for the kaon mass, which could be shown to be about 494 MeV/c^2.

(*b*) Corresponding to the decay mode of the neutral meson into two pions, we have a two-pion decay mode of the charged K. The particle decaying in this way was originally known as the θ-meson. The primary in this decay was first identified in emulsions as having the K-mass by observations on its track. The charged secondary could be identified as a π-meson, and was found to have a unique momentum in the cms of the decaying particle. As in the case of the $\pi \to \mu$ decay, this indicates that only one neutral particle can be present, although it was not possible in the early work to establish whether the neutral was a π^0 or a γ-ray. Increasing data provided a number of events in which a direct pair from π^0-decay (i.e. $\pi^0 \to e^+e^-\gamma$) was observed, thus removing the ambiguity.

For this decay, and for all other decays involving π^0-mesons or γ-rays, a substantial advance in technique was afforded by the development of bubble chambers containing liquids of short radiation length. In such chambers the probability of converting one or even both γ-rays from π^0-decay is high. Particularly useful in this respect have been studies carried out using bubble chambers containing liquid xenon, which has a radiation length of only 30 mm. For the decay under discussion, if the K-meson is brought to rest in a liquid-xenon chamber then in many cases the event will show two electron pairs from the decay, enabling a full reconstruction to be achieved.

(c) The last of the two-particle decays to be established was the process

$$K^+ \to \mu^+ \nu.$$

In fact, this mode turns out to be the most common decay channel for the K$^+$. The reason for the difficulty in establishing the nature of this decay was that the charged secondary has a rather high momentum, making it

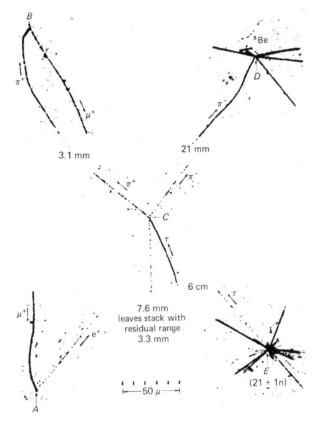

Fig. 5.2. Production of a τ-meson in a nuclear emulsion exposed to the cosmic radiation. The τ emerges from a 'star', slows down to rest and decays to three tracks. One of these is identified as a π$^+$ by its π → μ → e decay. Another is identified as a π$^-$ by its absorption by a nucleus in the emulsion to give another 'star' (Powell, Fowler and Perkins, 1959, p. 63).

more difficult to identify in a nuclear emulsion unless long tracks are available. The first events of this nature were identified, by means of a multi-plate cloud chamber, by the Ecole Polytechnique group in Paris. The cloud chamber used was, in fact, a double chamber in which the upper chamber was without plates but placed in a magnetic field. In this chamber the momentum of an incident particle could be measured. The lower chamber contained a series of metal plates, but no magnetic field was present. The multi-plate chamber gave the residual range for particles which stopped within it, and the combination of these data frequently gave the mass of the particle with precision adequate to identify it as a K-meson.

In the cases in which the secondary particle emitted by the stopped K was in such a direction as to pass through a number of plates, it was sometimes possible to determine its nature by means of observations on its residual range and rate of energy loss in passing through the plates. In addition, the fact that such charged secondaries were of uniformly high momentum suggested a two-body decay and detailed kinematic analysis indicated that the associated neutral particle should have very small or zero mass. The absence of associated showers in the plates established that the neutral particle could only be a neutrino.

A number of other decay modes have been established, the more frequent of which are

$$K^+ \rightarrow \pi^+\pi^0\pi^0 \quad \text{(expected partner to } \pi^+\pi^+\pi^- \text{ from charge independence)}$$
$$\mu^+\pi^0\nu$$
$$e^+\pi^0\nu$$
$$\pi^+\nu\gamma.$$

Various other rather rare decay modes, including four-particle decays such as $\pi^+\pi^-e^+\nu$, have also been observed.

The relative frequencies of the decay channels, usually referred to as the branching ratios, are summarised for the principal modes in appendix B.

5.3   Mass and lifetime for charged K-mesons

5.3.1   *Mass*

The most precise determination of the K-meson mass has been made by means of a reaction involving Σ-particles, which we have not yet discussed. This determination will be mentioned in a later section. An alternative method, which yielded an accurate value, consisted of a measurement of the momentum of the K-particles by magnetic analysis combined with a measurement of their energy from their range in a nuclear-emulsion stack. The arrangement is shown in fig. 5.3. A target placed in the internal circulating beam of the Bevatron yielded K-mesons, and other particles, which were allowed to pass through a set of magnetic quadrupole lenses (see chapter 1) and a bending magnet before entering an emulsion stack. The position at which they entered the stack was thus a function of the particle momentum. The momentum-position relationship was established by measuring the residual range of protons entering the stack over the region of interest. These measurements yielded a value for m of (493 ± 1) MeV/c^2.

5.3.2   *Lifetime*

A most precise value of the lifetime for the charged K-meson has been obtained by time of flight measurements at Brookhaven. An electrostatically (see chapter 1) separated K-meson beam was allowed to pass through scintillation counters, the distances between which were set at 15.75, 32.77 and 49.53 m. In addition to the electrostatic separation, the K-mesons were identified by means of two differential Cerenkov counters. Measurements were made of the attenuation of the K-meson flux at momenta of 1.6 and 2.0 GeV/c. A precise check of the geometric efficiency of the beam was obtained by using stable particles, protons or antiprotons, while the effect of the small amount of material in the beam was corrected for by studying the transmission as a function of the thickness of material, and extrapolating to zero absorber. A prime objective of these measurements was a precise comparison of the lifetime for the positive and negative particles, which could be obtained by changing the polarity of the bending magnet. The number of particles remaining after travelling a distance L is then given in terms of the mean lifetime by the expression

$$N = N_0 e^{-Lm_0/p\tau}$$

where N_0 is the original number of particles, m_0 the particle rest mass in MeV/c^2, p the particle momentum in MeV/c and L/τ is in units of the velocity of light. This relationship takes into account the dilatation of the particle lifetime due to its motion. The data obtained are consistent with a pure exponential decay to better than 1 % at the largest distance. The present best absolute lifetime of the K^+ is found to be 12.371 ± 0.0026 ns (1 ns $= 10^{-9}$ s). The ratio of the lifetimes τ_+ and τ_- for the positive and negative particles is found to be

$$\left(\frac{\tau_+}{\tau_-} - 1\right) = -0.0011 \pm 0.0009,$$

completely consistent with equality. We shall refer to this result later when we deal with invariance under the so-called *CPT* transformation.

5.4   Neutral K-mesons

(a) $K^0 \to \pi^+\pi^-$

We have already seen that the original observations of V-particles indicated the existence of a particle decaying into two charged π-mesons. Whenever the origin of such an event was observed it was found that the neutral line of flight was coplanar with the secondaries, indicating that no other neutral particles were present. The early data established that the mass of the primary was 492 ± 3 MeV/c^2, yielding a Q-value for the decay of 214 ± 3 MeV. The lifetime for these particles decaying in this way was best established in the early data using kaons produced and decaying in a 12 in (305 mm) propane bubble chamber. The momentum of the Ks was established from the direction and curvatures of the decay pions and used to determine the time of flight in each case. The distribution in time of flight was found to be closely exponential, at least for short times. The best value for the lifetime of the K^0 is $(0.8923 \pm 0.0022) \times 10^{-10}$ s.

(b) $K^0 \to \pi^0\pi^0$

This decay mode is to be expected, on the basis of charge independence, as an alternative to that involving two charged pions. As in previous situations where π^0-mesons are present, it is most clearly observed in the liquid-xenon bubble chamber, where in favourable circumstances electron pairs from all four decay γ-rays may be seen.

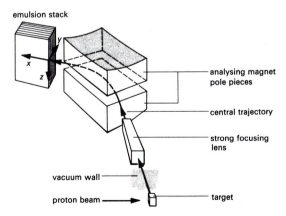

Fig. 5.3. Experimental arrangement for the determination of the kaon mass using magnetic analysis to measure the momentum and range in the emulsion to determine the energy (Peterson, 1957).

The ratio of the charged- and neutral-decay modes is found to be

$$\frac{K^0 \to \pi^+\pi^-}{K^0 \to \pi^0\pi^0} = \frac{69}{31}.$$

Even in the early observations a small percentage of so-called anomalous K-decays which could not be interpreted as decays into two pions were observed. We may anticipate the result of the Gell-Mann–Nishijima theory (which will be discussed in section 5.12) which predicted the existence of two different neutral kaons, having different lifetimes, of which the shorter-lived particle was expected to decay by two-pion modes while the longer-lived particle was expected to decay by three-particle modes.

Following this proposal, an experiment was carried out by Lande *et al.* at Brookhaven. A cloud chamber was set up, at a distance of 6 m from the target of the Cosmotron, in such a position that particles from the target could pass through a hole in the shielding mass into the chamber. A magnet placed in the line of flight swept all charged particles from the beam. At this distance and with primary protons having an energy of 3 GeV a completely negligible number of short-lived neutral kaons could reach the chamber. However, 23 V-events were observed, of which all but one were not coplanar with the line of flight from the target to the decay point. Also it was shown that none of the secondaries from these decays was a proton, while most of them were in fact lighter than K-mesons, that is, were pions, muons or electrons. This experiment thus established the existence of a long-lived neutral kaon, of lifetime of the order of 10^{-8} s, decaying into at least three particles. An alternative piece of evidence pointing to the existence of long-lived neutral kaons was provided by the application of another prediction of the Gell-Mann theory. This was that the K-particles must always be produced along with another so-called strange particle such as the Λ mentioned earlier in this chapter. In an experiment in which Λ-particles were produced by π-mesons in a propane bubble chamber, the K^0 was only observed to decay in the chamber in about 50 % (after correction) of all cases. This indicates that in the other 50 % of cases the K^0 lived for a time sufficiently long to escape from the chamber. By methods similar to those already described for the short-lived K-particles, the following decay modes have been identified:

$$K_L^0 \to \pi^\pm e^\mp \nu_e$$
$$\pi^\pm \mu^\mp \nu_\mu$$
$$\pi^+\pi^-\pi^0$$
$$\pi^0\pi^0\pi^0.$$

In addition, it has been found that a very small proportion of the long-lived neutral kaons decay into two pions. This will be discussed when the K^0-particles and their curious properties are treated more fully in chapter 8.

5.5   The Λ-hyperon

In section 5.1 we saw that the original observations of V-particles indicated the existence of a heavy particle, decaying into a proton and a π^--meson, in addition to the neutral K-meson which decays into two pions. This particle is known as the Λ-hyperon.

We may summarise the terminology as follows:

Nucleons are the proton and neutron, and antinucleons the antiproton and antineutron.

Hyperons were originally defined as particles heavier than the nucleons. However, as we shall see, this definition is no longer adequate and we must add the requirement that hyperons have baryon number equal to 1. The antihyperons are particles of mass greater than the nucleon with baryon number equal to -1. We shall see that, like the nucleons, hyperons are fermions and have half-integral spin.

Mesons are defined as strongly-interacting particles with baryon number 0, and we shall see that they are all bosons having integral spin.

All the above interact via the strong interactions and are known collectively as hadrons.

Leptons are particles subject only to the weak and electromagnetic interactions. The leptons turn out to be all fermions. In fact, as discussed in chapter 3, the only such particles known are the τ, the muon, the electron and the neutrinos and their antiparticles.

Returning to the Λ-hyperon, the decay mode first observed was

$$\Lambda \rightarrow p\pi^-$$

established by observations in cloud chambers and nuclear emulsions, which identified the decay tracks as a proton and a π^--meson. The two-body nature of the decay was determined by observations on the coplanarity of the neutral particle and the two charged tracks, in cases where the origin of the V could be observed. It should be noted that, in practice, the coplanarity measurements need to be very precise in order to exclude the existence of a third particle. In addition, the mass of the neutral could be established by determining the quantity.

$$(E_p + E_{\pi^-})^2 - (\mathbf{p}_p + \mathbf{p}_{\pi^-})^2$$

where E and $\mathbf{p}$ are the total energy and momentum of the particle. We shall have frequent cause to use this quantity, known as the *effective mass* of the two particles (see appendix A). The distribution of this quantity, for a typical set of V-particles produced in a bubble chamber experiment, is shown in fig. 5.4. The sharp peak gives a measurement of the mass, and provides additional evidence that there are only two decay particles. The mass is found to be $1115.60 \pm 0.05 \, \mathrm{MeV}/c^2$. The lifetime has been measured by plotting the distribution of the delay between production and decay, in the rest system of the Λ-particle, for events observed in a bubble chamber, where the production and decay of the particle were seen and the measurements allowed a good determination of the Λ-momentum,

Fig. 5.4. Distribution of the effective mass for a set of V^0-particles having effective mass near the Λ-mass. The data are from a sample of V^0 from K^-p interactions at $6 \, \mathrm{GeV}/c$.

necessary to make the transformation from the laboratory to the particle centre of mass. Such measurements yield a distribution well fitted by an exponential decay corresponding to a mean life of $(2.632 \pm 0.020) \times 10^{-10}$ s. On the basis of charge independence we should also expect the decay

$$\Lambda \to n\pi^0.$$

This process can be observed only by detecting the pairs from the π^0-decay γ-rays. In occasional cases where the Λ decays in the hydrogen bubble chamber, the neutron may collide with a proton in the liquid to produce a recoil track. From observations of events in which a Λ is expected to be present because of the existence of a K^0 (see section 5.12 concerning 'associated production'), such electron pairs, and sometimes proton recoils, have been observed. It might be possible that the Λ should decay via the mode

$$\Lambda \to n\gamma.$$

This was, however, excluded on a statistical basis by plotting the energy distribution of the hypothetical γ-rays for a number of events. For such a mode the distribution should be mono-energetic. Later work has detected such a decay with a very small branching ratio $((1.02 \pm 0.33) \times 10^{-3})$.

For the decay

$$\Lambda \to n\pi^0$$

we expect a flat distribution of γ-ray energy between 32 and 134 MeV (compare the process $\pi^- p \to \pi^0 n$ for absorption at rest; see section 2.5). Such a flat spectrum is in fact observed.

The ratio of the decay modes $(\Lambda \to p\pi^-)/(\Lambda \to n\pi^0)$ is found to be almost exactly 2. We shall see later that this ratio is an example of the rule that the isotopic spin changes by $\frac{1}{2}$ in weak-decay processes.

5.6 The $\Sigma^\pm$-hyperons

The first evidence for these particles came from tracks observed in nuclear emulsions exposed to cosmic rays (fig. 5.5). Observations on the grain density and multiple scattering of the tracks indicated masses of about 1200 MeV/c^2. The decay is into a single charged particle; with accumulation of data it became clear from the unique energy of the decay particle, in the decay centre of mass, that only one additional neutral particle could be present. By the usual methods three principal modes were identified:

$$\Sigma^+ \to p\pi^0$$
$$\Sigma^+ \to n\pi^+$$
$$\Sigma^- \to n\pi^-.$$

In addition, decay modes involving leptons and γ-rays are now known to occur with very low frequency. Examples of such modes are

$$\Sigma^+ \to \begin{cases} p\gamma \\ ne^+\nu_e \\ n\mu^+\nu_\mu \\ n\pi^+\gamma \\ \Lambda e^+\nu_e \end{cases} \qquad \Sigma^- \to \begin{cases} ne^-\nu_e \\ n\mu^-\nu_\mu \\ \Lambda e^-\nu_e \\ n\pi^-\gamma. \end{cases}$$

The branching ratio $(\Sigma^+ \to p\pi^0)/(\Sigma^+ \to n\pi^+)$ is found to be nearly equal to 1.

Fig. 5.5. An example of the production and decay of a Σ-particle in a nuclear emulsion. The Σ emerges from a large cosmic ray induced star. The Σ-track is continued on the right of the picture and decays at the lower right (Powell, Fowler and Perkins, 1959, p. 359).

The most precise values for the masses of the charged Σ-hyperons, and also of the K^--meson, have been obtained from range measurements in nuclear emulsions on the charged products of the decays $\Sigma^+ \to p\pi^0$ and $\Sigma^+ \to n\pi^+$ and the reactions

$$K^-p \to \Sigma^+\pi^- \qquad \textbf{5.1}$$
$$K^-p \to \Sigma^-\pi^+. \qquad \textbf{5.2}$$

The Σ^+-mass was obtained from the proton range in the decay $\Sigma^+ \to p\pi^0$. The K^--mass could then be obtained from the Σ^+-range in **5.1**. The mass difference between the particles of opposite charge could be obtained from the range difference for these particles in **5.1** and **5.2**. For events in which the K^- is captured at rest, and in which the Σ^+ decays at rest, these measurements yield the appropriate masses with high precision when account is taken of the small difference between the energy loss of slow negative and positive Σ-particles. Accurate masses are as follows:

$M_{\Sigma^+} = 1189.35 \pm 0.06 \, \text{MeV}/c^2$ (range measurement of proton in $\Sigma^+ \to p\pi^0$);

$M_{K^-} = 493.667 \pm 0.015 \, \text{MeV}/c^2$ (range of Σ^+ in reaction **5.1**);

$M_{\Sigma^-} = 1197.34 \pm 0.05 \, \text{MeV}/c^2$ (ranges of the pions in reactions **5.1** and **5.2**).

The lifetimes are found to be:

Σ^+: $(0.799 \pm 0.004) \times 10^{-10} \, \text{s}$;

Σ^-: $(1.479 \pm 0.011) \times 10^{-10} \, \text{s}$.

As might be expected, on the basis of charge independence and the existence of two channels for the decay of the positive Σ-hyperon and only one for the decay of the negative Σ-hyperon, the lifetime τ_{Σ^-} is found to be about twice τ_{Σ^+}. As with the Λ, the lifetimes are short, so that the distance between production and decay is normally short, except for particles of very high momentum which benefit from time dilatation. This feature made it difficult to establish accurate values for the mass from observations on the tracks in the early work.

5.7 The Σ^0-hyperon

It was a prediction of the Gell-Mann–Nishijima scheme (see section 5.12) that the hyperons should have isotopic spin equal to 1, implying an I-spin triplet with I_3 changing by one unit between members, and thus the existence of a neutral Σ-particle.

A consequence of this scheme was also that the decay

$$\Sigma^0 \to \Lambda\gamma$$

accessible to the neutral Σ, would not be inhibited by strangeness conservation as are the charged decays described in the previous section. It is thus possible for this decay to proceed via electromagnetic interactions, so that we might expect the lifetime to be in the region of 10^{-15} s characteristic of such processes. For such a lifetime the Σ^0 will travel only a rather short distance before decaying; this makes its identification difficult, particularly since it decays into two particles which are both uncharged. Such a decay thus corresponds to a neutral vertex in a bubble chamber.

The first convincing evidence for the Σ^0 was provided by an event of the following kind:

$$\pi^- p \rightarrow \Lambda K^0 + (\text{neutral}),$$

where the measurements of the missing momentum and energy identified the unobserved object as a γ-ray. Subsequently, events have been observed in which an electron pair from the γ-ray is produced along with the other particles. The event is then kinematically overdetermined and the Σ^0 mass may be calculated. This mass is found to be 1193 MeV/c^2. A more precise value may be obtained by measuring the $\Sigma^- - \Sigma^0$ mass difference. When a Σ^- stops in a hydrogen bubble chamber it may interact with a proton according to the processes

$$\Sigma^- p \rightarrow \Lambda n$$
$$\Sigma^- p \rightarrow \Sigma^0 n \rightarrow \Lambda \gamma n$$

if the Σ^- is heavier than the Σ^0. The first of these processes will yield monochromatic Λ-particles of 36.9 MeV. The second process will produce a continuous spectrum of Λ-particles. The $\Sigma^- - \Sigma^0$ mass difference can then be obtained from the upper and lower limits for the Λ-energy. The difference $M_{\Sigma^-} - M_{\Sigma^0}$ is found to be 4.9 MeV/c^2.

5.8   The Ξ^--hyperon

This particle is also known as the cascade hyperon, since it decays into a pion and a Λ-hyperon. In a cloud or bubble chamber the topology is that shown in fig. 5.6, where the Λ originates from the decay point of the negative particle and the charged member of the decay can be identified as a π^--meson,

$$\Xi^- \rightarrow \Lambda \pi^-.$$

Since both decay products may be observed, the mass may be determined with quite high precision and the best value currently available is 1321.32 ± 0.13 MeV/c^2. We note that this mass is such that decay to $\Sigma \pi$ is energetically just forbidden.

Fig. 5.6. An example of Ξ^--production and decay in the British 1.5 m bubble chamber exposed to a 6 GeV/c K$^-$-beam from the CERN proton synchrotron. The Ξ^- emerges as the lowest of four tracks from the interaction in the left-hand half of the picture and rapidly decays to a Λ and a π^-.

The lifetime obtained by methods similar to that for the Σ^- is found to be $(1.641 \pm 0.016) \times 10^{-10}$ s.

5.9   The Ξ^0-hyperon

Another prediction of the Gell-Mann–Nishijima scheme was that the cascade hyperon should be an I-spin doublet, thus suggesting the existence of a neutral cascade particle (compare the proton–neutron I-spin doublet). We might expect such a cascade particle to decay via the process

$$\Xi^0 \to \Lambda \pi^0$$

illustrated in fig. 5.7, where we have assumed production via the reaction

$$K^- p \to \Xi^0 K^0.$$

As with the Σ^0 the decay is an all-neutral vertex, rendering more difficult the identification and study of this particle. However, unlike the Σ^0, we might expect that the Ξ^0 should have a lifetime comparable with that of the Ξ^-, so that it may travel an observable distance before decay. Many examples of processes which can only be fitted by the production and decay of a neutral cascade particle have been observed. The best data can be obtained in cases where one, or even both, of the γ-rays from the π^0 in the decay produce electron pairs, as is the case quite frequently for the Ξ^0s decaying in a heavy-liquid bubble chamber. Even where this is not so, it may be possible to reconstruct the event. For instance, in the process illustrated in fig. 5.7 for incident K-mesons of known energy we may proceed as follows:

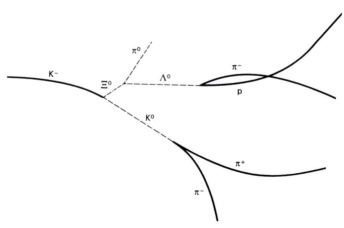

Fig. 5.7. Illustration of the formation and decay of a Ξ^0-particle.

(a) Attempt to identify the lower V from measurements on its decay tracks. This may enable it to be identified as a K^0-meson of known momentum.

(b) We may now test the hypothesis that only one other neutral particle is produced in the reaction, by calculating the magnitude and direction of the momentum necessary for conservation at the primary vertex. If we also apply energy conservation at this vertex we may calculate the mass of the neutral particles in the usual way.

(c) We also attempt to identify the particle responsible for the upper V in the figure, and in our example this may be identified as a Λ-hyperon whose momentum may be calculated in magnitude and direction.

(d) We may now test whether the calculated momentum vectors for the Ξ^0 and the Λ intersect in space. If this is found to be true, then for a two-particle decay of the Ξ^0 there will be a unique relationship between the emission angle and the momentum of the Λ. This relationship will provide a test of our hypothesis.

(e) In addition, if pairs from the π^0-decay γ-rays are observed yet a further check may be applied.

In practice, for such an event observed in a bubble chamber, the hypotheses for each charged and neutral vertex are tested individually and then an overall fit to the complete event is made.

The mass of the cascade zero is found to be (1314.9 ± 0.6) MeV/c^2 and its lifetime $(2.9 \pm 0.1) \times 10^{-10}$ s.

5.10 The Ω^--hyperon

This particle was predicted by Gell-Mann on the basis of SU(3) symmetry and its discovery provided one of the most notable triumphs of this theory (see chapter 10). The predicted mass was 1673 MeV/c^2 and the expected decay modes were

$$\Omega^- \rightarrow \begin{cases} \Xi^- \pi^0 \\ \Xi^0 \pi^- \\ \Lambda K^-. \end{cases}$$

The first example of the production of this particle was found in a bubble-chamber photograph at Brookhaven in 1964. The event is shown in fig. 5.8 and is interpreted as

$$K^- p \rightarrow \Omega^- K^+ K^0.$$

The incident beam consists of K^--mesons having a momentum of 5 GeV/c. The K^0 from the production vertex is not observed to decay in the chamber and the event could be fully analysed only because both γ-rays from the π^0, produced in the decay chain of the Ω^-, produced electron pairs in the bubble chamber. The decay was reconstructed according to the following scheme:

Many examples of Ω^--particles decaying in each of the three modes have now been observed and, indeed, a 'beam' of highly-relativistic strange baryons including Ω^- has been developed at CERN. The best values of the mass and lifetime are

$$M_{\Omega^-} = 1672.43 \pm 0.32\,\text{MeV}/c^2$$
$$\tau_{\Omega^-} = (0.822 \pm 0.012) \times 10^{-10}\,\text{s}.$$

The spin of the Ω^- has also been measured from the angular distribution of its decay products (see section 9.3) and found to be $\frac{3}{2}$.

Fig. 5.8. The first Ω^--particle to be observed (Brookhaven National Laboratory, 1964).

5.11   Strange-particle production and decay

We have seen that the lifetimes for the decay of the K-mesons, the
Λ-particle, the charged Σ-particles, the Ξ-particles and the Ω^- are all of the
order of 10^{-10} s, while the long-lived K^0 has a lifetime of about 10^{-8} s.
Following the discussion of chapter 4 it is clear that these lifetimes
correspond to decay by virtue of the weak interaction.

On the other hand, it became clear when the production of strange
particles was studied, using accelerator beams, that the production cross-
sections were quite large. For instance, it was found that at energies
somewhat above threshold the cross-section for Λ-production in π^-p
reactions was ~ 10 mb. The experiments which yielded this data were
performed using cloud chambers and nuclear emulsions, in the early
stages, and later predominantly by bubble chambers. The first accelerator-
produced strange particles were studied with the aid of the Brookhaven
3 GeV Cosmotron and later using the Berkeley 6 GeV Bevatron.

Again, referring to the discussion of chapter 4, we see that cross-
sections of this magnitude are characteristic of strong interactions.
Although there are uncertainties in the calculation of the coupling
constants, from either decay lifetimes or production cross-sections, the
difference in magnitude between the strong- and weak-interaction
coupling constants is so great that it would be extremely difficult to
account for the lifetimes of the strange particles if the decay proceeded by
means of a strong-interaction process. In fact there were some early
attempts to do this, based on the suggestion that the decays might be
inhibited by a large angular-momentum barrier if the strange particles had
high spins. This hypothesis was not particularly successful and was later
shown to be false when the spins of these particles were determined, and
found to be 0 for the K-mesons and $\frac{1}{2}$ for the Λ, Σ and Ξ.

The outstanding problem therefore was to account for the fact that
these particles were strongly produced but decayed via the weak
interaction. It was this behaviour which led to the appellation 'strange'.

5.12   Strangeness

An explanation of the strong-production, weak-decay anomaly
was offered in 1954 by Gell-Mann and Pais and independently by
Nishijima. This explanation proposed that there existed a quantum
number associated with the new particles which was conserved in the
production process, but not in the decay, which was therefore inhibited
relative to the production. It was thus necessary that in the production
process the new particles be produced in pairs having opposite values of

the new quantum number. Initially it could only be said that the experimental evidence did not contradict the hypothesis of 'associated production', but subsequent evidence demonstrated clearly that this hypothesis was indeed true.

The situation can be illustrated by analogy with the conservation of charge. In the field of the nucleus we can get electron-pair production by a γ-ray, i.e.

$$\gamma \rightarrow e^+ e^-,$$

but, of course, processes like $e \rightarrow \gamma$ are forbidden by charge conservation. Similarly, if we assign values of the new quantum number called strangeness (S) to all particles and if we have the selection rules:

for strong interactions strangeness is conserved and $\Delta S = 0$,
for weak interactions strangeness is not conserved and $|\Delta S| = 1$,

then the particles may be produced strongly in pairs, but can decay singly only by means of the weak processes. Strangeness differs from charge in that no force is known to be associated with this quantity.

The next problem is how to assign the strangeness quantum numbers. On a purely empirical basis we make some deductions from the observations that certain reactions take place. For instance, the process

$$\pi^- p \rightarrow \Lambda K^0$$

is observed in a bubble chamber as a stopping track with two associated Vs. For the π^- and the proton $S = 0$, so that the Λ and K^0 must have strangeness $\pm S$ (where S is greater than or equal to 1) although we cannot say which particle should have positive and which negative strangeness. If we accept the hypothesis that for weak processes $|\Delta S| = 1$, then the decay process $\Lambda \rightarrow p\pi^-$ shows that $|S| = 1$ for the Λ.

In order to make further progress we must note some other peculiarities of the strange particles. For nucleons and pions the centre of charge of the isotopic spin multiplet is equal to $B/2$, where B is the baryon number for the multiplet. Thus for the nucleon doublet the centre of charge is at $\frac{1}{2}(0+1) = +\frac{1}{2}$, for the antinucleons the centre of charge is at $-\frac{1}{2}$ and for the pions at 0, in all cases being given by $\frac{1}{2}B$.

For nucleons and pions the charge of any member of the multiplet is given by the expression

$$Q = I_3 + \tfrac{1}{2}B$$

where I_3 is the third component of the isotopic spin.

For the Λ, however, these relationships do not hold. Since all the evidence in familiar processes suggests that the baryon number is conserved we can say that the baryon number for the Λ is 1, since it decays into a proton and a pion. In addition, no other particles having the same mass as the Λ, but different charges, have been observed, suggesting that this particle is an isotopic spin singlet. Thus the centre of charge must be 0, whereas we might have expected it to be $+\frac{1}{2}$. We will find that this displacement of the centre of charge from its 'normal' value is characteristic of the strange particles. If we write $S = -1$ for the Λ, then the centre of charge will be at $\frac{1}{2}(B+S)$. If we replace B by $(B+S)$ in the formula for the charge we obtain

$$Q = I_3 + \tfrac{1}{2}(B+S). \qquad \textbf{5.3}$$

In this aspect the strangeness appears as a measure of the degree to which the centre of charge is displaced from its 'normal' value.

Turning to the Σ-particles, let us first suppose that only the Σ^+ and Σ^- are known, as was the case when the strangeness theory was proposed. From the decay we have $B = 1$ and the observations indicate that the centre of charge is at 0, so that $S = -1$ according to the hypothesis above. If only the charged Σs exist, they would be members of an isotopic spin doublet having $I = \frac{1}{2}$. However, according to **5.3**, this yields charges of $\pm\frac{1}{2}$ so that the scheme would be internally inconsistent. It was proposed in order to account for this difficulty that the Σs in fact formed an I-spin triplet, Σ^+, Σ^0, Σ^-. As before, $S = -1$, but now $I = 1$ and $I_3 = +1$, 0, -1 respectively and the charges are $+1$, 0 and -1 as required. The fact that the Σ^0 has open to it an electromagnetic decay mode

$$\Sigma^0 \to \Lambda\gamma$$

in which $\Delta S = 0$, which is not available to the charged Σs, suggested that it would decay with a lifetime $\sim 10^{-16}$ s, rendering its existence not immediately obvious. As pointed out in the previous chapter, this prediction of the Gell-Mann–Nishijima scheme has been amply confirmed by experiments.

Turning to the Ξ-particles, we recall that the decay

$$\Xi^- \to \Lambda\pi^-$$

is weak, suggesting $|\Delta S| = 1$ for this process, so that for the Ξ, $S = 0$ or -2. However, if $S = 0$ we might expect that the Ξ^- would decay overwhelmingly by the process

$$\Xi^- \to n\pi^-.$$

Since this is not so, we conclude that for the Ξ^-, $S = -2$. The centre of charge should therefore be at $-\frac{1}{2}$, implying the existence of a neutral partner for the Ξ^-. The formula **5.3** gives $I_3 = -\frac{1}{2}$ for the Ξ^- and $I_3 = +\frac{1}{2}$ for the proposed Ξ^0. For the Ξ^0 the expected decay is

$$\Xi^0 \to \Lambda\pi^0.$$

As described in the earlier section, this prediction of the scheme was also confirmed by experiment.

Note that the Σ^+ and Σ^- are not a particle-antiparticle pair like the π^+ and π^-. For the anti-Σ^+ ($\bar{\Sigma}^+$) we expect $Q = -1$, $B = -1$ and $I_3 = -1$ so that $S = +1$, illustrating the general result that particle and antiparticle have opposite strangeness.

We now examine how the Gell-Mann–Nishijima theory applies to the K-mesons. We start from the observation that the process

$$\pi^- p \to \Sigma^- K^+$$

is observed to occur. Therefore $S = +1$ for the K^+ and using **5.3** we obtain $I_3 = \frac{1}{2}$ for the K^+ consistent with the conservation of I_3. This I-spin assignment implies that the K^+ is one partner of an I-spin doublet, the other partner having $I_3 = -\frac{1}{2}$ and therefore charge 0. This partner is then apparently the K^0.

The question then arises as to what is the status of the K^-. It seems natural that the K^- should be the antiparticle of the K^+ so that it will have $S = -1$. This is the only self-consistent conclusion. To take the K^+, K^0, K^- as an I-spin triplet leads to an internal inconsistency.

The K^- should then form one half of an I-spin doublet, the other partner being a neutral K^0 which is the antiparticle of the K^0 associated with the K^+. The K^0 associated with the K^- is called the $\bar{K}^0$ and has opposite strangeness to the K^0, although these particles are in all other ways identical. The situation is thus slightly different from that of the π^0 which is in *every* way identical with its antiparticle. We will discuss further the peculiar properties of the K^0, $\bar{K}^0$ system in chapter 8.

An interesting consequence of the assignments of strangeness described above concerns the interaction cross-sections for K^-- and K^+-mesons. For interactions with nucleons, the baryon number in the initial state is $+1$. Since for $B = +1$ hyperons, Λ and Σ, the strangeness is -1, they can be formed readily even at low energies in K^-p interactions such as

$$K^- p \to \begin{cases} \Lambda\pi^0 \\ \Sigma^0\pi^0 \\ \Sigma^-\pi^+ \\ \Sigma^+\pi^-. \end{cases}$$

Reactions like this are not, however, possible for K^+-mesons. The fact that there are many more reaction channels open for K^- than for K^+-mesons implies that the mean free path for K$^-$ should be very much less than for K$^+$, as is observed experimentally.

5.13   Hypercharge

A convenient quantity in discussing elementary particles is the hypercharge Y, defined as the sum of the baryon number and the strangeness. In terms of the hypercharge we have the relations

Centre of charge $= \frac{1}{2}Y$, $Q = I_3 + \frac{1}{2}Y$.

6

Spin and parity of the K-mesons and non-conservation of parity in weak interactions

___

6.1 The τ–θ problem

As the experimental data accumulated it became clear that the masses of the K-mesons decaying into two pions and into three pions were identical, within small error limits. This observation suggested that the two-pion and three-pion decays were alternative decay modes of the same particle. Further weight was lent to this conclusion by the fact that the branching ratio of the two-pion and three-pion modes was found to be independent of the energy of the parent particle and also of its previous history, for instance, whether or not it had undergone scattering prior to decay. This implied that the parent particles of the two-pion and three-pion decays, as well as having the same mass, were identical in their interactions with nucleons. Thus all the evidence suggested that the so-called τ- and θ-particles were in fact different decay manifestations of only one K-meson.

The nature of the puzzle becomes clear only when one considers the possible spin-parity assignments to the τ and θ; we shall examine the possible assignments in the following sections.

6.2 Spin-parity for the θ-meson, $K \to \pi\pi$

Since the pions have zero spin, conservation of angular momentum demands that the spin of the K is equal to the relative orbital angular momentum L of the two pions. The parity of the di-pion is then $(-1)^L$. Thus the possible spin-parity assignments are

$$J^P = 0^+, 1^-, 2^+, 3^-, 4^+, \ldots \quad \text{('natural' spin-parity)}.$$

In addition, if we consider the decay mode of the neutral K-meson into two π^0-mesons it is clear that, since we are dealing with bosons, and since the space part of the wave function is symmetrical, the Pauli principle allows only symmetrical angular-momentum wave functions so that $J^P = 1^-, 3^-, \ldots$ are not allowed. Thus, assuming that the charged and neutral K-mesons decaying into two pions are different members of an I-spin multiplet and have the same spin-parity, then if parity is conserved in the decay the only possible J^P-assignments are those of even spin and even parity.

6.3 Spin-parity for the τ-meson, $K \to \pi\pi\pi$

One can treat the three-pion system most readily in terms of a di-pion, which in the case of the charged decay we take to consist of the two pions of like charge, plus an added third pion. We take the relative orbital angular momentum in the di-pion to be l, and of the third pion relative to the di-pion to be L (see fig. 6.1). The parity of the three-pion system is then

$$(-1)^3(-1)^l(-1)^L = -(-1)^L$$

since the di-pion must have even l for symmetry reasons.

The spin J of the three-pion system must lie in the interval

$$|L-l| \leqslant J \leqslant |L+l|.$$

The possible spin-parity assignments are then summarised in table 6.1. Comparing these assignments with those for the θ-meson, we see that the first possible J^P value common to both θ and τ is 2^+ for $l = 2$, $L = 1$. Higher angular momentum values could also give compatible J^P assignments, but we neglect them since decays from such states will be seriously inhibited by the high angular momentum barriers. Indeed, even

Fig. 6.1. Definition of the angular momenta in the charged three-pion system.

for the state mentioned, we might expect that the three-pion decay mode would be markedly inhibited for this reason.

The further study of this problem requires the determination of the spin-parity from an examination of the decay angular distributions for the meson. In order to see what form of angular distribution is to be expected for different J-assignments we write the angular part of the final three-pion wave function in the cms as a sum of partial waves specified by l and L. In order to examine the final state we define the following quantities:

(a) $\mathbf{q}$ is the relative linear momentum of the pions in the di-pion;

(b) $\mathbf{p}$ is the linear momentum of the third pion relative to the di-pion;

(c) θ is the angle between $\mathbf{p}$ and $\mathbf{q}$ so that $\cos \theta = (\mathbf{p} \cdot \mathbf{q})/pq$.

The relative proportions of the different angular-momentum states contributing to the final state will be fixed by the Clebsch–Gordan coefficients (see appendix B) but may also depend on the relative energies (proportional to p^2 and q^2) in the three-pion system. We specify this energy dependent weighting factor by $f_{L, l}(p^2, q^2)$. In the general case we can then write our final state wave function as

$$\sum_{L, l, m_L, m_l} c(L, l, J; m_L, m_l) \, Y_L^{m_L}(\theta_p, \phi_p) \, Y_l^{m_l}(\theta_q, \phi_q) f_{L, l}(p^2, q^2)$$

where with our coordinate system this reduces to

$$\sum_{L, l, m_l} c(L, l, J; 0, m_l) \, Y_l^{m_l}(\theta) f_{L, l}(p^2, q^2) \tag{6.1}$$

since $\cos \theta_p = 1$, $m_L = 0$ and $Y_L^{m_L} = $ a constant.

It remains to discuss the form of the function $f(p^2, q^2)$. In practice we resort to an approximation by assuming that the asymptotic solutions of the Schrödinger equation are valid. This is equivalent to the assumption that any momentum dependence arising from the internal structure of the τ-meson may be neglected. The approximation is good if the pion

Table 6.1

l	L	J^P
0	0	0^-
0	1	1^+
0	2	2^-
2	0	2^-
2	1	$1^+, 2^+, 3^+$
2	2	$0^-, 1^-, 2^-, 3^-, 4^-$

wavelengths are long compared with the τ-radius, so that there is a high probability that the pions lie 'outside' the τ (cf. the deuteron). With this approximation we can write

$$f = c_{L,l}\, p^L q^l \qquad\qquad \textbf{6.2}$$

where $c_{L,l}$ is a constant.

In order to obtain manageable (and realistic) formulae we also make the approximation of retaining only the lowest angular momentum values consistent with any given spin and parity, using the fact that the angular momentum barrier will inhibit the higher terms in comparison with the lower ones.

With these approximations we can write down the values of the square of the modulus of the matrix element for the decay process using **6.1** and **6.2**. These quantities are given in table 6.2. In order to obtain the final energy or momentum distributions we must multiply $|M|^2$ by the appropriate phase-space factor. For the angular distributions the transition probability is simply given by

$$\frac{\mathrm{d}N(\theta)}{\mathrm{d}(\cos\theta)} \propto I(\theta)$$

where $I(\theta)$ is the angular function in $|M|^2$, as given in the table.

It is convenient to parameterise the energy dependence of the decay in terms of $\varepsilon = E_{\pi^-}/E_{\mathrm{MAX}}$ where E_{MAX} is the maximum possible pion energy and is equal to two-thirds of the available cms energy. We may then write the phase-space factor as

$$\varepsilon^{\frac{1}{2}}(1-\varepsilon)^{\frac{1}{2}}\,\mathrm{d}\varepsilon\,\mathrm{d}(\cos\theta)$$

Table 6.2

J	P	L	l	$\lvert M\rvert^2 \propto$	$I(\varepsilon)$ not including phase space
0	+	not possible			
0	−	0	0	1	1
1	+	1	0	p^2	ε
1	−	2	2	$p^4 q^4 \sin^2\theta\cos^2\theta$	$\varepsilon^2(1-\varepsilon)^2$
2	+	1	2	$p^2 q^4 \sin^2\theta$	$\varepsilon(1-\varepsilon)$
2	−	$\begin{cases}2\\0\end{cases}$	$\begin{matrix}0\\2\end{matrix}$	depends on mixing	
3	+	$\begin{cases}3\\1\end{cases}$	$\begin{matrix}0\\2\end{matrix}$	depends on mixing	
3	−	2	2	$p^4 q^4 \sin^2\theta(5+3\cos^2\theta)$	$\varepsilon^2(1-\varepsilon)^2$

and expressing the pq dependence in terms of ε, as given in the table, we obtain for the energy distribution

$$\frac{\mathrm{d}N(\varepsilon)}{\mathrm{d}\varepsilon} = I(\varepsilon)\varepsilon^{\frac{1}{2}}(1-\varepsilon)^{\frac{1}{2}}$$

where $I(\varepsilon)$ is the energy-dependent part of the matrix element, given in the table. Experimentally the distribution in $\cos\theta$ is found to be nearly isotropic (see fig. 6.2) favouring J^{P} equal to 0^{-} or 1^{+}, but not 1^{-}. The best fit to the distribution in ε leads to $I(\varepsilon)$ being constant (that is, the energy distribution after extraction of the phase-space factor). This distribution thus indicated that J^{P} is 0^{-}. The conclusion of the analysis is thus that, assuming the τ and the θ to be the same particle, parity cannot be conserved in the decay of this K-meson.

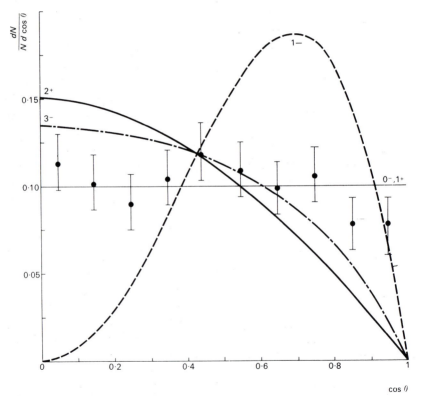

Fig. 6.2. The distribution of $\cos\theta$ (see text for definition of θ) for τ-decay. The curves show the calculated distributions for various spin-parity assignments. The approximately isotropic distribution agrees only with $J^{P} = 0^{-}$ or 1^{+} (Baldo-Ceolin *et al.*, 1957).

It is important to bear in mind that this analysis depends on certain assumptions:

(a) Validity of the 'small-radius' approximation used to obtain the form of f.

(b) The treatment has been non-relativistic, although any errors due to this approximation are likely to be small in view of the small Q-value of the decay.

(c) We have neglected the effects of any two-pion interactions in the final state. Other data on the two-pion system suggest that this assumption is justified at the energies in question.

(d) We have neglected the possibility that the observed distributions might be due to the mixing of, for instance, $L = 3$, $l = 0$ and $L = 1$, $l = 2$ states. That the mixing should be such as to produce distributions simulating the $L = 0$, $l = 0$ state seems sufficiently improbable that it may be disregarded.

6.4 Distribution in the Dalitz plot for the τ-decay

It is shown in appendix A.7 that the Dalitz plot has the property that, if the distribution of the energies in a three-particle system is determined only by the phase-space factor, then the Dalitz plot will be uniformly populated. It is of interest to investigate how the matrix-element dependence on p, q and θ, as discussed in the previous section, influences the distribution in the plot. We use the form of the plot where for each event the perpendicular distances from the point representing that event to the sides of the triangle are proportional to the kinetic energies of the mesons (see fig. 6.3). Thus PN is proportional to the kinetic energy of the 'unlike' meson, so that $PN \propto p^2$. It may then be shown that $PQ \propto q^2$ and $\cos \theta = GP/GH$. We recall that momentum conservation limits the points to within a circle inscribed in the triangle (non-relativistic case). An angular correlation between p and q is reflected by a variation in the density of points across the circle at any fixed value of PN. Points of constant $\cos \theta$ lie on an ellipse having major axis DO. On DO, $\theta = \pi/2$. On the circumference of the circle $\theta = 0$ so that the pions are collinear, a general result for the boundary of any form of Dalitz plot. Variations of density will also occur due to the terms dependent on p and q so that, for instance, the average density along lines parallel to the base of the triangle might be expected to vary with the Lth power of PN.

Special arguments concerning the distribution of points in the Dalitz plot may be made for limiting energies and angles. These arguments do not depend on the approximations made in the fuller treatment. For instance:

(a) A pion of zero energy is not possible if the spin-parity of the particle is odd–odd or even–even. Such a pion would need to be in an s-state, since otherwise it could not penetrate the centrifugal barrier. Thus for an odd-parity parent the remaining pions must be in a state of even angular momentum, resulting in an even value for the total spin. Similarly, for an even-parity parent the remaining pions must be in a state of odd angular momentum, giving total spin odd. Thus for J odd–odd or even–even, the density of points in the Dalitz plot should decrease towards the region corresponding to small pion momenta, i.e. the regions where the circle touches the triangle.

(b) A pion of unlike charge, with maximum energy, is not allowed if the spin-parity of the parent particle is odd–odd or even–even. In such a case, for instance, the two positive pions would have zero relative momentum so that the total spin will be the orbital angular momentum of the π^- relative to the di-pion. This is seen to exclude J^P odd–odd or even–even.

6.5   Other examples of parity non-conservation in weak decay processes

The explanation of the τ–θ anomaly as due to non-conservation of parity in the decay was proposed by Lee and Yang in 1956. At the same time they pointed out that no test of parity conservation in ordinary nuclear β-decay had ever been made and they suggested experiments to carry out such tests.

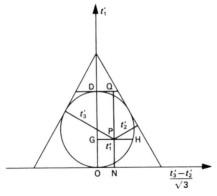

Fig. 6.3. The form of the Dalitz plot for three-pion decay. t_1', t_2', t_3' are the cms kinetic energies of the pions. The circle corresponds to the classical limit for momentum conservation.

The critical experiment which demonstrated conclusively that, in fact, parity was not conserved in nuclear β-decay was carried out immediately thereafter by Wu and collaborators (1957). In order to understand the principle underlying this experiment we note that the parity transformation reverses the direction of any momentum vector **p** but does not reverse the direction of angular momentum vectors **L**, since **L** is formed by a product of momentum and position vectors both of which change sign under the parity transformation. In the experiment of Wu, Ambler, Hayward, Hoppes and Hudson the distribution of β-particles emitted by an aligned radio-active source was examined. Consider electrons emitted in the same direction as the nuclear spin, or even into the hemisphere around the spin orientation. Applying the parity transformation to this situation the electron direction reverses, the spin direction remains unchanged and the electrons are now emitted opposite to the spin. Parity will only be conserved if the distribution of emitted electrons is symmetrical with respect to the spin orientation.

The apparatus used by Wu *et al.* is shown in fig. 6.4. The source used was ^{60}Co, which emits electrons of 0.312 MeV followed by a cascade of

Fig. 6.4. The cryostat and counters used by Wu *et al.* (1957) in the detection of parity violation in ^{60}Co decay.

two γ-rays of 1.19 and 1.32 MeV to give ^{60}Ni. In order to align the ^{60}Co nuclei the source was deposited in a crystal of cerium magnesium nitrate, which exerts a very strong internal magnetic field, and the thermal motions which would destroy the alignment were reduced to a minimum by cooling the crystal to 0.01 K by adiabatic demagnetisation. The ^{60}Co nuclei were aligned by applying an external magnetic field and the degree of alignment was measured by observing the anisotropy of the emitted γ-rays, which were detected by the sodium iodide crystals. When the cooling was complete, the magnetic field coils were removed and the decay electrons counted by means of an anthracene scintillator placed above the sample. The counting rate was measured as a function of time for nuclei with their spins aligned upwards and downwards. The results are shown in fig. 6.5 and show a clear anisotropy or correlation between the spin direction and the direction of emission, which decreases to zero as the sample warms up and the alignment disappears. Following our preliminary discussion, this observation indicates clear parity violation in the β-decay. This experiment is an example of a general method of detecting the effects of non-conservation of parity by looking for the effects on any process of terms, in the Hamiltonian for the system, which are not invariant under the parity transformation. Any pseudo-scalar term can produce such evidence of parity non-conservation. If **L** represents

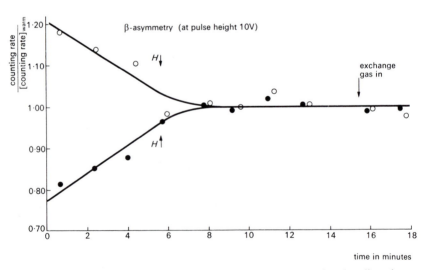

Fig. 6.5. The asymmetry in the β-ray counting rate for the aligned ^{60}Co nuclei in the experiment of Wu *et al.* (1957). A clear correlation between the β-direction and the alignment is observed, which decreases as the sample warms up and the alignment disappears.

the spin vector for the aligned nuclei then the quantity $\langle \mathbf{L} \rangle \cdot \mathbf{p}$ is a pseudo-scalar, where $\langle \mathbf{L} \rangle$ is the expectation value of $\mathbf{L}$. If parity is not conserved, then the intensity of the electrons may be a function of this quantity, leading to an electron distribution of the form

$$1 + x \cos \theta,$$

where we define $\cos \theta$ as $\mathbf{L} \cdot \mathbf{p} / |\mathbf{L} \cdot \mathbf{p}|$ and x is a constant.

Among other examples of parity non-conservation we may study the angular distribution of the Λ-decay. Taking the Λ-spin to be $\frac{1}{2}$ we see that the relative angular momentum of its decay products, the pion and the nucleon, can be 0 or 1. The states of zero or unit relative angular momentum will have odd or even parity respectively. If parity is conserved in the decay process only one of these states can be present, while if parity is not conserved we may have both states present and we may get interference between them. If we have Λ-particles which are polarised preferentially, either along or against the direction of the motion, then the effect of such interference may be observed in the angular distribution of the decay and it is, in fact, found that many processes resulting in the production of Λ-particles do yield such polarised Λs. In such a case, if θ is the angle between the Λ-direction and one of the decay particles in the decay cms, then interference between S-wave and P-wave decay processes will lead to a forward–backward anisotropy so that the angular distribution has the form

$$1 + \alpha \cos \theta,$$

where α is a constant. If the Λ-particles are not fully polarised, then the anisotropy is reduced and the distribution has the form

$$1 + \alpha P \cos \theta,$$

where P is the polarisation of the Λs, i.e.

$$P = \frac{N^+ - N^-}{N^+ + N^-}$$

where N^+ and N^- represent the numbers of Λs with spins oriented along and against the direction of motion. The asymmetry due to parity non-conservation in the decay has been observed in a number of experiments.

A fuller analysis shows that the proton emitted in the Λ-decay should be polarised if parity is not conserved. Such a polarisation has been observed. It implies that the decay is a function of the pseudo-scalar formed by the product of the proton spin and momentum.

A further consequence of parity non-conservation in β-decay processes involving neutrinos is that, for β-particles emitted even from non-aligned nuclei, the electrons are polarised such that their spin tends to lie in the same direction as the motion. This is sometimes expressed by saying that the helicity, which we may define as the cosine of the angle between the spin and the direction of motion, is equal to $+v/c \simeq +1$ for relativistic particles (v is the particle velocity).

Similar tests have demonstrated that parity is not conserved in the decay of the π and μ-mesons and, indeed, in no weak interaction process which has been investigated has parity conservation been found to hold good.

We shall return to this topic in more detail when considering the weak interaction in chapter 7.

7
Weak interactions: basic ideas

7.1   **Introduction**

We have already seen that there exists a class of interactions known as *weak* (chapter 4). The characteristic features of such interactions are:

(*a*) Long lifetimes for decay via weak processes ($\sim 10^{-10}$ s).

(*b*) Small cross-sections for interaction via weak interactions ($\sim 10^{-39}$ cm²) (except at very high energies).

(*c*) Leptons experience *only* weak and (if charged) electromagnetic interactions.

(*d*) Weak interactions do not conserve strangeness (chapter 5). Since strong interactions do conserve strangeness the strange particles decay weakly to non-strange particles with $|\Delta S| = 1$.

(*e*) Parity is not conserved in weak processes.

Despite the 'weakness' of these processes and the fact that in 'normal' matter they are in evidence only in the β-decay of radio-active nuclei, nevertheless they are of the greatest practical and theoretical importance. In main-sequence stars such as the Sun the thermonuclear reaction chain depends on the weak process

$$\text{pp} \rightarrow \text{d}\,\text{e}^{+}\nu_{e}$$

On the theoretical side it has recently been demonstrated that the weak and electromagnetic interactions are different aspects of the same fundamental process, lending new hope that ultimately all the interactions of matter may be unified. In this chapter the nature of the weak interaction is discussed in a more systematic manner than in the earlier parts of the book. We start with radioactive β-decay, the earliest studied weak process, related to which the terminology was established and the basic theory developed.

7.2   Neutron β-decay: phenomenology
Much of our knowledge of the basic β-decay processes

$$n \rightarrow pe^- \bar{\nu}_e$$
$$p \rightarrow ne^+ \nu_e$$
$$e^- p \rightarrow n\nu_e$$

is based on the β-decay of nuclei

$$(A, Z) \rightarrow (A, Z+1) + e^- + \bar{\nu}_e$$
$$(A, Z) \rightarrow (A, Z-1) + e^+ + \nu_e$$
$$(A, Z) + e^- \rightarrow (A, Z-1) + \nu_e.$$

For the nuclear processes the presence of the nucleons not participating directly in the process introduces some complications due to higher order processes. Nevertheless it is possible to extract the key features of the interaction relatively straightforwardly despite these difficulties.

In the so-called *allowed* transitions the outgoing particles carry away no *orbital* angular momentum. This is the most usual situation, as might be expected, since the outgoing particles normally have energies of the order of a few MeV. The wave function for the outgoing particles behaves as e^{ipr} and over the nuclear radius pr is easily seen to be $\sim 10^{-2}$ with $e^{ipr} \sim 1$. If e^{ipr} is expanded in terms of the orbital angular momentum l it is seen that $l = 0$ is dominant. Transitions with $l = 1$ are known as 'first forbidden' and those with $l = 2$ as 'second forbidden', and have lifetimes long compared with the allowed transitions to which we shall confine our attention.

Since e and ν have spin $\frac{1}{2}$ the possible changes in the nuclear spin are $\Delta \mathbf{J} = 0, 1$. Transitions with $\Delta \mathbf{J} = 0$ are known as *Fermi* transitions and with $\Delta \mathbf{J} = 1$ as *Gamow–Teller* transitions:

	Nuclear spin change	Lepton state
Fermi	$\Delta \mathbf{J} = 0$	singlet
Gamow–Teller	$\Delta \mathbf{J} = 1$	triplet
	$\Delta \mathbf{J}_z = 0, \pm 1$	

Since no orbital angular momentum is carried off there is no change in the nuclear parity and the nucleon configuration remains unchanged apart from the spin flip of the nucleon in the case of the Gamow–Teller transition.

Examples are observed of all three possible situations, Fermi, Gamow–Teller and mixed Fermi/Gamow–Teller transitions:

$$\text{Fermi}: 0^+ \to 0^+, \Delta \mathbf{J} = 0 \qquad \text{Gamow–Teller}: 1^+ \to 0^+, \Delta \mathbf{J} = 1$$

$$^{10}\text{C} \to {}^{10}\text{B}^*\text{e}^- \nu_e \qquad\qquad {}^{12}\text{B} \to {}^{12}\text{C}\,\text{e}^-\bar{\nu}_e$$

$$^{14}\text{O} \to {}^{14}\text{N}^*\text{e}^+ \nu_e$$

$$\text{Mixed}: \tfrac{1}{2}^+ \to \tfrac{1}{2}^+, \Delta \mathbf{J} = 0, 1$$

$$\text{n} \to \text{pe}^- \bar{\nu}_e$$

7.3 The matrix element for β-decay

In the original Fermi theory which accounts for many of the properties of the β-decay process it was taken to involve the interaction of four particles at a point:

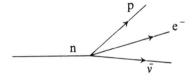

We can replace the process

$$\text{n} \to \text{pe}^- \bar{\nu}_e$$

by the more symmetrical situation

$$\nu + \text{n} \to \text{p} + \text{e}^-.$$

Fermi (1934) developed the first successful theory of the process in analogy with the well established theory for the generation of electromagnetic radiation where the amplitude is proportional to the electric current four-vector. For β-decay we have simultaneous transformation of $\text{n} \to \text{p}$ and $\nu_e \to \text{e}^-$, so the matrix element is written as proportional to the product of two 'currents'

$$M = Cj_{\text{lepton}} \cdot j_{\text{baryon}}$$

where C is the coupling constant setting the overall interaction strength. The 'currents' in this case are known as 'charged currents' since they involve transitions in which the electric charges of the participating particles change in contrast to the electromagnetic case involving absorption or emission of photons. The point interaction is also in contrast with the electromagnetic interaction mediated by photon exchange at ranges extending to infinity. We return to these questions in chapter 11 (11.2).

For the spin $\frac{1}{2}$ particles involved in β-decay the appropriate wave functions are the four-component spinors satisfying the Dirac equation. The matrix element has dimensions of an energy density (see section 4.2) and the lepton current, for example, has the form

$$j_{\text{lepton}} \propto \bar{\psi}_e O \psi_v$$

where O is a matrix operator transforming $v \to e$. Relativistic invariance places certain restrictions on the form of matrix element . Five covariant forms associated with the Dirac equation satisfy the requirements and in general we may write

$$M = \sum_i C_i(\bar{\psi}_p O_i \psi_n)(\bar{\psi}_e O_i \psi_v)$$

where, initially, we assume parity conservation.

The O_i may have scalar, vector, tensor, axial-vector or pseudo-scalar forms (S, V, T, A and P). The S and V interactions can only be associated with Fermi transitions. The T and A interactions can produce spin changes and are thus possible forms to account for Gamow–Teller transitions. The P interaction results in a factor of v/c in the matrix element where the nucleon velocity v in transitions is always $\ll c$ so that this interaction is certainly unimportant. Thus we can write the complete matrix element as

$$M = \sum_{i=S,V} C_i(\bar{\psi}_p O_i \psi_n)(\bar{\psi}_e O_i \psi_v) + \sum_{j=T,A} C_j(\bar{\psi}_p O_j \psi_n)(\bar{\psi}_e O_j \psi_v).$$

$$\text{Fermi} \qquad\qquad \text{Gamow–Teller} \qquad \textbf{7.1}$$

The Cs here are coupling constants for the different interactions. Now we can turn to the experimental evidence which enables us to take the first steps to distinguish the effective interactions.

(*a*) *Electron energy spectrum.* The detailed analysis involves some knowledge of the properties of the Dirac spinors and matrices which we do not treat in the present text (see, for instance, Halzen and Martin). We simply quote the result obtained by integrating the expression for the transition probability over the angles to obtain for electrons (n_-) and positrons (n_+)

$$\frac{dn_\mp}{dE_e} = \frac{p_e E_e}{2\pi^3}(E_0 - E_e)^2$$
$$\times \left[|M_F|^2(C_S^2 + C_V^2) + |M_{GT}|^2(C_T^2 + C_A^2) \right.$$
$$\left. \pm \frac{2m_e}{E_e}(|M_F|^2 C_S C_V + |M_{GT}|^2 C_T C_A) \right] \quad 7.2$$

where the M_F and M_{GT} are the Fermi and Gamow–Teller nuclear matrix elements and E_0 is the maximum possible electron energy.

Since pure Fermi and pure Gamow–Teller transitions are observed it is clear that we cannot have

$$C_S = C_V = 0 \quad \text{or} \quad C_A = C_T = 0.$$

An analysis of the detailed spectrum shapes in pure Fermi and pure Gamow–Teller transitions allows determinations of the ratios

$$\frac{C_S C_V}{C_S^2 + C_V^2} = 0.00 \pm 0.15, \quad \frac{C_T C_A}{C_T^2 + C_A^2} = 0.00 \pm 0.02.$$

Thus (Michel, 1957) these data suggest that either C_S or C_V and either C_A or C_T equal zero, i.e. there are only two effective couplings. We shall see that these are C_V and C_A.

(b) *Electron–neutrino correlations.* The correlations between electron and neutrino momenta $\mathbf{p}_e$ and $\mathbf{p}_\nu$ are given by

$$-(C_S^2 - C_V^2)\frac{\mathbf{p}_e \cdot \mathbf{p}_\nu}{E_e E_\nu} \quad \text{and} \quad \frac{1}{3}(C_T^2 - C_A^2)\frac{\mathbf{p}_e \cdot \mathbf{p}_\nu}{E_e E_\nu}$$

Fermi   Gamow–Teller

(the $\frac{1}{3}$ factor is due to the three possible orientations for the $J = 1$ total angular momentum of the leptons, thus reducing the correlations).

Writing

$$\frac{\mathbf{p}_e \cdot \mathbf{p}_\nu}{E_e E_\nu} = a_e \cos\theta_{e\nu}$$

the transition probability for a pure Fermi transition is proportional to

$$C_S^2(1 - a_e \cos\theta_{e\nu}) \quad \text{if } C_V = 0$$
$$C_V^2(1 + a_e \cos\theta_{e\nu}) \quad \text{if } C_S = 0$$

and for a pure Gamow–Teller transition

$$C_T^2(1 + \tfrac{1}{3}a_e \cos\theta_{e\nu}) \quad \text{if } C_A = 0$$
$$C_A^2(1 - \tfrac{1}{3}a_e \cos\theta_{e\nu}) \quad \text{if } C_T = 0.$$

The experiments measure the electron and nuclear recoil directions from which the neutrino direction can be inferred. The results provide unambiguous evidence for pure V, A couplings.

(c) *'ft values' and coupling constants.* With the above simplifications we write **7.2** as

$$\frac{\mathrm{d}n}{\mathrm{d}E_e} = \frac{p_e E_e}{2\pi^3}(E_0 - E_e)^2 [C_V^2 |M_F|^2 + C_A^2 |M_{GT}|^2].$$

The mean lifetime τ for the decay is obtained by integrating over energy:

$$n = \frac{1}{\tau} = \frac{1}{2\pi^3}[C_V^2 |M_F|^2 + C_A^2 |M_{GT}|^2] \int_{m_e}^{E_0} p_e E_e (E_0 - E_e)^2 \, \mathrm{d}E_e.$$

The integral is conventionally denoted by $m_e^5 f$ so that f is dimensionless. The ft value is then given by

$$(f\tau)^{-1} = \frac{m_e^5}{2\pi^3}[C_V^2 |M_F|^2 + C_A^2 |M_{GT}|^2]$$

and the value of the coupling constant may be obtained from measurements of $(f\tau)$. It is conventional to use the half life t rather than the mean life τ. Typical ft values for Fermi transitions are of the order of 3000 s, which yields a value of

$$C_V = 1.4 \times 10^{-49} \, \mathrm{erg \, cm^3} = \frac{10^{-5}}{M_p^2} \quad (\hbar = c = 1).$$

For allowed transitions the nuclear matrix elements $|M_F|^2 = |M_{GT}|^2 = 1$.

We have seen that we expect the bare neutron decay to be a mixed transition. Since we have three possible spin orientations in the Gamow–Teller case there is a weighting factor of 3 associated with the Gamow–Teller term and

$$\frac{1}{ft} = (C_V^2 + 3C_A^2)\frac{m_e^5}{2\pi^3 \ln 2} = (1080 \pm 16)^{-1} \, \mathrm{s}.$$

We can obtain the ratio of C_V/C_A by comparison with a pure vector or axial-vector decay, the conventional example being the (Fermi) decay of $^{14}O \rightarrow ^{14}N^*$. In this case the decay can come from two protons outside the ^{12}C core, thus requiring inclusion of an additional factor 2

$$\frac{(ft)^{14}O}{(ft)n} = \frac{3100 \pm 20}{1080 \pm 16} = \frac{C_V^2 + 3C_A^2}{2C_V^2}$$

yielding

$$\left|\frac{C_A}{C_V}\right| = 1.25 \pm 0.02$$

7.4   Parity non-conservation in β-decay

We have already seen that there is unambiguous experimental evidence for parity non-conservation in β-decay. Lee and Yang (1956) had pointed out before the experiment on ^{60}Co decay that the matrix element of equation **7.1** (which is a scalar) would require modification by addition of a pseudo-scalar part to require parity non-conservation in the interaction, so that C_i is replaced by

$$(C_i + C_i^1 \gamma_5) \frac{1}{\sqrt{2}}.$$

The factor of $1/\sqrt{2}$ is introduced to retain the numerical value of C_V given above. γ_5 in this expression is a matrix operator which ensures that the second term $C_i^1 \bar{\psi}_e O_i \gamma_5 \psi_v$ is *pseudo-scalar* ($\gamma_5^2 = 1$). The matrix element thus becomes

$$M = \sum_{V,A} \frac{1}{\sqrt{2}} (\bar{\psi}_p O_i \psi_n)[\bar{\psi}_e O_i (C_i + C_i^1 \gamma_5)\psi_v].$$

We may write the lepton bracket as

$$C_i \bar{\psi}_e O_i \psi_v + C_i^1 \bar{\psi}_e O_i \gamma_5 \psi_v$$
$$= \frac{(C_i + C_i^1)}{2} \bar{\psi}_e O_i (1 + \gamma_5)\psi_v + \frac{(C_i - C_i^1)}{2} \bar{\psi}_e O_i (1 - \gamma_5)\psi_v. \quad \textbf{7.3}$$

It follows from the Dirac equation applied to massless particles that the operator $(1 + \gamma_5)$ projects out positive helicity states for $\bar{v}$ and negative helicity states for v while the operator $(1 - \gamma_5)$ has the opposite effect:

	$\bar{v}$	v
$1 + \gamma_5$:	right-handed	left-handed
	positive helicity	negative helicity
$1 - \gamma_5$:	left-handed	right-handed
	negative helicity	positive helicity

In other words, the $(1 \pm \gamma_5)\psi$ are the positive and negative eigenstates of the helicity operator.

In fact we have already seen (section 6.5) that for leptons emitted along the z-axis in the ^{60}Co decay experiment of Wu *et al.* the particles are longitudinally spin-polarised. The actual helicity of the neutrino was unambiguously determined to be negative in an experiment by Goldhaber, Grodzins and Sunyar (1958). The process studied was electron capture in ^{152}Eu:

$$e^- + {}^{152}\text{Eu} \rightarrow {}^{152}\text{Sm}^* + \nu.$$

The excited state of Samarium decays by γ-emission to the ground state:

$$^{152}\text{Sm}^* \rightarrow {}^{152}\text{Sm} + \gamma.$$

The level diagram with the spins and parities of the states is shown in fig. 7.1(a). γ-rays from the ^{152}Sm* which are emitted along the direction of the recoiling ^{152}Sm* nucleus are selected by detecting the γs after resonant scattering on a second Sm target (fig. 7.1(b)). Only the γs along the line of the recoiling nucleus – i.e. opposite to the ν-direction – have enough energy to excite the resonance level and give resonance scattering.

From the spin values for the levels we see that the recoiling ^{152}Sm* must have its spin 1 aligned opposite to the electron and neutrino spins (fig. 7.1(c)). Thus the ^{152}Sm* has the same helicity as the ν so that the *helicities of the γ-ray and the neutrino must be the same.*

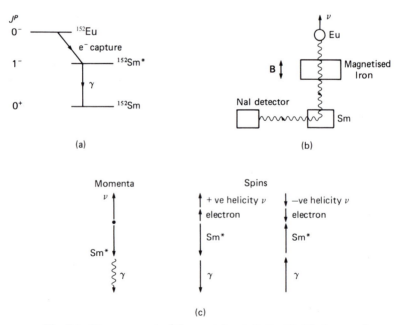

Fig. 7.1. Measurement of the neutrino helicity (Goldhaber *et al.*, 1958).

The γ-ray helicities were measured by passing the photons through iron which could be magnetised either parallel or antiparallel to the γ-direction. The scattering of the γs in the iron depends on the relative orientations of the γ-helicity and the magnetisation direction, since electrons with spin opposite to the γ-angular momentum can absorb this angular momentum by spin flip, whereas for an electron with spin parallel to the helicity there is no equivalent process. The absorption of the γs in the iron is therefore greater when magnetisation and helicity are opposite and this phenomenon allows a measurement of the γ-helicity. After allowing for known depolarising effects the data were found to be compatible with completely left-handed spin polarised neutrinos and $C_i = C_i^1$ in equation **7.3**.

Finally in this section we discuss briefly the relative signs of the V and A couplings. This parameter was measured in a study of the decay of polarised neutrons. When applied to this process, detailed development of equation **7.3** with insertion of appropriate operators and spinors shows that different spin configurations for the participating particles have amplitudes proportional to the combinations of coupling constants shown in fig. 7.2. Thus the amplitude of $(C_A + C_V)$ corresponds to electrons polarised opposite to and $(C_A - C_V)$ to electrons polarised parallel to, the neutron polarisation direction. In the experiment of Burgy *et al.* (1958) neutrons were polarised by reflection from magnetised cobalt mirrors, the polarisation direction being reversible by switching the direction of magnetisation. The link between polarisation and momentum (helicity) for the leptons then leads to different ratios of electrons along and opposite to the neutron spin for the $C_A + C_V$ and $C_A - C_V$ amplitudes. In the experiment both electron and proton were detected so that the momenta of all the particles were determined. The result gave

$$\frac{C_A}{C_V} = -1.26 \pm 0.02,$$

Fig. 7.2. Spin configurations and couplings in the decay of polarised neutrons.

consistent in magnitude with the number quoted earlier on the basis of ft values and in addition demonstrating that the couplings are of opposite sign or that the interaction is of the form 'V minus A'.

The properties of muon decay,

$$\mu \to \mathrm{e}\nu\bar{\nu}$$

muon capture

$$\mu^{-}\mathrm{p} \to \mathrm{n}\nu$$

and pion decays

$$\pi \to \mu\nu$$
$$\pi \to \mathrm{e}\nu$$
$$\pi^{-} \to \pi^{0}\mathrm{e}^{-}\bar{\nu}$$

may also be compared with the predictions for the V – A interaction. The experimental results are found to be in close agreement with the V – A expectations. For instance, the ratio $(\pi \to \mathrm{e}\nu)/(\pi \to \mu\nu)$ for purely axial coupling is found to be $(1.267 \pm 0.023) \times 10^{-4}$ compared with a prediction $\sim 1.28 \times 10^{-4}$ and the ratio $\pi^{\pm} \to \pi^{0}\mathrm{e}^{\pm}\nu/\pi^{\pm} \to \mu^{\pm}\nu = (1.02 \pm 0.07) \times 10^{-8}$ compared with the predicted value of $(1.06 \pm 0.02) \times 10^{-8}$. (The last result also involves the additional 'conserved vector current' hypothesis.)

In summary, all the data on strangeness conserving weak interactions is consistent with:

- (*i*) lepton conservation;
- (*ii*) 'two-component' neutrinos with left-handed neutrinos and right-handed antineutrinos;
- (*iii*) V – A interaction;
- (*iv*) universality of properties of muons and electrons (apart from mass).

7.5   Weak decays of strange particles

We can readily make some simple deductions concerning the weak decays of the strange particles at this stage. Other features of these decays, particularly concerning the processes involving leptons, are more readily discussed after treating the quark structure for hadrons which we will meet in chapter 10.

We first consider non-leptonic decays like

$$\Lambda \to \begin{cases} \mathrm{p}\pi^{-} \\ \mathrm{n}\pi^{0} \end{cases}.$$

We note that, since the nucleon and pion can only combine to give total isotopic spin $I = \frac{1}{2}$ or $\frac{3}{2}$ while the I-spin of the Λ is zero, therefore the I-spin cannot be conserved in this process. Gell-Mann suggested that the change in I-spin would be the minimum possible, i.e.

$$|\Delta I| = \tfrac{1}{2}.$$

This rule is more restrictive than the $|\Delta I_3| = \tfrac{1}{2}$ rule which follows from the relation

$$Q = I_3 + \tfrac{1}{2}(B+S)$$

coupled with $|\Delta S| = 1$ and $\Delta Q = \Delta B = 0$ which we have already met for the weak decays of strange particles. We can write I-spin wave functions for $(p\pi^-)$ and $(n\pi^0)$ in terms of the I-spin $\tfrac{1}{2}$ and $\tfrac{3}{2}$ wave functions using the Clebsch–Gordan coefficients (appendix B)

$$|p\pi^-\rangle = \sqrt{\tfrac{1}{3}}|\tfrac{3}{2}\rangle - \sqrt{\tfrac{2}{3}}|\tfrac{1}{2}\rangle$$
$$|n\pi^0\rangle = \sqrt{\tfrac{2}{3}}|\tfrac{3}{2}\rangle + \sqrt{\tfrac{1}{3}}|\tfrac{1}{2}\rangle.$$

These relations then give for the decay ratio

$$\frac{\Lambda \to p\pi^-}{\Lambda \to n\pi^0} = \begin{cases} 2:1 & \text{for } I = \tfrac{1}{2} \quad \text{i.e. } \Delta I = \tfrac{1}{2} \\ 1:2 & \text{for } I = \tfrac{3}{2} \quad \text{i.e. } \Delta I = \tfrac{3}{2} \end{cases}.$$

A small correction for the slightly different phase space available in the two decays changes the $I = \tfrac{1}{2}$ ratio from 2:1 to 1.90:1 to be compared with the experimentally measured $(1.80 \pm 0.03):1$. Some further small corrections for electromagnetic effects bring the measured and calculated numbers into good agreement.

Some decay ratios do, however, indicate the presence of a small contribution from $\Delta I = \tfrac{3}{2}$. For instance, the ratio for the decay of the I-spin singlet Ω^- to $\Xi\pi$ is analogous to that for the Λ so that we expect for $\Delta I = \tfrac{1}{2}$

$$\frac{\Omega^- \to \Xi^0\pi^-}{\Omega^- \to \Xi^-\pi^0} = 2$$

with some possible very small corrections. The measured value is 2.93 ± 0.33. The Σ-decays are more complicated in that even $\Delta I = \tfrac{1}{2}$ allows both $I = \tfrac{1}{2}$ and $I = \tfrac{3}{2}$ states since we start from $I = 1$ for the Σ. The detailed analysis also suggests some element of $\Delta I = \tfrac{3}{2}$.

We have already seen in chapter 5 that the strange particles have decay modes involving leptons such as for instance

$$\Lambda \to pe^-\nu$$
$$K^0 \to \pi^-e^+\nu.$$

The branching ratios for the leptonic modes of the hyperons are small, so that they are not easy to study. Very important advances in the study of such modes were made by producing low energy hyperons in very large numbers in a liquid hydrogen bubble chamber. In an experiment of this kind at CERN 10^7 K^--mesons were stopped in a hydrogen chamber. Candidates for leptonic decays of hyperons were selected by a crude measurement using fast rough digitisers connected on-line to a computer. Events passing the rough filter were studied in detail with measurements of high precision. Heavy liquid bubble chambers used in studies of this kind allow the detection of γ-rays and the easy identification of electrons.

As for the non-leptonic modes we have $|\Delta S| = 1$. So that we again have

$$\Delta Q = \tfrac{1}{2}\Delta S + \Delta I_3$$

but here the changes in Q, S and I_3 refer to the *strongly interacting particles* only, for which the original relationship holds. Since we now also have additional charged particles present, ΔQ need no longer be zero. Since ΔQ and ΔS can each equal ± 1 two possibilities apparently exist:

$$\Delta Q = \Delta S, \quad |\Delta I_3| = \tfrac{1}{2} \quad \text{and} \quad |\Delta I| \geqslant \tfrac{1}{2}$$
$$\Delta Q = -\Delta S, \quad |\Delta I_3| = \tfrac{3}{2} \quad \text{and} \quad |\Delta I| \geqslant \tfrac{3}{2}.$$

Note that although $|\Delta I| = \tfrac{1}{2}$ implies $\Delta Q = \Delta S$ the converse is not true.

In fact all the experimental evidence suggests that the '$\Delta Q = \Delta S$' rule is obeyed and that $\Delta Q = -\Delta S$ processes are at least strongly inhibited. Perhaps the best example concerns the Σ-decays

$$\Sigma^- \to n e^- \bar{\nu} \quad \Delta Q = \Delta S$$
$$\Sigma^+ \to n e^+ \nu \quad \Delta Q = -\Delta S.$$

The ratio of the branching ratio of the $\Delta Q = -\Delta S$ process to that for $\Delta Q = \Delta S$ is $< 5 \times 10^{-3}$.

8

Invariance under the *CP* and *T* operations, properties of K^0-mesons

8.1   The *TCP* theorem

The *TCP* theorem, derived in different forms by several workers (Schwinger in 1953, Luders in 1954 and Pauli in 1955), states that for locally interacting fields, a Lagrangian which is invariant under proper Lorentz transformations is invariant with respect to the combined operation *TCP*. By the operation *TCP* is meant the set of operations time reversal, charge conjugation (i.e. particle-antiparticle exchange) and the parity transformation, taken in any order. The proof applies only to the combined set of operations even though the theory may not be invariant under the individual operations *T*, *C* and *P*.

It is an obvious consequence of the *TCP* theorem that if an interaction is not invariant under one of the operations it must also fail to be invariant under the product of the other two.

We have already seen that for weak decays parity is not conserved, so that if these processes are to be invariant under time reversal they cannot be invariant under charge conjugation. A number of tests have demonstrated that this is in fact the case. For instance, in the decay of muons we have already mentioned that the decay electrons are fully polarised. If charge conjugation holds, we should expect that the helicity of the positron from μ^+-decay should be the same as that of the electron from μ^--decay, since these two processes are the charge conjugates of each other and since the only effect of the charge conjugation operator is to interchange particle and antiparticle. The helicities of the electrons and positrons from muon decay have been measured by observing the transmission through magnetised iron of the bremsstrahlung photons emitted by these particles. The transmission coefficient is a function of the photon helicity, which depends on the helicity of its parent particle. The result of these measurements indicates that, in fact, the positrons were fully right-handed, and the electrons fully left-handed, giving maximum violation of charge conjugation. Similar effects have been observed for both pion decay and nuclear β-decay.

A further interesting result concerning charge conjugation and parity can be derived with the aid of the *TCP* theorem. It can be shown that if the Hamiltonian responsible for a transition is invariant under charge conjugation, then the parity conserving and parity non-conserving final states cannot interfere with each other. Thus the observation of such interference effects, as for instance in the decay of the Λ-particle, demonstrated that not only is parity not conserved in this process, but neither is the process invariant under charge conjugation.

8.2   The K^0-particles and invariance under the operator CP

In order to discuss this problem we must consider the nature of the state functions for the K^0 and its antiparticle the $\bar{K}^0$. On the basis of our earlier discussion in chapter 6, we assume spin zero for the K^0s. In fact, for bosons which are different from their antiparticles the state function must be complex. Although this fact can only be proved by reference to the field theory for these particles, it may be rendered more plausible by reference to the modifications necessary to the Schrödinger equation for the inclusion of an electromagnetic field. In such a case we must replace the operators $\partial/\partial x$ by $(\partial/\partial x) - ieA$ and $\partial/\partial t$ by $(\partial/\partial t) + ieV$ where A and V are the vector and scalar potentials respectively. A consequence of this modification is that the wavefunctions representing the field must be complex. We therefore write the wave functions in the form

$$\phi = \phi_1 + i\phi_2$$

where ϕ_1 and ϕ_2 are real. In addition, the continuity equations require that particles represented by complex conjugate wave functions have opposite charge and current densities. Thus the antiparticle of that represented by the wave function ϕ will be represented by

$$\phi^* = \phi_1 - i\phi_2.$$

We may now use such wave functions for the kaon field and examine the effects of the charge conjugation operator. We know that

$$C\phi = \phi^*$$

so that $C\phi_1 = \phi_1$ and $C\phi_2 = -\phi_2$.

For particles such as the π^0 and the photon, which are the same as their antiparticles, $\phi_2 = 0$. Under charge conjugation, however, the wave function may still change sign since the eigenvalue of C may be ± 1. For charged particles and neutral particles which have distinct antiparticles, ϕ_2 is not equal to zero. Experiment shows that the K^0 and the $\bar{K}^0$ are certainly different, having opposite strangeness, so that they should be represented by complex wave functions. We note, however, that there is an important difference between the situation for the neutral K-mesons and that for other particle-antiparticle pairs. For charged particles, virtual transitions between particle and antiparticle are always forbidden by charge conservation. In addition, for baryons, such as for instance the neutron and the antineutron, virtual transitions between particle and antiparticle are forbidden by baryon conservation. The laws that forbid these transitions are true for all kinds of interactions. For the K^0 and the $\bar{K}^0$, on the other hand, transitions between particle and antiparticle are possible via the weak interaction (second order, $\Delta S = 2$ transitions) since they involve only violation of strangeness conservation. We may thus expect to find mixing of the neutral K-mesons, and this effect was predicted by Gell-Mann and Pais in 1954.

We write

$$\phi_{K^0} = \frac{1}{\sqrt{2}}(\phi_1 + i\phi_2)$$

and

$$\phi_{\bar{K}^0} = \frac{1}{\sqrt{2}}(\phi_1 - i\phi_2).$$

Although we do not expect invariance of the wave functions of the decaying particles under C and P separately, we do expect invariance under the operation CP. Applying this operator to the wave functions ϕ_{K^0} and $\phi_{\bar{K}^0}$ we can adjust the relative phase of the states to get

$$CP\phi_{K^0} = \phi_{\bar{K}^0} \quad \text{and} \quad CP\phi_{\bar{K}^0} = \phi_{K^0}.$$

Thus ϕ_{K^0} and $\phi_{\bar{K}^0}$ are not eigenfunctions of CP.

For the decay process we do not require eigenfunctions of the strangeness operator, and Gell-Mann and Pais proposed for the reasons described above that the appropriate wave functions should be the mixed wave functions ϕ_1 and ϕ_2

$$\phi_1 = \frac{1}{\sqrt{2}}(\phi_{K^0} + \phi_{\bar{K}^0}) \quad \text{and} \quad \phi_2 = \frac{-i}{\sqrt{2}}(\phi_{K^0} - \phi_{\bar{K}^0}).$$

These functions are indeed eigenfunctions of CP since

$$CP\phi_1 = \frac{1}{\sqrt{2}}(CP\phi_{K^0} + CP\phi_{\bar{K}^0}) = \frac{1}{\sqrt{2}}(\phi_{\bar{K}^0} + \phi_{K^0}) = \phi_1$$

and

$$CP\phi_2 = \frac{i}{\sqrt{2}}(\phi_{\bar{K}^0} - \phi_{K^0}) = -\phi_2.$$

The eigenvalues of ϕ_1 and ϕ_2 under CP are thus $+1$ and -1 respectively. These are the functions appropriate for the K^0-decays and the particles corresponding to which we might expect to exhibit different decay features. They are known as the K_1^0 and the K_2^0.

We may now examine the effect of the CP operation on the final state in the decay. For a two-pion system of relative orbital angular momentum L we have seen that the parity operator has eigenvalue $(-1)^L$. Application of the charge conjugation operator interchanges the positive and negative pions in the charged decay. As shown in Box 1 (using the usual property of the spherical harmonics) the eigenvalue of C is also $(-1)^L$. Thus the eigenvalue of the combined operation CP is $(-1)^{2L}$; i.e. the two-pion system has eigenvalue $+1$ under this operation. This means that to preserve invariance under the CP operation *only* the K_1^0 can decay to two pions.

Box 1 Effect of the CP operation on a $\pi^+\pi^-$ system with relative orbital angular momentum L

$$CP\left(\frac{\pi^+ \uparrow^L \pi^-}{}\right) \to C(-1)^L\left(\frac{\pi^+ \uparrow^L \pi^-}{}\right)$$

$$\to (-1)^L\left(\frac{\pi^- \uparrow^L \pi^+}{}\right)$$

$$\to (-1)^L\left(\frac{\pi^+ \quad \pi^-}{\downarrow_L}\right)$$

$$\to (-1)^L(-1)^L\left(\frac{\pi^+ \uparrow^L \pi^-}{}\right)$$

thus $\quad CP[\psi(\pi^+\pi^-)] = (-1)^{2L}\psi(\pi^+\pi^-)$

$$= +1\psi(\pi^+\pi^-).$$

In a similar way it may be shown that the eigenvalues for a three-pion system under the CP operation are $-(-1)^L$ where L is the orbital angular momentum quantum number of the neutral pion with respect to the di-pion (fig. 8.1 and Box 2). For the K_2^0 with eigenvalue -1 under CP only the three-pion or one-pion plus two-lepton modes of decay are permitted. Gell-Mann and Pais proposed that the lifetime for these three-particle decay modes, i.e. for the K_2^0, should be about a thousand times longer than for the K_1^0 on the basis of the decay rate for the similar process

$$K^+ \to \mu^+ \pi^0 \nu$$

for the charged K-meson.

Box 2 The effect of the CP operation on a $\pi^+\pi^-\pi^0$ system shown in fig. 8.1

$$CP[\psi(\pi^+\pi^-\pi^0)] \to (-1)^3(-1)^l(-1)^L C[\psi(\pi^+\pi^-\pi^0)]$$
$$\to -(-1)^{l+L} C[\phi(\pi^+\pi^-)]\alpha(\pi^0)$$
$$\to -(-1)^{l+L}(-1)^l \psi(\pi^+\pi^-\pi^0)$$
$$\to -(-1)^L \psi(\pi^+\pi^-\pi^0).$$

We should expect that the three-particle modes be inhibited relative to the two-particle decays simply due to the difference in available phase space. Thus the K_1^0 will not, in general, decay via the three-particle modes. For the $(\pi^0\pi^0\pi^0)$ combination the eigenvalue under CP can only be -1, so that K_1^0 decay to three neutral pions is forbidden by CP invariance.

8.3   The development of a K^0-beam

We shall see in the next section that even the postulate of invariance under the CP operation has turned out to be not exactly true. However, the violation is very small and the conclusions which we shall draw here, based on the results of the last section, are largely unaffected.

Fig. 8.1. Definition of angular momenta in the neutral kaon decay to three pions.

We shall consider the development of a beam of K^0-mesons. Let us suppose that the K^0s are generated in the process

$$\pi^- p \rightarrow \Lambda K^0.$$

In this strong interaction, strangeness is conserved and we have a pure K^0-state which we may regard as consisting of 50 % of K_1^0s and 50 % of K_2^0s. After about 10^{-9} s, as measured in the kaon rest system, nearly all the K_1^0s will have decayed into pion pairs. The intensity of the beam has been reduced to half the original intensity and it now consists of almost pure K_2^0-particles. A xenon bubble chamber has been used to show that 0.53 ± 0.05 of all K^0s decay via two-pion modes. This remaining beam of K_2^0s is no longer a state of pure strangeness, having equal strangeness $+1$ and strangeness -1 components. Thus, whereas the original K^0-beam could produce only reactions having a final state of strangeness $+1$, the stale beam can produce reactions of final state strangeness -1, such as

$$\bar{K}^0 n \rightarrow \Sigma^- \pi^+, \Sigma^+ \pi^-$$
$$\bar{K}^0 p \rightarrow \Lambda \pi^+$$

as well as the $S = +1$ reactions.

We recall that the cross-section for reactions involving strangeness -1 particles is much greater than that for strangeness $+1$, so that if the stale beam is allowed to pass through material then the $S = -1$ component is preferentially removed. Thus this process will again produce a preponderance of K^0 over $\bar{K}^0$ particles, the beam therefore again containing a proportion of K_1^0 which will decay by two-pion modes.

Experiments have, in fact demonstrated this regeneration property and have shown that a beam consisting initially only of K^0-mesons can be used after a suitable time to produce Λ- and Σ-particles as predicted by the Gell-Mann scheme.

Several authors have suggested an illustrative analogy between the behaviour of the K^0s and that of polarised light. Consider a circularly-polarised beam of light. This may be regarded as consisting of beams plane-polarised in orthogonal directions, which may be thought of as corresponding to the K_1^0 and K_2^0. If the beam is now passed into a bi-refringent material, one plane-polarised component will be removed leaving plane-polarised light. We take this as the analogue of the decay of the K_1^0. If the surviving component is now passed into a quarter-wave quartz plate then the plane of polarisation will be rotated and once more we have components in both the original planes of polarisation, a process which we compare with the regeneration of the K_1^0.

8.4 Time variation of the neutral kaons

We recall that for a stable stationary state solution of the wave equation, for a particle of mass m, the solution contains a phase factor e^{-imt}, where as usual we have taken units such that $\hbar = c = 1$ and m is in units of reciprocal time.

If the state is unstable, undergoing an exponential decay, then an additional phase factor $e^{-\frac{1}{2}\Gamma t}$ must be included, where Γ is the decay width, so that the mean lifetime is equal to Γ^{-1}. The combined phase-factor is often written as e^{-iMt}, where

$$M = m - \tfrac{1}{2}i\Gamma.$$

In these terms we can write the non-relativistic wave function for the K^0-state, after a proper time t, as

$$\psi(t) = \frac{1}{\sqrt{2}}(|K_1^0\rangle e^{-iM_1 t} + i|K_2^0\rangle e^{-iM_2 t}) \qquad \textbf{8.1}$$

where $M_1 = m_1 - \tfrac{1}{2}i\Gamma_1$ and $M_2 = m_2 - \tfrac{1}{2}i\Gamma_2$ and m_1 and m_2 are the K_1^0 and K_2^0 masses. As required, at time $t = 0$

$$\psi(0) = \frac{1}{\sqrt{2}}(|K_1^0\rangle + i|K_2^0\rangle).$$

It is clear from the expression **8.1** that $K_1^0 - K_2^0$ interference is possible and that the interference terms will involve the mass difference, $\Delta m = m_2 - m_1$, between the K_2^0 and the K_1^0. As time passes the K_1^0-component decreases relative to the K_2^0-component, because $\Gamma_1 \gg \Gamma_2$, and in addition its phase changes relative to the K_2^0 by an amount which is a function of Δm. This result is the key to a number of different interference effects and affords a method of measuring Δm.

Possibly the most obvious such interference effect is that of '*strangeness oscillation*'. Let us evaluate the intensity $|\psi(t)|^2$, where we write $\psi(t)$ explicitly in terms of $|K^0\rangle$ and $|\bar{K}^0\rangle$

$$\psi(t) = \tfrac{1}{2}(|K^0\rangle + |\bar{K}^0\rangle)e^{-iM_1 t} + \tfrac{1}{2}(|K^0\rangle - |\bar{K}^0\rangle)e^{-iM_2 t}.$$

Then multiplying by the complex conjugate quantity, and extracting the appropriate terms, we find for the intensity of the K^0-component

$$N(K^0) \propto \tfrac{1}{4}[e^{-\Gamma_1 t} + e^{-\Gamma_2 t} + 2\cos[(m_2 - m_1)t]e^{-\frac{1}{2}(\Gamma_1 + \Gamma_2)t}] \qquad \textbf{8.2}$$

and for the $\bar{K}^0$

$$N(\bar{K}^0) \propto \tfrac{1}{4}[e^{-\Gamma_1 t} + e^{-\Gamma_2 t} - 2\cos[(m_2 - m_1)t]e^{-\frac{1}{2}(\Gamma_1 + \Gamma_2)t}] \qquad \textbf{8.3}$$

As expected

$$N(K^0) + N(\bar{K}^0) \propto \tfrac{1}{2}(e^{-\Gamma_1 t} + e^{-\Gamma_2 t}).$$

For times short compared with $\tau_2 (= 1/\Gamma_2)$, we have the simpler expressions

$$N(K^0) \propto \tfrac{1}{4}[1 + e^{-\Gamma_1 t} + 2\cos(\Delta m \cdot t) e^{-\frac{1}{2}\Gamma_1 t}]$$
$$N(\bar{K}^0) \propto \tfrac{1}{4}[1 + e^{-\Gamma_1 t} - 2\cos(\Delta m \cdot t) e^{-\frac{1}{2}\Gamma_1 t}].$$

Thus in principle a study of the variations with time of the number of K^0-and $\bar{K}^0$-mesons in a beam consisting originally purely of K^0-particles ('strangeness oscillations'), affords a method of determining the mass difference Δm. Some examples of the effects to be expected for different values of Δm are shown in fig. 8.2. This figure illustrates the extraordinary sensitivity of the method, which makes it possible to measure a mass difference of as little as 10^{-6} eV/c^2 or 10^{-39} gm between the two K^0s. In practice it is possible to produce K^0-mesons in a bubble chamber and to study the proportion of $\bar{K}^0$ as a function of time, identifying such $S = -1$ particles by their production of hyperons. Several experiments of this kind have been performed yielding $\Delta m = (3.521 \pm 0.015) \times 10^{-6}$ eV/c^2. m_2 is found to be greater than m_1 in separate regeneration experiments.

8.5   Failure of invariance under *CP* in K^0-decay

Having analysed the K^0-decay phenomena on the basis of invariance under *CP*, we must now discuss the evidence provided by more recent experiments which shows quite clearly that there is a violation of *CP* invariance in these processes. The analysis in the preceding section is nevertheless a very good approximation to reality, since the degree of *CP*

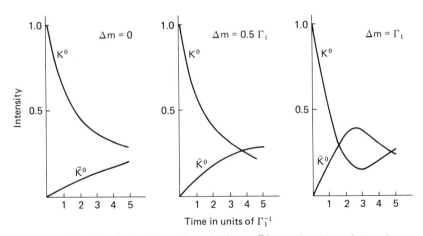

Fig. 8.2. Relative intensities of K^0 and $\bar{K}^0$ as a function of time, in a beam starting as pure K^0, for different values of Δm.

violation is very small. We may note that this is in sharp contrast to the violation of parity invariance in weak interactions, where the breaking is maximal.

The first experiment which demonstrated failure of CP invariance was carried out by Christenson, Cronin, Fitch and Turlay (1964), who detected a two-pion decay mode for the K_2^0 while studying regeneration phenomena. This development led to a change in nomenclature to K_S^0 (short-lived) and K_L^0 (long-lived) instead of K_1^0 ($CP = +1$) and K_2^0 ($CP = -1$).

The detection apparatus is shown in fig. 8.3. The experiment was carried out at the Brookhaven proton synchrotron where a beam from an internal target was selected at 30° to the direction of the 30 GeV internal circulating proton beam by means of a lead collimator. On the target side of the collimator was placed a 4 cm-thick block of lead which acted as a filter for γ-rays, while a bending magnet after the collimator swept charged particles from the beam. The detecting apparatus was placed behind a further collimator at a distance of 18 m from the target at which point only K_2^0-mesons and neutrons remain in the beam. Some γ-rays may also be present, although attenuated by the lead before the first collimator. The detecting apparatus consisted of a pair of symmetrically placed spark chamber spectrometers viewing a volume contained within a helium filled bag. Each of these spectrometers consisted of a pair of spark chambers, separated by a magnet, and triggered by scintillation and Cerenkov

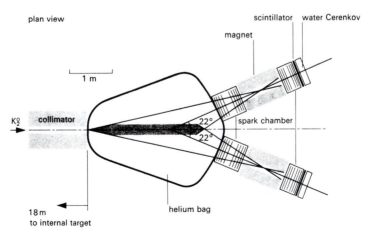

Fig. 8.3. Arrangement used by Christenson, Cronin, Fitch and Turlay (1964) for detection of the two-pion decay of the K^0-meson. The region seen by the two detection arms is heavily shaded.

counters, positioned as shown in the figure. The chambers are triggered only when a beam particle decays into charged particles of velocity $\gtrsim 0.75c$ passing into the two spark chamber assemblies.

$K^0 \to 2\pi$ decays were unambiguously detected in the following way. When two particles of opposite charge were detected in coincidence in the spark chambers the resultant momentum of the pair of particles was calculated, and also their effective mass, on the assumption that the particles were pions. The effective mass is given by

$$M_{1,2}^2 = (E_1 + E_2)^2 - (\mathbf{p}_1 + \mathbf{p}_2)^2$$
$$= M_1^2 + M_2^2 + 2(M_1^2 + p_1^2)^{\frac{1}{2}}(M_2^2 + p_2^2)^{\frac{1}{2}} - 2p_1 p_2.$$

The effective mass could then be studied as a function of the angle between the resultant two-particle momentum and the tightly collimated K_2^0 beam. Alternatively the distribution in this angle could be studied as a function of the effective mass, as is shown in fig. 8.4 for three regions of effective two-particle mass. These plots show quite clearly the existence of a long-lived K^0-meson *which decays into two pions*. A three-particle decay would not show this sharp correlation between K^0-mass and angle, while the density of helium in the apparatus was quite inadequate to give sufficient regeneration to explain the number of events observed.

Christenson, Cronin, Fitch and Turlay obtained

$$R = \frac{K_L^0 \to \pi^+ \pi^-}{K_L^0 \to \text{all charged}} = (2.0 \pm 0.4) \times 10^{-3}$$

and this result has been confirmed in a number of subsequent experiments using K_L^0-particles having a wide range of momenta.

The phenomenon is frequently described by the parameter $|\eta_{+-}|$ where

$$\eta_{+-} = \frac{\text{Amplitude}\,(K_L^0 \to \pi^+ \pi^-)}{\text{Amplitude}(K_S^0 \to \pi^+ \pi^-)}.$$

As is clear from our earlier discussion, the presence of such a two-pion decay mode implies that the K cannot be a pure eigenstate of the operator *CP*. It is natural to ask to what extent the all neutral decay

$$K^0 \to \pi^0 \pi^0$$

also occurs for the long lived kaon. Although such measurements are more difficult than for the charged system, values have been obtained. In an experiment at CERN all four γ-rays from the π^0s were detected by means of pair production in thick-plate spark chambers, with absolute calibration by comparison with the event rate from the K_1^0-mesons in a carbon target placed in the apparatus for this purpose. One difficulty of this method is to correct for events of the kind $K^0 \to 3\pi^0 \to 6\gamma$, where two γ-rays do not produce pairs and the remaining four simulate the $K^0 \to 2\pi$ decay. Such a background was corrected by means of a 'Monte Carlo' computer calculation, where fake events of the kind $K^0 \to 3\pi^0 \to 6\gamma$ were tested to determine in how many cases they might simulate a $K^0 \to 2\pi^0$ decay.

An experiment carried out at Princeton used a different technique, in which only one of the four γs was detected. It distinguished the $K^0 \to 2\pi^0$ decay from all other modes such as $K^0 \to 3\pi^0$, $K^0 \to \pi^+\pi^-\pi^0$, $K^0 \to \pi^0\pi^0\gamma$, $K^0 \to \pi^+\pi^-\gamma$, by exploiting the fact that only in this mode are γs produced with an energy greater than 170 MeV, in the K^0 cms. The energies of the

Fig. 8.4. Angular distribution near the forward direction (note scale) of the resultant two-particle momentum, for three ranges of two-particle effective mass m, in the decay of the long-lived K^0 (Christenson *et al.*, 1964).

γ-rays were measured by a spark-chamber magnetic spectrometer, but transformation to the K^0 cms demands a knowledge of the K^0 momentum. As in all these K^0 experiments, the K^0-beam has quite a wide spectrum of energies. However, in the Princeton 3 GeV accelerator the beam was divided into a series of short bursts and the K^0-momentum measured by the time of flight, as obtained from the delay between production and detection. The $K^0 \to 2\pi^0$ decay was then determined as a fraction of the known $K^0 \to 3\pi^0$ rate by comparison of the numbers of γ-rays above and below 170 MeV. For an accurate determination of the ratio of the amplitudes

$$\eta_{00} = \frac{\langle \pi^0\pi^0 | H | K_L^0 \rangle}{\langle \pi^0\pi^0 | H | K_S^0 \rangle}$$

from the observed γ-rays, allowance should also be made for the contribution of the modes other than $K^0 \to 3\pi^0$ to the γ-spectrum below 165 MeV.

Since both K_S^0 and K_L^0 can decay to two pions we expect interference effects in the two-pion intensity as a function of time in a way analogous to the situation we have discussed above for K_1^0 and K_2^0. Starting with a pure K^0-beam, the two-pion intensity is then given by a modification of equation **8.2**:

$$N_{2\pi}(t) \propto [e^{-\Gamma_S t} + |\eta_{+-}|^2 e^{-\Gamma_L t} + 2|\eta_{+-}| \cos(\Delta m \cdot t + \phi_{+-}) e^{-\frac{1}{2}(\Gamma_S + \Gamma_L)t}].$$

Δm is the $K_L^0 - K_S^0$ mass difference and ϕ_{+-} is the phase difference between the $K_S^0 \to \pi^+\pi^-$ and $K_L^0 \to \pi^+\pi^-$ amplitudes. Such interference effects have been measured for both $\pi^+\pi^-$ and $\pi^0\pi^0$ decay modes to yield values for ϕ_{+-} and ϕ_{00}. The complex quantities η_{+-} and η_{00} can be written

$$\eta_{+-} = |\eta_{+-}|e^{i\phi_{+-}} \quad \text{and} \quad \eta_{00} = |\eta_{00}|e^{i\phi_{00}}.$$

The best values presently available for the decay parameters are

$$|\eta_{+-}| = (2.266 \pm 0.018) \times 10^{-3} \quad |\eta_{00}| = (2.245 \pm 0.019) \times 10^{-3}$$
$$\phi_{+-} = (44.6 \pm 1.2)° \quad \phi_{00} = (54 \pm 5)°.$$

The phenomena of *CP* violation in K^0 decay can be described in terms of a variety of parameters. In a formal approach it is natural to introduce parameters related to the K^0–$\bar{K}^0$ mixing in the formalism of the 'mass matrix' (see section 10.6). Such a parameter ε, measures the orthogonality of the K_L^0 and K_S^0 states where

$$\langle K_L^0 | K_S^0 \rangle = 2 \operatorname{Re} \varepsilon$$

so that if CP invariance were exact, the states were truly orthogonal and K_L^0 did not decay to two pions then ε would be equal to zero.

In principle the two pion final state can have I spin equal to 0 or 2 (Bose symmetry forbids $I = 1$) so that we consider two complex decay amplitudes:

$$\langle I = 0|T|\mathrm{K}\rangle = A\,e^{i\delta_0}$$
$$\langle I = 2|T|\mathrm{K}\rangle = A\,e^{i\delta_2}$$

If we define

$$\varepsilon_0 = \frac{\langle I = 0|T|\mathrm{K}_{\mathrm{L}}^0\rangle}{\langle I = 0|T|\mathrm{K}_{\mathrm{S}}^0\rangle} \quad \text{and} \quad \varepsilon_2 = \frac{\langle I = 2|T|\mathrm{K}_{\mathrm{L}}^0\rangle}{\langle I = 2|T|\mathrm{K}_{\mathrm{S}}^0\rangle}$$

it can be shown that

$$\varepsilon_0 = \varepsilon \quad \text{and} \quad \varepsilon_2 = \frac{i\,\mathrm{Im}\,A_2}{\sqrt{2}\,A_0}\,e^{i(\delta_2 - \delta_0)}$$

where δ_0 and δ_2 are the $\pi\pi$ scattering phase shifts at the K mass. ε_2 is conventionally known as ε'. Assuming *CPT* invariance ε and ε' are related to η_{+-} and η_{00} by

$$\eta_{+-} = \varepsilon + \varepsilon' \quad \text{and} \quad \eta_{00} = \varepsilon - 2\varepsilon' \qquad \textbf{8.4}$$

If the $\Delta I = \frac{1}{2}$ rule holds then only the $I = 0$ two-pion state is allowed so that $\varepsilon' = 0$ and $\eta_{+-} = \eta_{00}$.

ε and ε' are complex quantities but the phase of ε can be determined from rather general theoretical considerations in terms of measured quantities including Δm, the $\mathrm{K}_{\mathrm{L}}^0 - \mathrm{K}_{\mathrm{S}}^0$ mass difference and τ_{S}, the $\mathrm{K}_{\mathrm{S}}^0$ mean life. It turns out to be close to the superweak value of 44^0. The phase of ε' is equal to $\delta_2 - \delta_0 + \frac{\pi}{2}$ and inserting the values of δ_0 and δ_2 from an analysis of $\pi\pi$ phase shifts we obtain a value of $(48 \pm 8)^0$ so that ε and ε' have *equal* phase shifts to within 10–15^0. Using the result 8.4 and the well founded approximation that $\varepsilon' \ll \varepsilon$ we can then write the modulus of the ratio ε'/ε:

$$\left|\frac{\varepsilon'}{\varepsilon}\right| = \mathrm{Re}\left(\frac{\varepsilon'}{\varepsilon}\right) \doteq \frac{1}{3}\left(1 - \left|\frac{\eta_{00}}{\eta_{+-}}\right|\right)$$

and

$$|\varepsilon| = \left|\eta_{+-}\left[1 - \mathrm{Re}\left(\frac{\varepsilon'}{\varepsilon}\right)\right]\right|$$

The origin of the *CP* violation (and the implicit *T* violation) is not presently understood. Since the phenomenon is potentially a clue to fundamental aspects of particle physics it has attracted substantial theoretical and experimental attention. Of the several theoretical models originally proposed two have so far survived the comparison with increasing experimental data. The '*superweak* model' proposed by

Wolfenstein in 1964 postulates a new $S = 2$ CP violating interaction with a coupling weaker than conventional second order weak processes. The superweak model predicts ε' identically equal to zero so that $\eta_{+-}/\eta_{00} = 1$. In addition it requires that

$$\phi_{+-} = \phi_{00} = \tan^{-1}\left(\frac{2\Delta m\tau_{\mathrm{S}}}{h}\right) \quad \text{and} \quad \mathrm{Re}\,\varepsilon = |\eta_{+-}|\left[1+\left(\frac{2\Delta m\tau_{\mathrm{S}}}{h}\right)^2\right]^{-\frac{1}{2}}$$

Substituting the best measured values for these quantities we obtain predictions

$$\phi_{+-} = \phi_{00} = (46.67 \pm 0.13)^0 \quad \text{and} \quad \mathrm{Re}\,\varepsilon = (1.639 \pm 0.014) \times 10^{-3}$$

which may be compared with the current best measurements:

$$\phi_{+-} = (44.6 \pm 1.2)^0 \quad \phi_{00} = (54 \pm 5)^0 \quad \mathrm{Re}\,\varepsilon = (1.630 \pm 0.083) \times 10^{-3}$$

Another class of models, sometimes known as the '*milliweak* models' require a CP violating part $\sim 10^{-3}$ in the weak interaction. When CP violation was discovered, weak interaction theory provided no place for such a phenomenon. If, as we now believe, (see chapter 12 and in particular section 12.8) there exist *six* basic quarks then there is room in the theory for a residual phase factor which, if it exists, will necessarily give rise to a small CP violation $\sim (1-3) \times 10^{-3}$, the value depending on the mass of the top quark and on certain other 'Standard Model' parameters. This 'milliweak' model was first proposed by Kobayashi and Maskawa in 1972.

The determination of $|\varepsilon'/\varepsilon|$ to the required level of one part in 10^3 or better via measurements of η_{+-} and η_{00} is a very demanding experiment. Major experiments are, at the time of writing, in progress at CERN and at Fermi National Accelerator Laboratory (FNAL) and two of these have reported results from analyses of the first parts of their data. These experiments measure (among other quantities) the double ratio

$$R = \frac{\eta_{00}}{\eta_{+-}} = \frac{\Gamma(\mathrm{K^0_L} \to \pi^0\pi^0)}{\Gamma(\mathrm{K^0_S} \to \pi^0\pi^0)} \times \frac{\Gamma(\mathrm{K^0_S} \to \pi^+\pi^-)}{\Gamma(\mathrm{K^0_L} \to \pi^+\pi^-)}$$

Since the two-pion modes for the $\mathrm{K^0_L}$ have branching ratios of only $(1-2) \times 10^{-3}$ it is critical to minimise background arising from other channels and to understand any residual backgrounds to very high precision by means of Monte Carlo simulation.

In the CERN NA31 experiment both charged and neutral decays are recorded simultaneously. Measurements are made first in a $\mathrm{K^0_L}$ beam generated from a target bombarded by a beam of 450 GeV protons from the CERN SPS. The kaon decay products are detected and identified at a distance of about 100 m from the production target by a system of drift chambers and calorimeters which measure the particle directions and energies and distinguish electrons, pions and muons. The same system is then used for K_S^0 by moving the primary target downstream towards the detector.

In the FNAL experiment half of the neutral K_L^0 beam is intercepted by a regenerator to give K_S^0. The K_L^0 and K_S^0 are measured simultaneously but only the charged *or* neutral mode is detected at any one time. A measure of the experimental problem is that in the FNAL data the numbers of events in the signal channels were

$$K_L^0 \rightarrow \pi^+\pi^- \quad 43 \times 10^3 \qquad K_L^0 \rightarrow \pi^0\pi^0 \quad 52 \times 10^3$$
$$K_S^0 \rightarrow \pi^+\pi^- \quad 179 \times 10^3 \qquad K_S^0 \rightarrow \pi^0\pi^0 \quad 201 \times 10^3$$

to be distinguished against a background of

$$K_L^0 \rightarrow \pi^{\pm}e^{\mp}\nu \quad 18 \times 10^6 \qquad K_L^0 \rightarrow \pi^0\pi^0\pi^0 \quad 10 \times 10^6$$

Both experiments make extensive comparisons of all measured distributions with expectation from Monte Carlo models to ensure that the geometry and efficiency of the apparatus is fully understood.

The results of these two measurements are presently (January 1990) apparently in disagreement (though only at the level of two standard deviations). The CERN NA31 experiment finds for $\left|\frac{\varepsilon'}{\varepsilon}\right|$ a value of $+(3.2 \pm 1.0) \times 10^{-3}$ while the FNAL experiment obtains $-(0.5 \pm 1.5) \times 10^{-3}$.

At present therefore the FNAL experiment is consistent with the superweak model whereas the CERN result is in disagreement with superweak at the level of three standard deviations. Both these experiments have further data to analyse while at least one other experiment, which will measure all four channels (K_L^0, K_S^0 charged and neutral) at the same time, will shortly take data. The results of these high precision studies will hopefully lead to a better understanding of the fundamental origin of the CP phenomena.

8.6   *CPT* invariance

CPT invariance implies the equality of the masses, magnetic moments and lifetimes of particles and their antiparticles. It may be shown that the mass difference between K^0 and $\bar{K}^0$ is less than or equal to the K_1^0–K_2^0 mass difference (the masses may be related by means of the mass matrix, described in section 10.6). Thus, following our earlier discussion (section 8.4), it is clear that the best test in respect of masses is the comparison of the K_1^0- and K_2^0-masses, where the difference $\Delta m/\Gamma_1$ was shown to be ~ 0.5 yielding $\Delta m/m \sim 10^{-14}$. Writing for complete *CPT* conservation

$$\langle K^0 | H_{St} + H_\gamma + H_{WK} | K^0 \rangle = \langle \bar{K}^0 | H_{St} + H_\gamma + H_{WK} | \bar{K}^0 \rangle$$

the above result implies invariance of H_{St} to $\sim 10^{-14}$, H_γ to $\sim 10^{-12}$ and H_{WK} to $\sim 10^{-8}$ (strangeness-conserving non-leptonic part of H_{WK}).

The magnetic moments of both e^+/e^- and μ^+/μ^- have been measured to very high precision. The results are generally expressed in terms of the g factor where the magnetic moment is written as

$$\boldsymbol{\mu} = g\mu_{\mathrm{B}}\mathbf{s}$$

where μ_{B} is the Bohr magneton and $\mathbf{s}$ is the spin. Comparison of the values for particle and antiparticle yields

$$(g_{e^+} - g_{e^-})/\text{average} = -(0.5 \pm 2.1) \times 10^{-12}$$
$$(g_{\mu^+} - g_{\mu^-})/\text{average} = -(2.6 \pm 1.6) \times 10^{-8}$$

A comparison of particle and antiparticle lifetimes yields (see section 5.3 for kaons)

$$\mu: \Delta\tau/\tau = (0.03 \pm 0.08) \times 10^{-3}$$
$$\pi: \Delta\tau/\tau = (0.5 \pm 0.7) \times 10^{-3}$$
$$K: \Delta\tau/\tau = (0.11 \pm 0.09) \times 10^{-3},$$

again in good agreement with *CPT* (including the leptonic part of H_{WK}) but to a lower degree of precision.

We thus conclude that all the experimental evidence supports invariance under *CPT* and that the limits to which such invariance has been tested are very good.

8.7   *C* and *P* invariance in strong interactions

For strong interactions it is possible to test *P* violation by looking for effects due to pseudo-scalar terms in low energy nuclear processes, or in nucleon–nucleon scattering. An example of such a parity non-conserving term is a longitudinal polarisation in, for instance, proton–neutron scattering. A search for such polarisation has been made in an experiment in which a beryllium target was bombarded by protons of 380 MeV. Neutrons emerging from the target in the line of the incident proton beam and having an energy greater than 350 MeV were passed into a solenoid such that their spin was rotated by the magnetic field to convert any longitudinal polarisation into transverse polarisation. The magnitude of the resulting transverse polarisation was determined by allowing the neutrons to scatter on a hydrogen target and measuring the up–down asymmetry in the neutron–proton scattering. For unpolarised neutrons no such asymmetry exists. No asymmetry was observed, and the limits of the measurement indicated that any parity non-conserving amplitude in the initial interaction must be less than about 10^{-3} of the parity-conserving amplitude. Likewise, no other evidence for parity non-conservation in strong processes has been found.

A more direct test has been made by seeking for parity non-conserving transitions in the decay of excited states of nuclei. For instance, an excited state ^{20}Ne with spin parity 1^+ may be formed by bombardment of ^{19}F by protons of appropriate energy. Such a state could decay to the ground state of ^{16}O by α-particle emission only if parity is not conserved in the transition, since ^{16}O has spin parity 0^+ and the α-particle spin is also zero. This transition is not observed and by this method a limit of about 10^{-6} on the relative size of the parity non-conserving amplitude has been obtained.

The best method of testing C invariance in strong interactions is to compare the angular and energy distributions of π^+ and π^-, K^+ and K^-, or K^0 and $\bar{K}^0$ particles in proton–antiproton annihilation into mesons. If the process is invariant under C, then the angular and energy distributions for the particle and the antiparticle should be identical. Such studies have established a limit of about 10^{-4} for the relative amplitudes for the C non-invariant and C invariant processes.

8.8   *C, P* and *T* invariance in electromagnetic interactions

As we shall discuss in chapter 11, weak and electromagnetic interactions are now understood to be different aspects of the same basic interaction, so that we may expect to find subtle interference effects which give rise to small parity violations in 'electromagnetic' processes such as electromagnetic transitions in atoms. We shall return to this subject in chapter 11. In this section we deal with some tests of C, P and T invariance in electromagnetic interactions.

One such test, involving invariance under both the parity transformation and the operation of time reversal in electromagnetic processes, is the existence or otherwise of an electric-dipole moment in the neutron. We can show that the existence of such a moment implies lack of invariance under either P or T and hence, if the CPT theorem is true, also under C. The Hamiltonian for the interactions between the magnetic- and electric-dipole moments for such a particle, with an electromagnetic field may be written in the form

$$H_1 = \rho_m \, \boldsymbol{\sigma} \cdot \mathbf{H} + \rho_e \, \boldsymbol{\sigma} \cdot \mathbf{E}$$

where ρ_m and ρ_e are the magnitudes of the magnetic- and electric-dipole moments, σ is the spin vector and **H** and **E** are the magnetic and electric field vectors. A consideration of **H** and **E** shows that **H** is even under the parity transformation and **E** odd, while **H** is odd under time reversal and **E** is even under this operation. Also, the spin vector σ is even under P and odd under T, i.e. it behaves like the magnetic field in this respect. Thus the contribution to the Hamiltonian due to the magnetic-dipole moment is invariant under the parity operation and under time reversal, while the contribution due to the electric-dipole moment changes sign under both operations. This means that the electric-dipole moment of the neutron must be zero unless the electromagnetic interaction is not invariant under time reversal and space inversion. Thus the observation of such a moment would imply failure of time reversal invariance and parity invariance.

If we write the electric-dipole moment of the neutron in the form

$$d_n = eL_n$$

where e is the charge on the electron and L_n is a length, then the best limit is presently that of an experiment at the ILL laboratory at Grenoble which yields $L = -(3 \pm 4) \times 10^{-26}$ cm. Unfortunately, it is difficult to know how to calculate properly the expected value of the quantity L_n, but we may note that the 'standard model' with six quark flavours (see chapter 12) predicts $L \sim 10^{-28}$ cm.

The best-studied test of C invariance in electromagnetic processes involves the decays of the η^0-meson. This meson will be discussed in some detail in the following chapter. For the present we merely state certain of its properties as follows: mass $= 549$ MeV/c^2, narrow 'width' indicating a lifetime consistent with electromagnetic decay, and decay modes

$$\eta^0 \to \begin{cases} \pi^+\pi^-\pi^0 \\ \pi^0\pi^0\pi^0 \\ \pi^+\pi^-\gamma \\ \pi^0\gamma\gamma \\ \gamma\gamma. \end{cases}$$

Even the three-pion decays probably proceed by means of electromagnetic processes (see chapter 9) although no γ-rays are actually produced. They are found to be of similar intensity to those decays involving γ-rays.

If the decay $\eta \to \pi^+\pi^-\pi^0$ is invariant under charge conjugation then the π^+ and π^- must behave completely alike in the final state, so that in a plot of the kind shown in fig. 8.5 (see also appendix A) we expect complete symmetry between the right- and left-hand halves of the diagram. Another way of testing the symmetry is to define a parameter

$$A = \frac{N_+ - N_-}{N_+ + N_-}$$

where we have N_+ events in which the cms energy of the π^+ is greater than that of the π^- and N_- events in which the opposite is the case. For invariance under C, the parameter A must clearly equal zero.

Several experiments have tested C violation in this way, to various degrees of accuracy. The η^0-particles may be made for instance in the reaction

$$\pi^+ d \rightarrow p\eta^0(p)$$

where the π^+ has reacted with the neutron in the deuterium and the proton has acted as a 'spectator'. The result of such an experiment, carried out by studying interactions of 0.82 GeV/c-mesons in a deuterium filled bubble chamber, is shown in fig. 8.5. Each point on the diagram represents an event of the type shown above, where the η decays by $\eta \rightarrow \pi^+\pi^-\pi^0$. Such events were identified by observation of a short recoil proton, corresponding to the spectator recoil, followed by kinematic fitting. Each

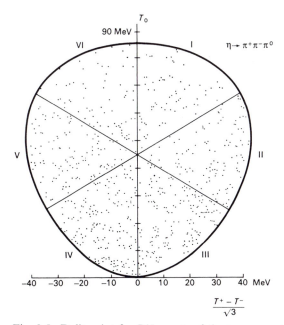

Fig. 8.5. Dalitz plot for 765 events of the type $\eta \rightarrow \pi^+\pi^-\pi^0$ in the experiment of Larribe *et al.* (1966). T_0, T_+ and T_- are the kinetic energies of the π^0, π^+, π^--mesons respectively in the η-cms.

event was also examined to ensure that the bubble densities of the tracks were in agreement with the values predicted by the kinematic fit. This last process often resolved ambiguities between possible hypotheses, in particular resolving a number of cases where a fit could be achieved on the assumption that either of the positive tracks was the proton. In fig. 8.5 there are represented 765 events. In this experiment, however, there were 21 000 events measured; 17 000 of these fitted the reaction

$$\pi^+ d \to p\pi^+ \pi^- (p)$$

while several hundred events fitted the other processes. The number finally included in the analysis was further reduced by imposing strict criteria on the volume of the chamber used for measurement, on the minimum length for the recoil proton and on the mass accepted for the η-meson. These criteria were necessary to avoid any bias of the data due to, for instance, badly measured events which might affect the π^+ and π^- energies differently. The final value obtained for A in this experiment was

$$A = -0.048 \pm 0.036.$$

The advantage of the use of a bubble chamber for this experiment is

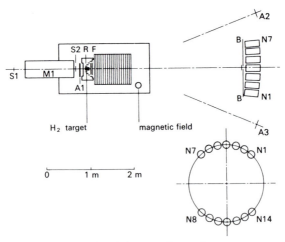

Fig. 8.6. Schematic of the experimental arrangement of Cnops *et al.* (1966) to study the $\eta \to \pi^+ \pi^- \pi^0$ decay. The neutrons from the production process $\pi^- p \to n\eta$ were detected in the counters N1–N14. The counters F (with hole for beam) ensured that at least one charged particle entered the spark chambers. The counters R and A1 in anti-coincidence ensured that no charged particles emerged except into the spark chambers. Counts in S1 and S2 together with no count in B ensured a beam particle interaction in the hydrogen target. M1 is a magnet to compensate the beam deflection in the main magnet surrounding the spark chambers.

that the possibility of biases which might produce a difference between π^+ and π^- is very small. That this is so is due to the observation of the interaction vertex itself and the essentially '4π-geometry' in which events are observed having all angles for the outgoing tracks. The disadvantage is that it is difficult to obtain a very large sample of events so that the statistical error remains appreciable.

A much lower statistical error may be achieved by means of a spark chamber-counter experiment, although in this case the danger of systematic errors is greater. The arrangement of an experiment of this type is shown in fig. 8.6. Here the η-particles were produced in a liquid hydrogen target by incident π^--mesons of 0.713 GeV/c by means of the process

$$\pi^- p \rightarrow n\eta.$$

The spark chambers detect the π^+- and π^--mesons from the η-decay to $\pi^+\pi^-\pi^0$, while the neutrons are detected by a system of scintillation counters having their axes directed at the target. The spark chambers were triggered by a coincidence-anticoincidence arrangement which ensured that a beam particle had interacted in the target, that no charged particle had emerged other than into the spark chambers, that at least one charged particle had entered the chambers and that a neutral particle had entered the neutron counters in a specified time interval. The neutron time of flight was measured to yield a value for its momentum. In order to avoid any asymmetries in the result due to asymmetries in the magnetic field or the optical system, the field was reversed during the experiment. After kinematic fitting and application of certain selection criteria, the final sample in this experiment yielded 10 665 events and the final value obtained for A was

$$A = (0.3 \pm 1.1)\%$$

The best current values for the left–right asymmetries are

$$\pi^+\pi^-\pi^0 \quad A = (0.12 \pm 0.17)\%$$
$$\pi^+\pi^-\gamma \quad A = (0.88 \pm 0.40)\%.$$

In summary, there is no evidence for C non-invariance in either strong or electromagnetic processes at the present level of sensitivity of the tests.

9

Strongly-decaying resonances

9.1 Introduction

In the course of the foregoing chapters we have studied unstable particles which decayed via the weak and electromagnetic interactions. In particular, in dealing with strangeness, it was shown that the strange particle decays had lifetimes generally in the region of 10^{-8}–10^{-10} s, characteristic of the weak interaction. We also have examples of *electromagnetic* decay processes involving γ-rays, such as

$$\pi^0 \to \gamma\gamma$$
$$\Sigma^0 \to \Lambda\gamma$$

for which the lifetimes are 10^{-15}–10^{-16} s. It is then natural to ask whether there are particles which decay via the strong interaction. Since we know from consideration of the cross-section magnitudes that the strong interaction is some 10^{14} times stronger than the weak interaction, we might expect lifetimes for strong-decay processes of about 10^{-23} s.

If particles decaying via strong interactions exist they may therefore be expected to have lifetimes such that, even if travelling near the speed of light, the distance they travel before decay is so short as to make it quite impossible to distinguish production and decay points. Thus in seeking such particles we cannot for instance expect to see a track in a bubble chamber but must rely on indirect evidence. Applying the uncertainty relationship in the form

$$\Delta E \cdot \Delta t \simeq \hbar \qquad\qquad \textbf{9.1}$$

to possible particles with lifetimes (i.e. Δt) $\tau \sim 10^{-23}$ s, it is clear that the width or uncertainty in mass may be appreciable. For $\tau \simeq 10^{-23}$ s, ΔE is about 100 MeV. This means that for such a particle we might expect to find masses varying by energies of this order.

9.2 The Δ^{++} (1232): formation experiments

The discussion of isotopic spin in chapter 2 revealed that there was a particularly strong interaction between π-mesons and nucleons in the state of total I-spin $I = \frac{3}{2}$ when the incident pion kinetic energy was about 180 MeV. Experimentally, this result is deduced from a substantial peak in the $\pi^+ p$ elastic scattering cross-section (as well as the values of the $\pi^- p$ elastic and charge exchange cross-sections). We may interpret this strong interaction at a particular incident energy as being equivalent to the formation of a well defined particle, resonance, or nucleon isobar, of very short life, according to a diagram of the form shown in fig. 9.1 The resonance then acts as a short-lived particle having $I = \frac{3}{2}$, $I_3 = +\frac{3}{2}$ and charge $Q = +2$. The mass is the total centre-of-mass energy at which the resonance peak occurs. For $T_\pi = 180$ MeV we have $P = 288$ MeV/c so that

$$E^* = \sqrt{(E^2 - p^2)} = \{(0.18 + 0.14 + 0.94)^2 - (0.288)^2\}\,\text{GeV}$$
$$= 1.23\,\text{GeV} = M.$$

Looking again at fig. 2.14 we see that the width of the resonance peak is about 120 MeV, so that according to the uncertainty principle argument the lifetime must indeed be about 10^{-23} s, characteristic of strong decay.

Such a method of studying resonance particles is known as a 'formation' experiment. For such an experiment the cross-section for a given process, like πp elastic scattering, is studied as a function of incoming particle momentum. In the simplest case the formation of an intermediate state resonance particle is indicated by a peak in the total cross-section.

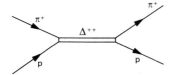

Fig. 9.1. Formation of the $\Delta(1232)$ in the s-channel in $\pi^+ p$ interaction.

This kind of process is sometimes referred to as the appearance of a resonance in the '*s*-channel', where the symbol is used for the total-energy squared in the cms.

9.3   Resonance spin and parity: the Δ (1232)

It is not the intention to develop here the full theory of scattering in terms of partial waves, which may be found in any standard text on quantum mechanics. Rather, we present a brief treatment designed to bring out the principal features.

In a study of a resonance in the *s*-channel, such as the Δ^{++} in $\pi^+ p$ scattering, the differential scattering cross-section will be a function of the angular momentum of the intermediate state, i.e. the resonance spin.

Since the pion spin is zero, the angular momenta to be combined in the initial state are the proton spin and the orbital angular momentum l ($\hbar = 1$ throughout). In the initial state the proton spin can be 'up' or 'down' so that the angular momentum J of the state is $J = l \pm \frac{1}{2}$. We note that the proton spin may be 'flipped' in the interaction. In this case, since J is conserved, a change $\Delta s = \pm 1$ in the *z*-component of the proton spin implies $\Delta l_z = 1$ since $\Delta j_z = 0$. We make our *z*-direction along the line of flight of the pion and proton in the cms so that $l_z = 0$ in the initial state, and for spin flip $l_z(\text{final}) = 1$.

The scattered amplitude $f(\theta, \phi)$ can then be written as a sum over the associated Legendre polynomials $Y_{l,m}$ defined in the usual way as

$$Y_{l,m}(\theta, \phi) = C_{l,m} P_l^m(\cos \theta) e^{im\phi}$$

where θ is the usual polar scattering angle, ϕ is the azimuthal angle and C is a function of m and l. For a polarised target, a situation in which more protons are aligned in one direction than in the other may be attained by means of a magnetic field acting on a suitable molecule at very low temperature. For such a target the ϕ-distribution may be anisotropic, but otherwise the angular distribution will always be isotropic in the azimuthal angle. Thus, integrating over ϕ, we write

$$f(\theta) \propto \sum_l C(l, m, E^*) P_l^m(\cos \theta)$$

where E^* is the total centre-of-mass energy and the Cs are weighting factors. The spin-flip terms have $m = \pm 1$, while the no-flip terms have $m = 0$. Thus we see that spin-flip and no-spin-flip interactions give different θ-distributions.

If $l = 0$ (*s*-wave scattering), so that the intermediate state has spin $\frac{1}{2}$, then only $Y_{0,0} = 1/\sqrt{(4\pi)}$ can enter, so that the angular distribution is isotropic.

If $l = 1$, then $J = \frac{1}{2}$ or $\frac{3}{2}$, i.e. we can have $p_{\frac{1}{2}}$ or $p_{\frac{3}{2}}$ states. Each of these states can include spin-flip and no-flip terms. Let us write the orbital angular momentum wave function for the pion–nucleon combination as $\psi(l, m)$ and the nucleon spin wave function as $\chi(s, s_z)$. The $\psi(l, m)$ are simply proportional to our $Y_{l, m}$. We arbitrarily choose the initial proton spin 'up', $S_z = +\frac{1}{2}$, since whichever choice we may make is irrelevant, the flip or no-flip being the significant factor. For this choice the initial state may be written as proportional to

$$\psi(l, 0)\chi(\tfrac{1}{2}, +\tfrac{1}{2}).$$

The final state with $J = \frac{3}{2}$, $J_z = \frac{1}{2}$ is then

$$\phi(\tfrac{3}{2}, \tfrac{1}{2}) = \sqrt{\tfrac{1}{3}}\psi(1, 1)\chi(\tfrac{1}{2}, -\tfrac{1}{2}) + \sqrt{\tfrac{2}{3}}\psi(1, 0)\chi(\tfrac{1}{2}, \tfrac{1}{2})$$

where we have used the Clebsch–Gordan coefficients for combination of angular momenta 1 and $\frac{1}{2}$. Also

$$\phi(\tfrac{1}{2}, \tfrac{1}{2}) = \sqrt{\tfrac{2}{3}}\psi(1, 1)\chi(\tfrac{1}{2}, -\tfrac{1}{2}) - \sqrt{\tfrac{1}{3}}\psi(1, 0)\chi(\tfrac{1}{2}, \tfrac{1}{2}).$$

Thus both these states involve $Y_{1,1}$ and $Y_{1,0}$ terms but in different ways. A pure $p_{\frac{1}{2}}$-state would give an amplitude proportional to

$$\alpha\sqrt{\frac{3}{4\pi}}\cos\theta + \beta\sqrt{2}\sqrt{\frac{3}{8\pi}}\sin\theta e^{i\phi}$$

where we have written α and β for $\chi(\frac{1}{2}, \frac{1}{2})$ and $\chi(\frac{1}{2}, -\frac{1}{2})$ respectively. The pure $p_{\frac{3}{2}}$-state has amplitude proportional to

$$\alpha\sqrt{2}\sqrt{\frac{3}{4\pi}}\cos\theta - \beta\sqrt{\frac{3}{8\pi}}\sin\theta e^{i\phi}.$$

The spin-flip and no-flip amplitudes are orthonormal and do not interfere, so that we add the squares of the amplitudes to get for the $p_{\frac{1}{2}}$ state

$$\cos^2\theta + \sin^2\theta \rightarrow \text{isotropy},$$

and for the $p_{\frac{3}{2}}$ state

$$2\cos^2\theta + \tfrac{1}{2}\sin^2\theta \propto 1 + 3\cos^2\theta,$$

where we have dropped the nucleon spin wave functions.

Our conclusion is that for a pure spin state decaying via the strong interaction we expect an isotropic angular distribution for a spin-$\frac{1}{2}$ state, and a distribution proportional to $1 + 3\cos^2\theta$ for a spin-$\frac{3}{2}$ state. In fact, at the $\Delta(1232)$ resonance in $\pi^+ p$ elastic scattering, the differential cross-section has a form very close to $1 + 3\cos^2\theta$. Some deviation from such a distribution can arise from the existence of a background of events which have not passed through the intermediate resonance state.

Since the resonance state is $p_{\frac{3}{2}}$ the parity is

$$P(\pi)P(p)(-1)^l = +1.$$

In some cases a useful clue to the spin of a resonance state may be obtained from the total cross-section. The partial-wave analysis leads to an expression for the total elastic cross-section

$$\sigma_T = \frac{2\pi}{k^2} \sum_l (2j+1) \sin^2 \delta_l$$

where $\hbar k$ is the momentum of the incoming particle, $j\hbar$ is the total angular momentum and δ_l is the 'phase shift' for the wave with orbital angular momentum $l\hbar$. Suppose that the interaction takes place predominantly in a single state of well defined angular momentum $l\hbar$, as in the case of an s-channel resonance. In that case there is only one important term in the summation and

$$\sigma_T = \frac{2\pi}{k^2} (2j+1) \sin^2 \delta_l.$$

The maximum value of σ_T occurs for $\delta_l = \pi/2$. Thus the maximum value of the total cross-section for resonance in a given partial wave is

$$\frac{2\pi}{k^2}(2j+1)$$

and for $j = \frac{3}{2}$ we have an upper limit of $8\pi/k^2$ for the elastic scattering via an intermediate state of spin $\frac{3}{2}$. The limit $8\pi/k^2$ is drawn on fig. 9.2 and it is seen that the cross-section does indeed reach just this limit at the resonance energy.

9.4   'Production' experiments

The Δ^{++} (1232) was the first resonance found in a formation experiment, but its status as a strongly-decaying very short lived particle was only recognised when it was also found in 'production' in studies of the reaction $pp \rightarrow pn\pi^+$.

Since there are three particles in the final state, each one may have a range of momenta within the limits of energy and momentum conservation. As we have seen in other applications, the transition probability can be written as

$$T = \frac{2\pi}{\hbar}|M|^2 \frac{\mathrm{d}N}{\mathrm{d}E}.$$

If $|M|^2$ is not a function of the individual particle momenta, then the momentum distribution of pion, proton and neutron will each range throughout the values allowed by energy and momentum conservation, with probabilities set by the density of states or 'phase space' factor dN/dE. The dependence of dN/dE on the momenta is derived, for some simple cases, in appendix A. If, on the other hand, there is a strong interaction between two of the particles for certain momenta, this would be reflected by a strong dependence of M on these quantities.

In particular, suppose that the π^+ and the proton form a $\Delta(1232)$ which lives for 10^{-23} s before decaying. In such a case the primary reaction is, in fact, a two-body process.

$$pp \rightarrow n\Delta^{++} \rightarrow np\pi^+,$$

as illustrated in fig. 9.3. In the cms, however, the momenta of the product particles in a two-body process are uniquely constrained, so that we should expect a unique momentum for the neutron instead of the phase-space distribution. In practice, since the $\Delta(1232)$ has a substantial width or spread in mass, the neutron momentum might be expected to have a corresponding spread. The distribution of the neutron momentum in this

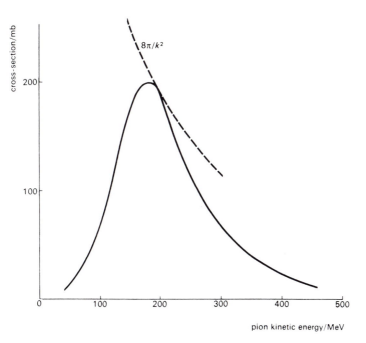

Fig. 9.2. The π^+p scattering cross-section showing the maximum cross-section $8\pi/k^2$ for scattering in the $j = \frac{3}{2}$ state.

reaction for protons of 2.8 GeV/c is shown in fig. 9.4 The disagreement with the phase-space distribution is obvious and it is qualitatively equally clear that the distribution might be fitted by a phase-space contribution plus a peak centred at a momentum of 0.80 GeV/c. The reader can check that this momentum corresponds to recoil of the neutron against a particle of mass 1236 MeV/c^2.

Fig. 9.3. Illustration of the process pp → nΔ^{++} in the cms with subsequent decay of the Δ to pπ^+.

Fig. 9.4. cms distribution of the neutrons from the reaction pp → pnπ^+ at 2.81 GeV/c. The continuous curve is that expected if the distribution is governed by phase space only. The dotted curve is that expected for $\Delta(1232)$ only. The dashed curve includes both this resonance and an $I = \frac{1}{2}$ N^* (Fickinger *et al.*, 1962).

An alternative way of studying resonances in production experiments is by an examination of the distribution of the 'effective mass' of a group of particles. Using the general relativistic relation between total energy E, momentum $\mathbf{p}$ and mass m, $E^2 - \mathbf{p}^2 = m^2$, we have for the effective mass of a group of particles

$$m^2_{1, \ldots, i} = \left(\sum_{n=1}^{i} E_n \right)^2 - \left(\sum_{n=1}^{i} \mathbf{p}_n \right)^2.$$

We may calculate the expected distribution of m^2 if the matrix element does not depend on these quantities, i.e. if the transition probability depends only on the density of states or phase-space factor. For the three-particle case, such phase-space calculations are presented in appendix A. The phase-space distribution can be calculated analytically for up to four particles in the final state, but for larger numbers the integrals cannot be evaluated directly. However, a recursion relation may be used to relate the phase space for n bodies to that for $n-1$ bodies or a 'Monte Carlo' calculation may be carried out in which artificial events are generated in a random way, but with energy and momentum conservation imposed. In all cases the phase-space distribution is smooth without any narrow peaks. If, however, a short-lived resonance exists in the mass combination $m_{1, \ldots, i}$, then the distribution will show a peak at this mass.

For an unstable particle decaying according to an exponential law, the shape of the distribution in mass takes what is known as a Breit–Wigner peak. For a state decaying exponentially with a mean life τ, we may write a wave function where the usual time-dependent part is multiplied by an exponential factor

$$\psi(t) = e^{-t/2\tau} e^{-iE_R t/\hbar}$$

where E_R is the resonance energy. What we need is the wave function in terms of energy, so we take the Fourier transform and get

$$\phi(E) = \int_0^\infty \psi(t) e^{iEt/\hbar} \, dt$$

$$= \int_0^\infty e^{-\frac{1t}{2\tau} - it(E_R - E)/\hbar} \, dt$$

$$= \frac{-i\hbar}{(E_R - E) - \frac{1}{2}i\hbar\tau^{-1}}.$$

Writing the uncertainty relation **9.1** in terms of the width of the peak Γ and the mean life as

$$\Gamma\tau \simeq \hbar$$

we have

$$\phi(E) = \frac{-i\hbar}{(E_R - E) - \frac{1}{2}i\Gamma}.$$ **9.2**

Thus we may expect a distribution

$$|\phi(E)|^2 \propto \frac{1}{(E_R - E)^2 + \frac{1}{4}\Gamma^2}.$$

At $E = E_R$ we have the maximum $\propto \Gamma^{-2}$ while at $E = E_R \pm \frac{1}{2}\Gamma$ we get a value half that at maximum, in line with the usual definition of Γ as the full width at half-maximum.

The Breit–Wigner amplitude of **9.2** when expressed in the form

$$\phi(E) \propto \frac{1}{(E_R - E)^2 + \frac{1}{4}\Gamma^2}\left[(E_R - E) + \frac{i\Gamma}{2}\right]$$

can be conveniently represented as a function of energy on the Argand diagram. In the Argand diagram the equation of a circle of centre $(0, \frac{1}{2}i)$ and radius $\frac{1}{2}$ is

$$x^2 + (y - \frac{1}{2})^2 = \frac{1}{4}$$

and we may see that the Breit–Wigner amplitude sweeps out such a circle by writing it in the form

$$\phi(E) = \frac{2}{\Gamma}(x + iy)$$

where

$$x = \frac{(E_R - E)\frac{1}{2}\Gamma}{(E_R - E)^2 + \frac{1}{4}\Gamma^2}$$

and

$$y = \frac{\frac{1}{2}\Gamma\frac{1}{2}\Gamma}{(E_R - E)^2 + \frac{1}{4}\Gamma^2}$$

(see fig. 9.5). At A, $E = E_R - \frac{1}{2}\Gamma$, at B, $E = E_R$ and at C, $E = E_R + \frac{1}{2}\Gamma$. Thus as E increases through the resonance the amplitude vector **OP** sweeps out a counter-clockwise circle.

The passage of the vector counter-clockwise through the purely imaginary value at B is a necessary condition for resonance, although recent studies have shown that causes other than resonances may also result in this kind of behaviour. The phase shift may be shown to be half the angle OMP so that at resonance $\delta = \frac{1}{2}\pi$.

The amplitude passes rapidly through the upper half of the circle compared with the regions O to A and C to O. The velocity $d\theta/dE$ is a maximum at *B*, and this condition is often useful in identifying the resonance energy in a case where the resonance is superimposed on a non-resonant background. In such an event the amplitude will no longer describe a simple circle but the circle will be superimposed on a background variation with results as shown in fig. 9.6. A further modification to the simple theory outlined above is necessary if the resonance is not purely 'elastic'. Such a situation may arise if a resonance has several decay channels. Suppose, for instance, that in a formation experiment of pions on protons one can produce a resonance which can decay both by the 'elastic' channel, but also into a nucleon and two pions or a Λ^0 plus K^0

$$\pi^- p \to \Delta \to \begin{cases} \pi^- p & (a) \\ \pi^- p \pi^0 & (b) \\ \Lambda^0 K^0 & (c). \end{cases}$$ **9.3**

The full width for the process **9.3** is then the sum of the three partial widths $\Gamma(a)$, $\Gamma(b)$ and $\Gamma(c)$. In such a case, the elastic amplitude sweeps out a path within the 'unitary circle' rather than along it. If the resonant amplitude is less than the radius of the unitary circle then the phase shift at resonance is 0 rather than $\frac{1}{2}\pi$.

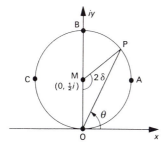

Fig. 9.5. Argand diagram for a Breit–Wigner resonance.

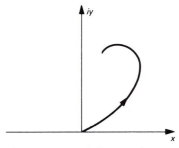

Fig. 9.6. Argand diagram for a resonance superimposed on a non-resonant background.

9.5 The Dalitz plot

A particularly useful technique in the study of resonances in production experiments involving three particles in the final state is the Dalitz plot, discussed in detail in appendix A and already introduced in the discussion of the τ-decay (see section 8.4) to which it was first applied.

We may illustrate the application of this technique particularly well by studying, for instance, the process

$$K^+p \rightarrow K^0\pi^+p. \qquad\qquad \textbf{9.4}$$

This process has been extensively studied in bubble chambers over a wide range of incident energies. A separated beam of K-mesons is produced, by means of the techniques described in chapter 1, and directed into the chamber. Reactions of the kind **9.4** have the form shown in fig. 9.7 and may be identified by measurement, reconstruction in space and kinematic analysis. A Dalitz plot can then be constructed by plotting the kinetic energies of any two particles one against the other, or, as has become more usual, by plotting the squares of the effective masses of two pairs of particles.

In fig. 9.8 is shown the Dalitz plot for this process with $m^2(K^0\pi^+)$

Fig. 9.7. A reaction of the type $K^+p \rightarrow K^0\pi^+p$, using 10 GeV/$c$ kaons in a hydrogen bubble chamber. The second beam track from the bottom is seen to interact yielding two charged tracks and a V^0 in the forward direction.

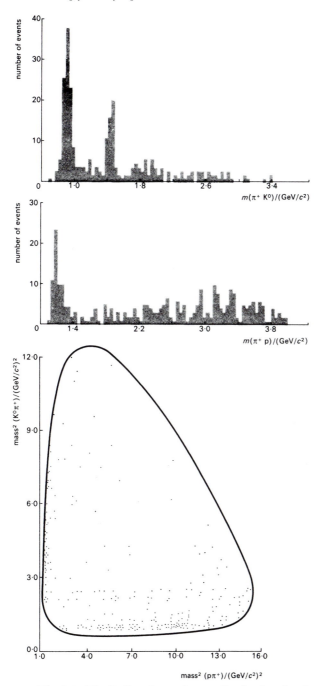

Fig. 9.8. The Dalitz plot and mass distribution for the reaction $K^+p \rightarrow K^0\pi^+p$ at 10 GeV/c. The $\Delta(1232)$ and $K^*(892)$ and $K^*(1415)$ are clearly visible.

plotted against $m^2(p\pi^+)$ for an incident kaon momentum of $10\ \mathrm{GeV}/c$. From the evidence already presented for the existence of the $\Delta^{++}(1232)$ we should expect a concentration in a band centred at $(1.232)^2$ on the $p\pi^+$-axis and, indeed, such a band is observed. We might also expect some uniform phase-space background, but at this energy the background is seen to be quite small.

Our discussion to date, however, has not suggested the existence of the excited meson states, two of which are clearly visible in fig. 9.8 as bands centred at $(0.892)^2$ and $(1.415)^2$ on the $K^0\pi^+$-axis. These are known as the $K^{*+}(892)$ and $K^{*+}(1415)$ resonances. They have strangeness $+1$, baryon number zero, and decay into $K^0\pi^+$, into $K^+\pi^0$ and in the case of the $K^*(1415)$ also into the $K\pi\pi$ final state.

We see that the resonance bands on the plot may overlap. In such an event we have two different amplitudes (for instance those for $K^*(892)$ and $\Delta(1232)$) leading to the same final state. This gives the possibility of interference, yielding an excess or deficit of particles in the overlap region over what might be expected by a simple addition of bands. A good example of this effect is shown in fig. 9.9, where results from the same process are presented but for lower incident momenta. The accessible kinematic region is of course different, the $K^*(1415)$ is not accessible, the $K^*(892)$ and $\Delta(1232)$ bands overlap and the overlap region shows constructive interference.

The Dalitz plot, as such, may only be used for a three-particle final state. However, a modification of the plot is often useful for a larger number of final particles. For instance, consider the reaction

$$K^+p \to p\pi^+K^+\pi^-.$$

We expect to find resonances in the $p\pi^+$ and $K^+\pi^-$ systems, so that it is natural to examine the structures in this process by plotting $m(p\pi^+)$ against $m(K^+\pi^-)$. In this case the boundary of the kinematically allowed region is a triangle as shown in fig. 9.10. The Δ and $K^*(892)$ bands show clearly and there is a marked interference in the overlap of the bands. It should be remembered that for this type of plot a phase-space background will not yield a uniform distribution over the allowed region. Nevertheless, the marked bands cannot be due to phase-space variations, since calculation shows that the phase-space density varies smoothly over the plot, and, indeed, only changes rapidly near the edges of the allowed region on which it falls to zero.

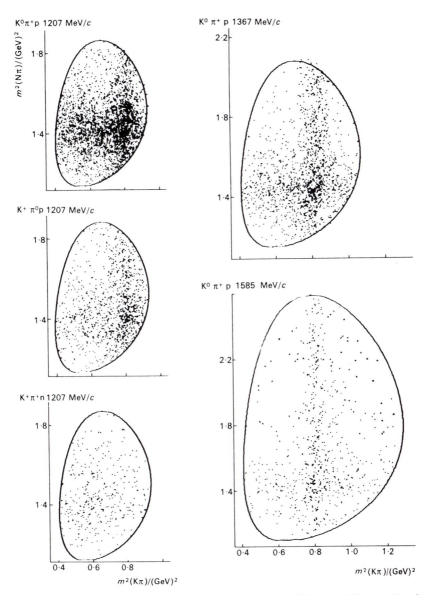

Fig. 9.9. Dalitz plots for the processes $K^+p \to K^0p\pi^+$ and $K^+p \to K^+p\pi^0$ at various momenta showing interference between $K^*(892)^+$ and $\Delta(1232)^{++}$. A plot for the process $K^+p \to K^+n\pi^+$ is shown for comparison. Note that in this case there is no K^*-band and the Δ is very weak.

9.6   Resonances with $B = 1$ and $S = 0$ (Δ and N)

The Δ^{++} (1232), which we have discussed as an example in the preceding sections, is perhaps the most readily produced of all the strangeness zero baryonic resonances. We have seen that it has isotopic spin $I = \frac{3}{2}$, $I_3 = \frac{3}{2}$. This means that it is an I-spin quadruplet ($N = 2I + 1$) and that we might expect to find charge states $+$, 0, $-$ as well as $++$.

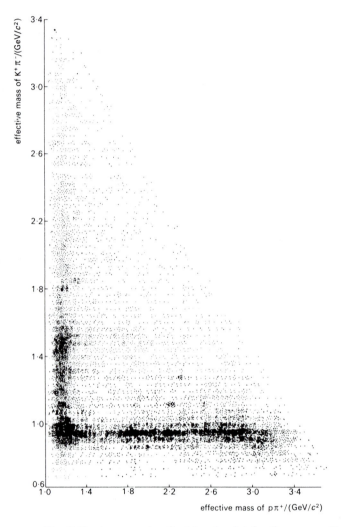

Fig. 9.10. An example of a triangle plot for the process $K^+p \rightarrow K^+p\pi^+\pi^-$ for 10 GeV/c kaons. The plot, representing 8331 events, shows strong horizontal and vertical bands due to the K*(892) (decaying to $K^+\pi^-$) and Δ^{++}(1232) resonances respectively.

These states are found in formation experiments, and they are also seen in production.

Many other Δ or N resonances have been found, nearly all in formation experiments with pions on nucleons. Most of these resonances are not very evident as actual peaks in cross-sections, but may appear, for instance, as enhancements in particular terms in, say, the elastic-scattering angular distribution, when a resonance of a particular spin is evidenced as an enhancement in the appropriate partial wave. In practice, the study of many of these resonances has been by means of phase-shift analysis in which the angular distributions are analysed in terms of the amplitudes of the partial waves. The complex amplitudes may then exhibit resonances according to the criteria we have already discussed. The $\Delta(1232)$ is, in many experiments, more evident than the other baryon resonances not only because it is often strongly produced but also because of its low Q-value. Thus it occurs near the lower end of phase space. In formation experiments one also has the advantage, for $I = \frac{3}{2}$ resonances, of having available a pure $I = \frac{3}{2}$ initial state. The study of the higher mass resonances is always further complicated by the presence of the 'tails' of those of lower mass.

A nomenclature which has become commonly accepted is that $I = \frac{1}{2}$, $S = 0$ resonances are called Ns, and $I = \frac{3}{2}$, $S = 0$ resonances are called Δs. There is no convincing evidence for the existence of resonances with $I > \frac{3}{2}$. We shall return to the classification of these particles, and the proposed reason for the absence of higher I-spins, in the following chapter. The known resonances with $B = 1$ and $S = 0$ are listed in appendix B. Δs have been clearly identified with spins up to $\frac{11}{2}$ and masses up to 2420 GeV/c^2.

9.7 Resonances with $B = 1$ and $S = -1$ (Λ and Σ)

The study of these resonances has in general been by different means from the strangeness zero particles. There are several reasons why this is so:

(*a*) For strangeness zero resonances access can be had to both the $I = \frac{3}{2}$ and $I = \frac{1}{2}$ states by pion–nucleon elastic scattering. This has meant that high statistics elastic or charge exchange scattering experiments, using counters, have been possible. For this reason formation experiments of high precision have been made yielding the data for detailed phase-shift analyses.

(*b*) For $|S| = 1$ resonances one requires in formation experiments an incident K-meson beam. Also, in production experiments the cross-sections for formation of final states containing strange particles are so much higher for incident kaons than for other incident projectiles that most of the work has been done with K-meson beams.

Since separated beams of K-mesons (see chapter 1) came later than pion beams, and are always much less intense, formation experiments are more difficult. More important, however, these resonances have, in general, large branching ratios into channels other than KN. Indeed, since the threshold for decay into KN is about 1440 MeV/c^2, this channel is not accessible to the lower mass $S = -1$ resonances.

These factors have led to much of the study of the strange particle resonances being carried out in bubble chambers, both for production and formation experiments. In general, a separated beam of K-mesons was allowed to enter the chamber. In formation experiments all channels were studied as a function of incident momentum, batches of photographs being taken at 20–100 MeV/c intervals. In a production experiment, mass distributions may be studied at a single momentum. In either case the total number of photographs required is large, frequently exceeding half a million in a single production experiment, and ten to fifty thousand per momentum interval in formation.

The standard nomenclature now used for the strange baryon resonances uses the symbol Λ for states with $S = -1$ and $I = 0$ (isotopic spin singlets), Σ for $S = -1$ and $I = 1$ (isotopic spin triplet), Ξ for $S = -2$ and $I = \frac{1}{2}$ and Ω for $S = -3$ and $I = 0$, the mass being given in brackets in each case.

We shall illustrate the methods of study of the strangeness -1 resonances by two examples, the first being the first such resonance discovered, and being a simple case, and the second involving a more elaborate analysis.

The $\Sigma(1385)$ is copiously produced in K$^-$p interactions over a wide range of incident momenta. The simplest process leading to production of this resonance is

$$\mathrm{K^-p} \to \Lambda\pi^+\pi^-, \qquad\qquad \textbf{9.5}$$

which appears in a bubble chamber as an event with two charged prongs and, if the Λ decays via pπ^-, a neutral vee. Application of energy and momentum conservation at the production and Λ-decay vertices, combined with observation of the track densities, can usually distinguish this process from the reaction of similar topology K$^-$p $\to$ K^{0}pπ^-. A Dalitz plot for the reaction **9.5** in which $M^2(\Lambda\pi^+)$ is plotted against $M^2(\Lambda\pi^-)$ shows two bands at masses of 1382 MeV/c^2 corresponding to formation of the so-called $\Sigma(1385)$. (Although the best mass known for this particle is 1382 MeV/c^2, it is still conventionally known by the mass from earlier measurements.)

We note that since $B = +1$ and $S = -1$, then $Q = I_3 = \pm 1$ for the $\Sigma(1385)^{\pm}$. No higher charge manifestations of the $\Sigma(1385)$ have been found, so that we can write $I = 1$. The I-spins of the π- and Λ-decay products are 1 and 0 and we should expect the three charge states and decay modes

$$\Sigma(1385)^+ \to \Lambda\pi^+, \quad \Sigma(1385)^0 \to \Lambda\pi^0, \quad \Sigma(1385)^- \to \Lambda\pi^-.$$

The decay $\Sigma(1385)^+ \to \Sigma\pi$ is also possible, but the branching ratio for this decay is found to be only $12 \pm 2\%$.

The spin of the $\Sigma(1385)$ has been measured to be $\frac{3}{2}$ from a study of the decay angular distribution. The most useful distribution is that of the cosine of the angle between the normal to the production plane and the direction of one or other of the decay particles from the $\Sigma(1385)$, as measured in the $\Sigma(1385)$ centre of mass. Any anisotropy in this distribution indicates a polarisation of the $\Sigma(1385)$, since correlation information is carried from the production to the decay vertex, but the details of the spin analysis will not be treated here.

The analysis of the $\Lambda(1520)$ (Watson, Ferro-Luzzi and Tripp, 1963) characteristics illustrates a number of important principles, and this experiment was a classic of its kind. The study was a formation experiment with K^--mesons on protons, the incident momenta ranging from 250–520 MeV/c. The runs were made using a liquid hydrogen bubble chamber. We note first that at a mass of 1520 MeV/c^2 the resonance has open to it KN, $\Sigma\pi$, $\Lambda\pi$, $\Lambda\pi\pi$, and even $\Sigma\pi\pi$ decay channels, of which the last is rather near threshold and will be neglected. We shall consider, therefore, the processes:

$$K^-p \to \begin{cases} K^-p & (a) \\ \bar{K}^0 n & (b) \\ \Sigma^{\pm}\pi^{\mp} & (c) \\ \Sigma^0\pi^0 & (d) \\ \Lambda\pi^0 & (e) \\ \Lambda\pi^+\pi^- & (f) \\ \Lambda\pi^0\pi^0 & (g). \end{cases}$$

The cross-sections, as a function of incident kaon momentum, for the above processes (a)–(f) are shown in fig. 9.11. From this data we can determine the I-spin of the $\Lambda(1520)$ (corresponding to incident K-meson momentum 390 MeV/c). The I-spin of the final states may be obtained with the aid of the appropriate Clebsch–Gordan coefficients. For the kaon

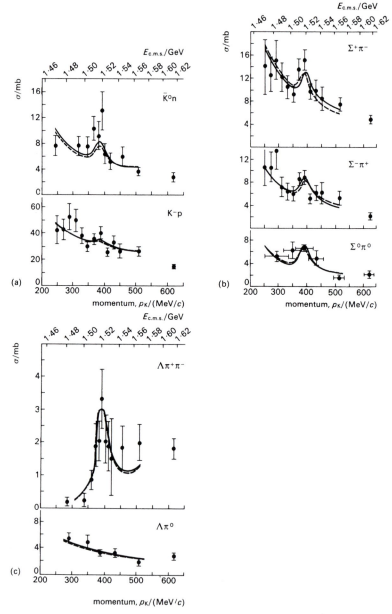

Fig. 9.11. Cross-sections as a function of momentum for: (a) K⁻p charge exchange and elastic scattering; (b) $\Sigma^+\pi^-$, $\Sigma^-\pi^+$ and $\Sigma^0\pi^0$ production; (c) $\Lambda\pi^+\pi^-$ and $\Lambda\pi^0$ production. The solid line corresponds to the best fit of all cross-sections, angular distributions and polarisations for negative $K N \Sigma$-parity while the dashed line corresponds to the best fit for positive $K N \Sigma$-parity.

nucleon final states we have in the usual notation, where the first term in the brackets refers to I_3 for the nucleon and the second to I_3 for the kaon,

$$\psi_{I=0} = \sqrt{\tfrac{1}{2}}(\tfrac{1}{2}, -\tfrac{1}{2}) - \sqrt{\tfrac{1}{2}}(-\tfrac{1}{2}, \tfrac{1}{2})$$
$$\psi_{I=1} = \sqrt{\tfrac{1}{2}}(\tfrac{1}{2}, -\tfrac{1}{2}) + \sqrt{\tfrac{1}{2}}(-\tfrac{1}{2}, \tfrac{1}{2})$$

so that

$$\phi(p, K^-) = \sqrt{\tfrac{1}{2}}(\psi_{I=1} + \psi_{I=0})$$
$$\phi(n, \bar{K}^0) = \sqrt{\tfrac{1}{2}}(\psi_{I=1} - \psi_{I=0}).$$

Also the $\Lambda\pi^0$ state is a pure $I = 1$ state, while the $\Sigma^0\pi^0$ and $\Lambda\pi^0\pi^0$ states formed in this process are pure $I = 0$ states. The student can readily verify these statements and write the appropriate wave functions to show that the $\Lambda\pi^+\pi^-$ and $\Sigma^\pm\pi^\mp$ states are $I = 0$, $I = 1$ mixtures. It is obvious from the data that the rather clear presence of the resonance in the $\Sigma^0\pi^0$ and $\Lambda\pi^+\pi^-$ states, and its absence in $\Lambda\pi^0$, fix the I-spin as zero.

We can make some further quantitative checks on this conclusion. If we consider the $\Sigma\pi$ channels we have ratios

	$\Sigma^+\pi^-$	$\Sigma^0\pi^0$	$\Sigma^-\pi^+$
$I = 0$	1	1	1
$I = 1$	1	0	1.

For the $\Lambda\pi\pi$ states we have

	$\Lambda\pi^+\pi^-$	$\Lambda\pi^0\pi^0$
$I = 0$	2	1
$I = 1$	1	0.

The $\Lambda\pi^0\pi^0$ channel should account for most of the unfitted V-zero events at these low momenta and the experimental ratio for resonance decay in these modes is in fact found to be $\sim \tfrac{1}{2}$.

Finally, we find that in production experiments of the kind

$$K^-p \to \Lambda(1520) + \text{pions},$$

where there is no restriction on the Λ charge, the resonance is still produced only in the neutral state.

We can now deduce the resonance spin, with some degree of certainty, from qualitative deductions from the differential cross-sections for the elastic and charge-exchange scattering, which are shown in fig. 9.12. Although the number of events at any energy is relatively small (~ 160 charge exchange events at 390 MeV/c) the data clearly allow certain useful conclusions. In both the elastic and charge-exchange processes the 390 MeV/c distributions show a strong $\cos^2 \theta$ dependence, which is absent above and below this momentum. Since no term higher than $\cos^{2J} \theta$ can occur, where J is the resonance spin, we see that $J > \frac{1}{2}$. The spin must of course be half-integral, since decay is into, for instance, KN. Also there is no need for $\cos^4 \theta$ in any distribution so that $J \geqslant \frac{5}{2}$ is unlikely, although not absolutely excluded. If $J = \frac{3}{2}$ then we could have either a $P_{\frac{3}{2}}$ or a $D_{\frac{3}{2}}$

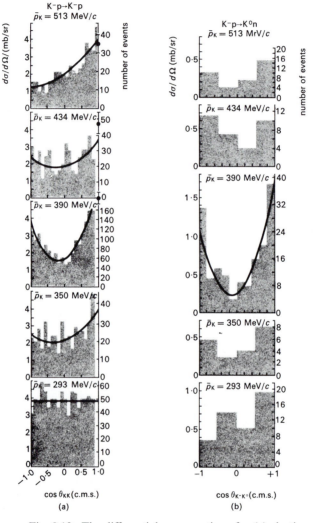

Fig. 9.12. The differential cross-sections for (a) elastic and (b) charge exchange, K^- p-scattering at various momenta in the vicinity of $\Lambda(1520)$, which is centred at 394 GeV/c with half-width at half-maximum corresponding to ± 21 MeV/c K^- laboratory momentum (Watson, Ferro-Luzzi and Tripp, 1963).

state ($J^P = \frac{3}{2}^+$ or $\frac{3}{2}^-$). The angular distributions for decay of these states are the same (this is the well-known Minami ambiguity). However, it turns out that in this case the ambiguity can be resolved by considering the way in which the interference with the background changes as a function of momentum.

In order to do this fully we require to be able to write down the relations between the coefficients of the terms in the angular distribution

$$\frac{\mathrm{d}\sigma}{\mathrm{d}\Omega} \propto \sum_n A_n \cos^n \theta$$

and the amplitudes for scattering in the various states. This is beyond the scope of the present text so that we must make some unsubstantiated assertions. At incident momenta less than $250 \, \mathrm{MeV}/c$ the angular distributions are isotropic, consistent with pure s-wave scattering, as one might expect at low momenta. At resonance the large $\cos^2\theta$ term is present, but the odd-power $\cos\theta$ interference term which shows as a forward–backward asymmetry, is not large. In this region we expect the continued existence of the s-wave background plus the resonance. However, the amplitude–coefficient relationships referred to above show that S–P$_3$ interference leads to a a large $\cos \theta$ term but no $\cos^2\theta$ term, although the P$_3$ alone will still yield such a term. On the other hand, as S–D$_3$ interference develops it gives a fast increase in the $\cos^2\theta$ term, a decrease in the isotropic term and no $\cos\theta$ term; these are all in agreement with the observations, which are indeed quantitatively quite well fitted by a D$_3$ resonance ($J^P = \frac{3}{2}^-$) with an s-wave background.

The presently known Λ and Σ resonances and their properties are listed in appendix B.

9.8   Resonances with $B = 1$ and $S = -2$ (Ξ)

Having seen that there exist resonances with $B = 1$, and $S = 0$, -1 and knowing also of the existence of the Ξ-particles with $B = 1$ and $S = -2$ it should be no surprise to find that there are also 'excited states' of the Ξ-particle or Ξ resonances. These we might expect to decay via $\Xi\pi$, $\Lambda\bar{K}$, $\Sigma\bar{K}$ or even more complicated cascades. It is clear that, due to the comparative rarity of Ξ-particles or reactions involving Λ and K or Ξ and K in the final state, the study of Ξ has been beset by the problem of statistics. Massive K$^-$p bubble-chamber experiments (5×10^6 pictures in the CERN 2 m hydrogen chamber) have eased this problem for the lower mass Ξs but for higher mass Ξs it still exists.

As an example of Ξs we take the $\Xi(1530)$, which was found in production in the process (Pjerrou *et al.*, 1962; Bertanza *et al.*, 1962).

$$K^-p \to \Xi K$$
$$\quad \downarrow$$
$$\quad \to \Xi\pi.$$

The ratio of Ξ-decays to $\Xi^0\pi^-$, $\Xi^-\pi^0$, $\Xi^-\pi^+$ and $\Xi^0\pi^0$ gives the *I*-spin

$$\frac{\Xi^0\pi^-}{\Xi^-\pi^0} = \frac{\Xi^-\pi^+}{\Xi^0\pi^0} = \begin{cases} 2 & \text{for} I_\Xi = \frac{1}{2} \\ \frac{1}{2} & \text{for} I_\Xi = \frac{3}{2} \end{cases}.$$

The result strongly favours $I = \frac{1}{2}$.

The decay angular distributions in such experiments establish $J^P = \frac{3}{2}^+$. $J^P = \frac{3}{2}^+$ allows this resonance to fit well into one of the multiplets of SU(3), which will be discussed in the following chapter.

An unusual feature of this resonance is its narrow width of $9.1 \pm 0.5\,\text{MeV}/c^2$. The presently known Ξ-particles are listed in appendix B.

9.9   Meson resonances: quantum numbers, *J, I, P, C*

The study of the meson resonances has uncovered a rich world of particles of varied properties. All the early particles were discovered in production experiments using bubble chambers. More recently, much work in this area has been done using a variety of 'spectrometers' – arrangements of electronic detectors with magnetic analysis for particle momenta. Such systems can be triggered by selecting certain characteristics of desired events and can thus operate with high beam fluxes to give sensitivity to processes with very small cross-sections.

Formation experiments for meson resonances require the production of an *s*-channel state with baryon number zero so that only through $\bar{p}p$ collisions can meson resonances manifest themselves in formation. This implies that only meson resonances with masses greater than $2\,\text{GeV}/c^2$ can be studied in this way.

The mesons are bosons, that is, integral spin particles, for which the baryon number is zero. Thus for mesons $Q = I_3 + \frac{1}{2}S$ and *Y*, the hypercharge, $= S$. The mesons thus have integral *I*-spin if $S = 0$ and half-integral *I*-spin if $S = \pm 1$, as exemplifed in the pions and kaons which we have already studied.

For the strongly decaying resonances we expect conservation of *I* as well as *J*, so that we can often draw conclusions about the quantum numbers of a meson resonance decaying into two particles from the nature of the final state. For decay into two spinless particles, the generalised Pauli principle (section 3.7) demands that if the particles are identical, e.g. $\pi^0\pi^0$, then the relative orbital angular momentum L must be even. If the two decay products are not identical, but members of the same I-spin multiplet then

$(L + I)$ must be even,

or, in other words, L and I must be even or odd together.

The above deductions from the generalised Pauli principle follow from the fact that the parity of the space wave function for two members of the same multiplet is $(-1)^L$. If J is the resonance spin, then particles having parity $(-1)^J$ are said to have natural spin-parity and those having parity $(-1)^{J+1}$, unnatural spin-parity. Decay into two pseudo-scalar mesons is only possible for natural-parity resonances. In this connection we note that, unlike the baryons, meson particle and antiparticle have the *same* parity.

We have already used the *charge-conjugation* operator C in studying the K^0, $\bar{K}^0$ system (section 8.2). For a state to be an eigenfunction of C it must be electrically neutral, since in changing particle to antiparticle C reverses the charge. Equally, B and S must also be zero. In the decay of the K_1^0 we saw that the operations C and P were equivalent. This is similarly true for decay into K^+K^- and $K^0\bar{K}^0$ so that for these decays I, J, P and C must be all even or all odd. Also, as we have seen for the π^0-meson (2.8), decay to $\gamma\gamma$ implies even parity under charge conjugation.

We can also draw some simple conclusions for decay into K_1^0 and K_2^0 combinations. From our previous analysis of the $K^0\bar{K}^0$ system it is clear that

$$CP|K_1^0K_1^0\rangle = CP|K_2^0K_2^0\rangle = (-1)^J$$
$$CP|K_1^0K_2^0\rangle = -(-1)^J.$$

However, since C and P are equivalent for these systems they must be even under the CP operation, so that for K^0 decay

even-spin resonances can decay only to $K_1^0K_1^0$ or $K_2^0K_2^0$,
odd-spin resonances can decay only to $K_1^0K_2^0$.

9.10   Meson resonances: *G*-parity

When several conservation laws operate for the same system it is sometimes possible to obtain new quantum numbers and selection rules by combining the original ones. The new conservation law may reveal features not evident in the originals.

We recall that I-spin invariance holds good only for strong interactions. Charge-conjugation invariance holds for strong interactions and also for electromagnetic interactions. Only for strong interactions can we combine charge conjugation and I-spin invariance to obtain a new selection rule.

If the charge of a system is not zero it cannot be an eigenfunction of C. However, if $B = S = 0$ the effect of C is simply to reverse the charge. Thus a system with $B = S = 0$ will be an eigenfunction of the combined operators CR where R is the charge-inversion operator. The R operator can be written in terms of the I-spin operators in that to invert the charge one needs to invert I_3, i.e. rotate the system by π about the I_2-axis. Thus $R = e^{i\pi I_2}$. The G operator is

$$G = CR = Ce^{i\pi I_2}.$$

We may study the effect of the C, R and G operators on π-mesons by writing the wave functions for the pions in terms of the I-spin components of the boson field. In these terms we write

$$\psi_{\pi^+} = \frac{1}{\sqrt 2}(\phi_1 - i\phi_2)$$

$$\psi_{\pi^-} = \frac{1}{\sqrt 2}(\phi_1 + i\phi_2)$$

$$\psi_{\pi^0} = \phi_3$$

where ϕ_1, ϕ_2, ϕ_3 are the I_1, I_2, I_3 components of the field. Although we cannot prove these relations within the scope of the present text, we may note at least that they satisfy the requirements concerning antiparticles and complex conjugation, discussed in connection with the $K^0\bar{K}^0$ system.

When R is applied to the ϕs we see that, since it produces a rotation about the I_2-axis, ϕ_2 remains unchanged while the others reverse sign:

$$R\phi_1 = -\phi_1, \quad R\phi_2 = \phi_2, \quad R\phi_3 = -\phi_3$$

since $C(\pi^+) = \pi^-$ and $C(\pi^0) = \pi^0$ we see that

$$C\phi_1 = \phi_1, \quad C\phi_2 = -\phi_2, \quad C\phi_3 = \phi_3.$$

Thus

$$G\phi_1 = -\phi_1, \quad G\phi_2 = -\phi_2, \quad G\phi_3 = -\phi_3$$

and $G(\pi) = -\pi$. Thus the pion wave functions are eigenfunctions of G with eigenvalue of G-parity -1.

We can derive the G-parity of an I-spin multiplet by considering its neutral member, since the G-parity is the same for all members of the multiplet. This may be seen to be so from the following argument. We can construct operators from the I-spin operators which will produce transformations between the members of an I-spin multiplet (cf. the raising and lowering operators in the quantum theory of angular momentum). G, however, may be shown to commute with all the I-spin operators $[G, I] = 0$. Now suppose that we have a given G-parity for the neutral member of a multiplet, e.g. $G(\pi^0) = -1(\pi^0)$, and that we make an operator I^+ from I_1, I_2, I_3 which has the effect of increasing the value of the third component by one unit so that

$$I^+(\pi^0) = x\pi^+.$$

Now apply the operator G

$$G(\pi^+) = G\left(\frac{1}{x}\right)I^+(\pi^0)$$

$$= \frac{1}{x}I^+G(\pi^0)$$

$$= \left(\frac{1}{x}\right)(-1)I^+(\pi^0)$$

$$= -1(\pi^+)$$

where we have used the fact that G commutes with the I-spin operators from which I^+ is constructed.

The fact that the G-parity of a multiplet can be determined from that of its neutral member allows us to derive a useful relation between G-parity and I-spin. As with ordinary angular momentum states, we can represent the isotopic spin wave function for a system, or particle, by a spherical harmonic $Y_I^{I_3}(\cos \theta)$, where θ is the angle which the I-spin vector makes with the I_3-axis. For the neutral member of a multiplet, with $B = S = 0$, I_3 is also zero, so that the isotopic spin wave function can be written $Y_I^0(\cos \theta)$. The R-operator (rotation by π about the I_2-axis) changes θ to $\theta + \pi$, which results in multiplying the wave function by $(-1)^I$. Thus for non-strange mesons

$$G = C(-1)^I.$$

If the meson decays *strongly* into $\pi\pi$ or KK, then the C-parity is given by $(-1)^I$ and for this case the G-parity is given by

$$G = (-1)^{I+l}. \qquad\qquad \textbf{9.6}$$

9.11 Meson resonances: spin and parity

In the following chapter we shall see that the mesons and meson resonances fall into multiplets of nine members (nonets), all of which have the same spin and parity, and that these multiplets are accounted for by the 'SU(3)' symmetry group. It will thus be convenient for us to discuss the experimental evidence concerning the meson resonances by looking first at the *pseudo-scalar meson* nonet ($J^P = 0^-$) and then, in turn, at the *vector mesons* ($J^P = 1^-$) and *tensor mesons* ($J^P = 2^+$).

9.11.1 Pseudo-scalar mesons

We have already studied the isotopic spin triplet of pions and the K^+, K^0 and $\bar{K}^0$, K^- I-spin doublets, all of which have $J^P = 0^-$. The other members of the nonet are the η and the η' mesons.

In fig. 9.13 is shown the effective mass distribution of the $\pi^+\pi^-\pi^0$ combination in the reaction in which the η was first observed

$$\pi^+ d \rightarrow pp\pi^+\pi^-\pi^0$$

at a momentum of 1.23 GeV/c. Instead of the smooth phase-space

Fig. 9.13. The first observation of the η-meson. The plot shows the effective mass of $\pi^+\pi^-\pi^0$ from the reaction $\pi^+ d \rightarrow pp\pi^+\pi^-\pi^0$. The upper peak corresponds to the ω-meson and the lower to the η-meson. The smooth points correspond to the calculated phase-space distribution from the total average (p, 3π) cms energy (1850 MeV) (Pevsner *et al.*, 1961).

distribution the spectrum shows two sharp peaks at masses of 550 and 780 MeV/c^2. The upper of these is known as the ω-meson and the lower as the η-meson. Both of these resonances have also been seen in many other reactions. No charged η has ever been observed, so that the I-spin is zero. The η is also found to have neutral decay modes

$$\eta \rightarrow \gamma\gamma$$
$$\eta \rightarrow \pi^0\pi^0\pi^0$$

which account for 71% of all decays. The $\pi^+\pi^-\pi^0$ mode has a branching ratio of 24% and the only sizeable channel apart from those mentioned above is $\eta \rightarrow \pi^+\pi^-\gamma$ (4.9%).

The γ-modes are clearly electromagnetic decays and the fact that these have an appreciable branching fraction suggests that the strong decay modes must be inhibited by some conservation law and that the the width should be $\lesssim 1$ keV, characteristic of electromagnetic processes.

The decay to $\gamma\gamma$ limits the possible spin to 0 or 2 (see section 2.8 concerning the spin of the π^0). For even spin parity the η could decay strongly to two pions, so that the parity must be odd. The distribution in the Dalitz plot (cf. τ-decay) is fitted by $J^P = 0^-$. Since the η must be odd under CP (cf. K_1^0, K_2^0 discussion, section 8.2) it is even under C (as can also be inferred from the $\gamma\gamma$-decay) and the G-parity is given by

$$G = +(-1)^0 = +.$$

But the G-parity of the three-pion final state is $(-1)^3 = -$, so that G-parity is not conserved in the decay. This accounts for the inhibition of the three-pion modes. The most precise value of the η-width is obtained from the decay rate for the purely electromagnetic mode $\eta \rightarrow \gamma\gamma$ (see section 4.3). The fitted branching ratio for this mode is then used to obtain the total width of 0.83 ± 0.12 keV/c^2.

The η' meson is also narrow (0.28 ± 0.10 MeV/c^2), with mass 958 MeV/c^2 and decay modes

$$\eta' \rightarrow \pi^+\pi^-\eta,$$
$$\rightarrow \pi^+\pi^-\gamma,$$
$$\rightarrow \pi^0\pi^0\gamma.$$

The quantum numbers are

$$IJ^{PCG} = 0, 0, ^{-++}.$$

The properties of the pseudo-scalar mesons are summarised in appendix B.

9.11.2 *Vector mesons*

Unlike the pseudo-scalars, the vector meson nonet includes no semi-stable ($\tau \sim 10^{-10}$ s) particles but has only strongly decaying members. These are the

$$\rho^{\pm 0}, \quad \omega^0, \quad K^*(892)^{\pm 0}, \quad \bar{K}^*(892)^0, \quad \phi^0.$$

It is not appropriate to this text to describe in detail the evidence and the arguments used in determining the quantum numbers and other properties of all these resonances. We shall rather present some examples and list the properties in appendix B.

In fig. 9.13 we have already seen examples of ω-production in the mass combination ($\pi^+\pi^-\pi^0$). No charged ω has been seen, so we expect $I = 0$. $G = -1$, so $C = -1$ (strong decay) and L for the di-pion ($\pi^+\pi^-$) is odd (fig. 9.14). For the I-spin wave function we get $I = 0$ by combining an $I = 1$, $\pi^+\pi^-$ di-pion with $I = 1$ for the π^0. For the di-pion the $I = 1$ system is antisymmetric (for $I_3 = 0$, $I = 1$, we have

$$\frac{1}{\sqrt{2}}(\pi^+\pi^- - \pi^-\pi^+)$$

– see Clebsch–Gordan coefficients), so for the three-pion system the I-spin wave function is also antisymmetric. Since overall symmetry is required by Bose–Einstein statistics, the space part of the wave function must also be antisymmetric. Since C is odd so is L. The assignments are given in table 9.1.

As before, $\mathbf{q}$ is the momentum of either π in the di-pion cms and $\mathbf{p}$ is the

Table 9.1

J^P for ω	L	l	Parity of space w.f.	Dependence on momenta
0^-	1	1	$+$	$\mathbf{p} \cdot \mathbf{q}$ – zero on medians
1^+	1	0	$-$	$\mathbf{q}$ – zero at $q = 0$
1^-	1	1	$+$	$\mathbf{p} \times \mathbf{q}$ – zero on boundary $(\mathbf{p} \parallel \mathbf{q})$

Fig. 9.14 Definition of angular momenta for ω-decay.

momentum of the π^0 relative to the di-pion in the three-pion cms. The dependences in the last column lead to rather distinctive distributions in the Dalitz plot, which allow determination of $J^P = -1$.

The ρ-meson is a broad resonance ($\Gamma = 153 \pm 2$ MeV/c^2, $m = 770 \pm 3$ MeV/c^2) found to decay into $\pi^+\pi^0$, $\pi^+\pi^-$, $\pi^-\pi^0$, and is thus seen to be likely to have I-spin $= 1$. This hypothesis is confirmed by measurement of the ratio of ρ^-/ρ^0 production in the reactions

$$\pi^-p \to \begin{cases} \pi^0\pi^-p \\ \pi^+\pi^-n \end{cases}.$$

In these two reactions, and in the process

$$\pi^+p \to \pi^+\pi^0p$$

ρ production is copious, the momentum transfer to the proton is small and the other features of the reactions suggest that they proceed via one-pion exchange (fig. 9.15). In this case the ratio of ρ^-/ρ^0 production in the first two reactions will be (see appendix B for C–G coefficients) $\frac{1}{2}$, 0, $\frac{2}{9}$, for $I(\rho) = 1$, 0 and 2, respectively. Experimentally, the ratio is found to be 0.5.

Figure 9.15 illustrates a method by which we can (indirectly) study π–π scattering at the upper vertex, although not 'on the mass shell' (i.e. with energy and momentum conserved) since the exchanged pion is virtual. However, an apparently successful recipe exists which allows extrapolation to the actual physical situation. Many analyses of ρ-production using these reactions have studied the distribution of $\cos \theta$, where θ is the scattering angle in the cms of the di-pion. For the simplest model, with $J(\rho) = 1$, this distribution should be pure $\cos^2 \theta$. Initial and final state effects may modify the distribution by addition of an isotropic term to give

$$A + B \cos^2 \theta,$$

but the distribution should still be symmetrical about 90°. For the charged ρ, the angular distributions below and above the resonance are asymmetric, as would be expected if the resonance amplitude interferes with a background which is itself changing only slowly. At resonance the angular distribution becomes symmetric (fig. 9.16). For the ρ^0, however,

Fig. 9.15. π–π scattering by ρ-production in $\pi p \to \rho p$.

the asymmetry persists in the resonance and does not change sign passing through the resonance. The most plausible explanation of this asymmetry occurring only in the neutral state is that there may exist an $I = 0$ interaction, almost in phase with the ρ-amplitude. Since an $I = 1$ resonance cannot decay into $\pi^0\pi^0$ (consider the appropriate C–G coefficients), while an $I = 0$ resonance can decay in this way, the $\pi^0\pi^0$ channel is a good one in which to study the $I = 0$ interaction (sometimes known as the ε^0). However, study of this channel is difficult since it involves detection of the γ-rays from the π^0 decays.

The $K^*(892)$ resonances appear strongly in almost every reaction in which they can be produced. For instance, fig. 9.8 shows the $K\pi$ mass spectrum from the process

$$K^+p \rightarrow K^0p\pi^+$$

for incident kaons of momentum 10 GeV/c. The distribution shows peaks at 890 and 1400 MeV/c^2. By studying reactions such as

at incident kaon momentum high enough to produce the K^*, but not so high that the secondary K^+ and p are frequently indistinguishable in the fit or in ionisation, one may determine the ratio

$$\frac{K^{*+} \rightarrow K^0\pi^+}{K^{*+} \rightarrow K^+\pi^0}.$$

This ratio is found to be 2.0, as expected for $I = \frac{1}{2}$ ($I = \frac{3}{2}$ gives a ratio of 0.5). The decay angular distributions fix J as 1 so the parity is -1. The strangeness is $+1$ for K^{*+}, K^{*0}, and -1 for $\bar{K}^{*0}$ and K^{*-}, so G is not a good quantum number.

The ϕ-meson is observed to decay into $K\bar{K}$ only in neutral states and no charged ϕ has ever been observed, though if it existed such an object could readily be produced in, for instance,

$$K^-p \rightarrow \Sigma^+K^0K^-.$$

The mass of the ϕ is 1020 MeV/c^2 and it is found to be very narrow (4.41 ± 0.05 MeV/c^2). An example of ϕs produced in the processes

$$\pi^-p \rightarrow K^+K^-K^+K^- \ldots$$

where the trigger demanded a combination of signals in Cerenkov and scintillation counters characteristic of the production of two pairs of oppositely charged kaons is shown in fig. 9.17. The production of pairs of ϕs is markedly more abundant than would be expected on the basis of a random superposition.

Decay angular distributions for ϕ establish the spin-parity as 1^-. From **9.6** we see that $G = -1$.

9.11.3 *The tensor nonet*

The tensor nonet is a group of mesons all of spin-parity 2^+. Its members are the $a_2(1320)$, $f_2(1270)$, $f'_2(1525)$ and $K_2^*(1430)$ resonances (originally known as the $A_2(1320)$, $f(1270)$, $f'(1525)$ and $K^*(1400)$).

The f_2-meson decays into two pions and is seen only in the neutral state. The ratio

$$\frac{f_2 \to \pi^0 \pi^0}{f_2 \to \pi^+ \pi^-}$$

is in agreement with 0.5, so $I = 0$, $G = +1$, J^P must be $0^+, 2^+, 4^+, \ldots$ and

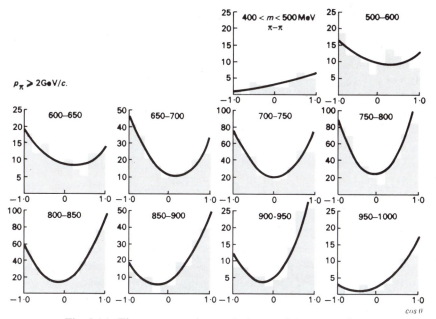

Fig. 9.16. The π–π scattering angle (see text) in $\pi^- p \to \pi^0 \pi^- p$ as a function of $\pi^0 \pi^-$ mass. Only events with four momentum squared less than $5(m_\pi)^2$ have been used to ensure 'peripheral' rho production corresponding to the diagram of fig. 9.15. The forward–backward asymmetry is seen to disappear at 750 MeV/c^2, corresponding to the ρ-mass (Walker *et al.*, 1967).

$C = +1$. The distribution of $\cos \theta$ (θ is the $\pi\pi$ scattering angle, as in the discussion of the ρ-meson) is symmetrically forward–backward peaked, so 0^+ is excluded while 2^+ gives a reasonable fit. A convincing analysis of the spin-parity for particles decaying to two pions is illustrated in fig. 9.18. In fig. 9.18 (a) shows the $\pi^+\pi^-$ mass distribution found in a study of the process

$$\pi^- p \rightarrow \pi^+ \pi^- n$$

at 6 GeV/c (Crennell *et al.*, 1968). The mass distribution shows clear peaks corresponding to the ρ-meson and the f-meson (1264 GeV/c^2) and also a small peak corresponding to a meson of even higher mass (1650 GeV/c^2) known as the ρ_3 (originally named g) meson. The $\pi^-\pi^-$ scattering angular distribution for events with four-momentum transfer less than 1.0 (GeV/c)^2 was fitted to the distribution

$$A_n P_n \cos \theta$$

and the even coefficients are plotted as a function of $m(\pi^+\pi^-)$ in (b), (c) and (d) of fig. 9.18. The highest coefficient to be expected is A_{2J} and we see that for the ρ-region ($J = 1$) only A_2 is found, while for the f-region both A_2 and A_4 are substantial while A_6 is absent, indicating $J = 2$. Although A_6

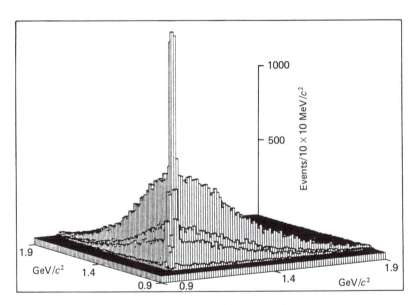

Fig. 9.17. $M(K^+K^-)/M(K^+K^-)$ for the reaction $\pi^- p \rightarrow$ $K^+K^-K^+K^-\ldots$. The sharp peak at the overlap of the ϕ-bands indicates correlated ϕ-production (Booth *et al.*, 1984).

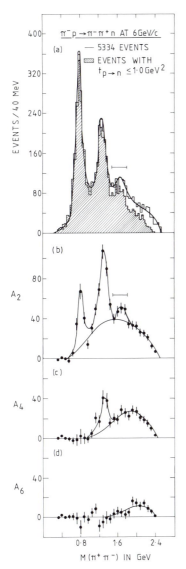

Fig. 9.18. (a) shows the effective mass distribution for $\pi^+\pi^-$ from the reaction $\pi^- p \rightarrow \pi^+\pi^- n$ at 6 GeV/c. The peaks due to the ρ^0, f_2^0 and ρ_3^0 mesons are visible. (b), (c) and (d) show the variation with $m(\pi^+\pi^-)$ of the coefficients A_n when the π–π scattering distribution is fitted to $A_n P_n(\cos\theta)$. The presence of A_4 and absence of A_6 for the f_2^0 mass suggests $J = 2$ (Crennell *et al.*, 1968).

is also present for the ρ_3, the decay is very asymmetric (substantial odd-A coefficients) and it is not possible to determine J for the ρ_3 from this distribution.

The f_2'-meson appears as a peak in the $K\bar{K}$ mass spectrum in reactions like

$$K^-p \to \Lambda K_1^0 K_1^0$$

at a mass of 1525 ± 5 MeV/c^2. The $K_1^0 K_1^0$ mode means that this meson is even under charge conjugation. A good illustration of this point was presented by the discoverers of the f' (Barnes *et al.*, 1965). For any $K^0\bar{K}^0$-system one can derive the proportion of events with one (N_1) or with two (N_2) visible K^0-decays under the assumptions:

(a) that there is no resonance so that the K^0 and $\bar{K}^0$ decay uncorrelated;

(b) that the K^0, $\bar{K}^0$ arise from the decay of a resonance with $C = +1$; and

(c) that the K^0, $\bar{K}^0$ arise from the decay of a resonance with $C = -1$.

The result is shown as a function of $m^2(K^0\bar{K}^0)$ in fig. 9.19. The reader should check the expected values of the ratio as an exercise.

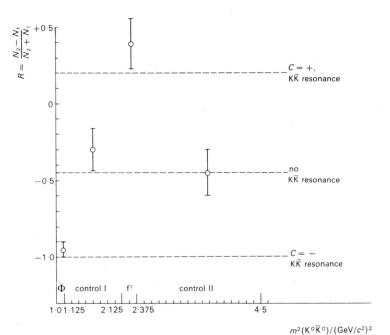

Fig. 9.19. The ratio R (see text) as a function of $m^2(K^0\bar{K}^0)$ for the process $K^-p \to \Lambda K^0\bar{K}^0$. In the f'-region the ratio reaches ~ 0.2 as expected for a resonance with $C = +1$ (Barnes *et al.*, 1965).

The $K_2^*(1430)$-meson has already been mentioned, in passing, in the discussion of the $K^*(892)$ and fig. 9.8. Its decay angular distributions establish fairly clearly that $J^P = 2^+$ and the branching to different charge states gives unambiguously that $I = \frac{1}{2}$.

The a_2-meson, the isotopic spin triplet of the nonet, has been seen in many bubble chamber experiments as a peak at 1318 MeV/c^2 in the $\pi\rho$ mass distribution. It is also observed to decay into $K_1^0 K_1^0$, $K^- K_1^0$, and $\eta\pi$, $\omega\pi\pi$. All the data is consistent with $I = 1$, $C = +1$. The $K_1^0 K_1^0$ mode requires $J^P = 0^+$, 2^+,.... For $a_2 \rightarrow \pi\rho$, however, $J^P = 0^+$ is forbidden, so that 2^+ is the lowest allowed assignment. The decay angular distributions also favour this assignment.

The properties of the $J^P = 2^+$-mesons are summarised in appendix B.

9.11.4 *Other meson resonances*

The three nonets ($J^P = 0^-$, 1^-, 2^+) which are described above are well established and formed a crucial element of the data on which the quark model was based. There are in addition a number of other meson resonances, some of which have well-established quantum numbers, and there is evidence for nonets with $J^{PC} = 1^{++}$, 1^{+-} and 0^{++}, as well as for higher spin resonances such as the ρ_3 mentioned above, the $\omega_3(1670)$ and the $K_3^*(1780)$ all with $J^P = 3^-$.

A list of the established states is given in appendix B. A comprehensive account of the current state of knowledge on resonance particles is given in the *Review of Particle Properties*, published every one or two years by Physics Letters in Europe and also available from CERN and from Lawrence Berkeley Laboratory in the U.S.A.

10

SU(3) and the quark model: classification and dynamic probes

10.1   Introduction

The discovery of such a wealth of apparently 'elementary' particles stimulated new activity in the search for a pattern amongst them, as a first step towards the understanding of their nature. The discovery of such a pattern is analogous to, for instance, the discovery of the Rydberg formula in atomic spectroscopy. The Bohr atom finally provided an explanation of the formula, and we shall see that the quark model provides an explanation of the symmetry pattern of the elementary particles.

We have already become familiar with the limited symmetry of isotopic spin multiplets. In that case we grouped together particles which were the same except for properties associated with the electric charge. The degeneracy of the multiplet is removed by the symmetry-breaking Coulomb interaction. Alternatively, we can regard the members of the multiplet as states linked by rotations in isotopic spin space and we can define a group of rotation operators which enable us to step from one state to another.

The Coulomb interaction is not strong compared with the so called 'strong' interactions, and the symmetry breaking to which it gives rise is small. For instance, the masses of particles in the same isotopic spin multiplet differ only by at most a few per cent. In order to extend the

216

symmetry, to group larger numbers of particles together, we must recognise the existence of much stronger symmetry breaking forces since the mass differences between, say, *I*-spin multiplets, are substantial, even compared with the particle masses themselves. Nevertheless it turns out to be true that the symmetry, though broken, remains in many ways very useful.

10.2   Baryon and meson multiplets

In attempting to group together different *I*-spin multiplets we may seek to group particles having the same baryon number, spin and parity, but allow the strangeness (or equivalently the hypercharge) to vary within the multiplet. This hypothesis has the merit of success over others, such as allowing spin and parity to vary within a multiplet and demanding the same strangeness, which might *a priori* appear equally reasonable.

For *I*-spin multiplets, we may represent the members of the multiplet as points spaced at unit intervals on the I_3-axis. The raising and lowering operators allow steps to the right and left along the axis. For instance, for the $\Delta(1232)$ multiplet we have the configuration shown in fig. 10.1. In order to extend the classification to include other *I*-spin multiplets of the same J^P and B we need to move from a one- to a two-dimensional diagram, where the axes are I_3 and Y. We first take the $J^P = \frac{1}{2}^+$ baryons, of which eight were known when the classification was first proposed. The original $\frac{1}{2}^+$ octet consists of p, n, Σ^+, Σ^0, Σ^-, Λ^0, Ξ^0 and Ξ^-; two *I*-spin doublets, a triplet and a singlet. The diagram for this octet is shown in fig. 10.2. The octet forms a hexagonal pattern on the Y–I_3 diagram and it is clear that in order to make transitions between states we need operators which effect diagonal steps as well as the step operators within the *I*-spin multiplet.

Many other particles also readily fall into similar patterns. In particular, for the pseudo-scalar, vector and tensor mesons we find nonets which are shown on the Y–I_3 plot in fig. 10.3. In the meson nonets, we notice that, unlike the baryon case, particle and antiparticle appear in the same multiplet, since for both particle and antiparticle B is the same.

10.3   Symmetry groups

Having noted the existence of these multiplets we may seek a classification to describe them. The SU(3) group was proposed for this purpose in 1961 by Gell-Mann and independently by Ne'emann.

Some of the formal properties of groups are summarised below, though the reader wishing to pursue this aspect of the subject more deeply is recommended to consult one of the several excellent texts such as *Lie Groups for Pedestrians* (Lipkin, 1965). In this section we use some of these properties to bring out the principal physical ideas. The idea of examining

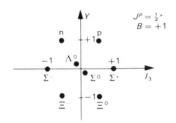

Fig. 10.1. The members of the $\Delta(1232)$ isotopic spin multiplet plotted on the I_3 axis.

Fig. 10.2. The $J^P = \frac{1}{2}^+$, $B = +1$ octet plotted on the Y–I_3 plane.

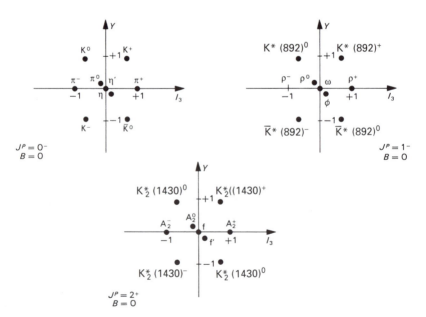

Fig. 10.3. The pseudo-scalar, vector and tensor meson nonets plotted on the Y–I_3 plane.

the symmetry of any system by 'rotating' it in the appropriate space is a familiar one. We accomplish this by means of 'rotation' operators, in the most general sense of the term 'rotation'. The set of rotation operators is said to form a symmetry group. For instance, on functions of geometric coordinates we have the translation, space-rotation and inversion groups. These operators act on the wave functions.

The most familiar example to compare with the operators we shall require is the space-rotation group. These operators act on the angular coordinates of the wave function $\psi(\theta_1, \theta_2, \theta_3) = \psi(\theta)$. They have the form

$$R(\theta) = e^{i\theta \cdot J}.$$

Here, the θ are the rotation angles and J the angular momentum. These operators form an infinite group.

Now, if we have a limited space of states such that under the operators of the group each state acted on by an operator of the group transforms into another state in this space, then we have an invariant subspace of states. For such a set of states, an operator of the group cannot connect a state within the set to a state outside. For instance, the set of spherical harmonic functions $Y_{0,0}$, $Y_{1,1}$, $Y_{1,0}$, $Y_{1,-1}$ form a four dimensional invariant subspace of the rotation group. The subspace is said to be irreducible if it contains no smaller subspace. Thus, pursuing the above example, we see that:

$Y_{0,0}$ is an irreducible singlet,

$Y_{1,1}$, $Y_{1,0}$, $Y_{1,-1}$ is an irreducible triplet.

The conventional way of expressing this division is to write

$$(Y_{0,0}, Y_{1,1}, Y_{1,0}, Y_{1,-1}) = Y_{0,0} \oplus (Y_{1,1}, Y_{1,0}, Y_{1,-1}).$$

An irreducible invariant subspace is known as a multiplet of the operator group.

A general theorem can be proved which states that if the Hamiltonian for the system commutes with the operators of the symmetry group then the multiplets are sets of degenerate eigenstates of the system. We know that for the elementary particles the states are not degenerate, so that the multiplet is split. This is due to additional terms in the Hamiltonian which do not commute with the group operators. An example of terms which cause splitting for the rotation group is, for instance, the $L \cdot S$ coupling which splits the levels in atomic spectra.

In the space rotation group there is an infinite number of operators, but we can deal with this infinite group in a simple way because all the operators can be expressed in terms of only three basic operators, J_1, J_2 and J_3. If $R(\theta)$ depends analytically on θ, at least near $\theta = 0$ and thus near $R(0) = 1$, then we can relate the Js to R by means of the relation

$$iJ_k = \left[\frac{\delta R(\theta)}{\delta \theta_k} \right]_{\theta=0}.$$

A Lie group is a continuous group composed of the operators $U(\alpha) = U(\alpha_1, \ldots, \alpha_n)$ which depends analytically on all the n parameters and for which $U(0) = 1$. As in the particular case above, we can extract basic operators

$$iG_i = \frac{\delta U}{\delta \alpha_i}\Big|_{\alpha=0}$$

which are called the generators of the group but which are not themselves members of the group. The dimension of the group is given by the number of the generators.

We have already seen that the larger multiplets for particles contain the *I*-spin multiplets. Suppose, therefore, that we enlarge a Lie group by adding more coordinates for the group to transform. In this case we have to add more operators to the group.

An idea of fundamental importance in the present application of group theory is that of the basic or elementary multiplet. For a Lie group of rank *l* there exist *l* such elementary multiplets, and from these can be built up more complicated multiplets by taking repeated 'products' between the basic multiplets. A simple example from the rotation group is the triplet deuterium ground state consisting of spin $\frac{1}{2}$ neutron and proton. The spin $\frac{1}{2}$ particles are members of two-member multiplets or doublets of the rotational group for angular momentum $\frac{1}{2}$ (the generators of which are the Pauli spin matrices). The product of two such doublets then yields triplet and singlet multiplets according to $2 \otimes 2 = 3 \oplus 1$. This kind of decomposition of the product of two multiplets is a process which is familiar under another guise in angular momentum theory as the Clebsch–Gordan expansion

$$Y_{l_1 m_1}(\theta, \phi)\, Y_{l_2 m_2}(\theta, \phi) = \sum_{L=|l_1-l_2|}^{l_1+l_2} \sum_{M=-L}^{+L} C_{m_1 m_2 M l_1 l_2 L} Y_{LM}(\theta, \phi).$$

The form of the decomposition or the values of the Clebsch–Gordan coefficients must be worked out from the commutation relations of the generators.

Thus for the Simple Unitary (Lie) group of order 3, SU(3), the elementary multiplet is a triplet, and the multiplet products can be shown to decompose as, for instance,

$$3 \otimes 3 \otimes 3 = 10 \oplus 8 \oplus 8 \oplus 1.$$

10.4   The SU(3) classification and the quark model

For SU(2) there are only two basic states, that is, the basic multiplet is a doublet such as the p, n I-spin multiplet. In order to include states with non-zero hypercharge, it is necessary to move to a larger group of order 3, such as the group SU(3) for which the basic multiplet is a triplet. The state to be added to the SU(2) doublet is then a state with non-zero hypercharge and $I = 0$. The strangeness and isotopic spin quantum numbers of the triplet are then the same as p, n, Λ.

An early model in which it was indeed proposed that all the known particles were built from the physical proton, neutron and lambda particles was proposed by Sakata in 1956. However, this model quite rapidly ran into some problems, such as the parity of the Σ, which was on this model composed of $N\bar{N}\Lambda$ (e.g. $\Sigma^+ = p\bar{n}\Lambda$) in mutual S-states, so that $J^P = \frac{1}{2}^-$ and not $\frac{1}{2}^+$ as is experimentally found to be the case. The basic SU(3) multiplet is now referred to as the SU(2) doublet u, d plus the non-zero hypercharge state s. These states are now identified with the *quarks* from which all the particles are built. It has often been stressed that the SU(3) classification does not depend on the actual existence of particles corresponding to the basic multiplet. We shall return to the question of existence of the quarks below. For the present we merely use a basic triplet of states, equivalent under the strong interaction, to construct the pattern of the observed particles.

In order to find the patterns we require the set of operators forming the group SU(3) which will transform the basic triplet states. For SU(3) the operators may be represented by a group of unimodular, unitary 3×3 matrices in a similar way to the Pauli spin matrices for SU(2). There are $3 \times 3 - 1$ such independent traceless matrices. However, the proper treatment of these eight operators requires a more detailed knowledge of group theory than can be presented here.

We rather follow the treatment of Lipkin (1965) to derive the operators in terms of bilinear products of the creation and annihilation operators for u, d and s. These operators are denoted $a_u^\dagger, a_d^\dagger, a_s^\dagger$ and a_u, a_d, a_s respectively.

The technique is best illustrated by application first to the simplest case of the I-spin doublet u, d. Only bilinear products which do not change the baryon number can be permitted, so that in this case there are four such products:

$a_u^\dagger a_d$ – changes d to u $\equiv$ step up operator $= \tau_+$;
$a_d^\dagger a_u$ – changes u to d $\equiv$ step down operator $= \tau_-$;
$a_u^\dagger a_u$ – counts 'ups' (i.e. the number of protons is given by the eigenvalue);
$a_d^\dagger a_d$ – counts 'downs'.

In order to include states with $S \neq 0$ we include the $a_s^\dagger, a_s$, obtaining:

$a_s^\dagger a_u$ – changes u to s $= C_-$;
$a_s^\dagger a_d$ – changes d to s $= C_+$;
$a_u^\dagger a_s$ – changes s to u $= B_+$;
$a_d^\dagger a_s$ – changes s to d $= B_-$.

Together with the first two we thus have six step operators:

τ_+ and τ_- leave Y unchanged and change I_3 by ± 1;
B_+ and B_- change Y by $+1$ and I_3 by $\pm\frac{1}{2}$;
C_+ and C_- change Y by -1 and I_3 by $\pm\frac{1}{2}$.

The operations are illustrated in fig. 10.4. Such a set of step operators satisfies the rule that operation on any state transforms it into another state of the multiplet or annihilates it. These operators require a hexagonal type symmetry for all the multiplet diagrams. This is due to the fact that instead of the single I_3 symmetry axis in SU(2) there are now three symmetry axes at 120° to each other.

Along the 'I-spin' axis the hypercharge is constant but the charge varies. The step operators for the basic multiplet change d $\leftrightarrow$ u. Along the $B_- - C_+$ axis the charge Q remains constant, Y varies and the step operators change d $\leftrightarrow$ s. This is known as the 'U-spin' axis. If invariance under I_3

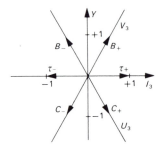

Fig. 10.4. Step operators in the Y–I_3 plane.

changes was exact then the properties of a system would be independent of its charge. If invariance under *U*-spin was true the properties would be independent of the hypercharge. The third axis, corresponding to u↔s transformations, has no such simple interpretation. It is sometimes known as the *V*-spin axis.

As has already been mentioned, in SU(3) there are eight independent operators of which we have discussed the six step operators formed from bilinear combinations of the creation and annihilation operators for u, d, s. It is convenient to choose the other two operators as

$$\tau_0 = \tfrac{1}{2}(a_u^\dagger a_u - a_d^\dagger a_d)$$
$$N = \tfrac{1}{3}(a_u^\dagger a_u + a_d^\dagger a_d - 2a_s^\dagger a_s).$$

By writing the charge, strangeness and baryon number operators which count these quantities as Q, S and B we see that

$$Q = \tfrac{1}{3}(2a_u^\dagger a_u - a_d^\dagger a_d - a_s^\dagger a_s)$$
$$S = -a_s^\dagger a_s$$
$$B = \tfrac{1}{3}(a_u^\dagger a_u + a_d^\dagger a_d + a_s^\dagger a_s)$$

so that using $Q = I_3 + \tfrac{1}{2}(B + S)$ we have that operation with τ_0 gives the I_3 eigenvalue. Also

$$N = B + S = Y.$$

None of these operators changes B.

For these eight operators the basic triplet and antitriplet are shown in fig. 10.5. The operators transform the basic states into each other. By combining these basic triplets nine states are obtained, which can be generated from the vacuum by the products of creation and annihilation operators. This is shown in fig. 10.6, where we have omitted the states at the centre generated by the remaining three products or by operators such as τ_0, B and N. The nine states split into an octet plus a singlet. The singlet has the quantum numbers of the vacuum, while the octet also has two states with $I_3 = Y = 0$. As far as the SU(3) algebra is concerned these two

Fig. 10.5. The basic SU(3) triplet (u, d, s) and antitriplet (ū, d̄, s̄).

octet states are degenerate, but we note that the four $Y = 0$ states must include the $I_3 = 0$ member of the I-spin triplet. Thus the two states at the origin have $I = 1$ and $I = 0$.

From this analysis we see that in the simplest model all the mesons ($B = 0$) can be constructed from quark–antiquark (q$\bar{\text{q}}$) pairs.

In a similar way the baryons and antibaryons may be constructed from three quark (qqq) or three antiquark ($\bar{\text{q}}\bar{\text{q}}\bar{\text{q}}$) combinations. We shall return to this aspect of the model in section 10.7.

A consequence of the nature of the step operators is that the multiplets are all hexagonal lattices in the Y–I_3 plane, having the following general properties:

(*a*) All diagrams are symmetrical about the Y-axis ($+I_3 \leftrightarrow -I_3$) corresponding to such transformations as d $\leftrightarrow$ u and also about axes corresponding to u $\leftrightarrow$ s and d $\leftrightarrow$ s transformations.

(*b*) The multiplicity of points at any coordinate increases by one at each 'ring' as one moves in from the boundary, until one arrives at a point, or triangle, inside which triangle it remains constant.

(*c*) Charge conjugation changes the sign of Y and I_3. Thus the charge-conjugate multiplet is the original reflected through the origin.

SU(3) gives, in addition to the multiplets we have already discussed, others, such as those formed by combining two octets or three triplets:

$$8 \otimes 8 = 27 \oplus 10 \oplus \overline{10} \oplus 8 \oplus 8 \oplus 1$$
$$3 \otimes 3 \otimes 3 = 10 \oplus 8 \oplus 8 \oplus 1.$$

On the Y–I_3 diagram the decuplet and 27-plet have the forms shown in fig. 10.7, where the quark content of the states in the decuplet is also indicated.

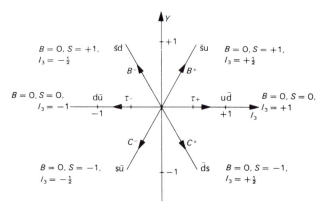

Fig. 10.6. States obtained by combining the basic triplet and antitriplet.

10.5 Prediction of the Ω^-

We might expect the u- and d-quarks to be very similar in mass since they are members of the same I-spin doublet and also since we know that the neutron and proton built from different combinations of u- and d-quarks are very close in mass. The symmetry breaking which is evident in the SU(3) multiplet may, however, be manifest in a difference in mass between the u- and d-quarks, on the one hand, and the s-quark. We shall see that other effects arising from quark–quark interactions can also give rise to mass differences in their combinations but for the $J = \frac{3}{2}$ decuplet we may note that the situation is particularly simple in that

(*a*) no space on the diagram is occupied by more than one particle, so that possible 'mixing' problems are avoided;

(*b*) as we move down from one I-spin multiplet to another the number of s-quarks increases at each step.

If in this case the mass differences between I-spin multiplets are simply due to the difference in mass between the u-, d- and s-quarks then we expect an *equal mass spacing* rule.

In 1962 there were known only nine members of the $J^P = \frac{3}{2}^+$-multiplet with the space at the bottom vacant (fig. 10.8). The Ω^- proposed by Gell-Mann (and independently by Ne'eman) as the missing member has the quantum numbers:

$$B = +1, \quad J^P = \tfrac{3}{2}^+, \quad Y = -2, \quad S = -3, \quad I = 0, \quad I_3 = 0,$$
$$Q = -1.$$

Applying the equal mass spacing rule we have:

$$m_\Omega - m_{\Xi(1530)} = m_{\Xi(1530)} - m_{\Sigma(1385)}$$
$$= m_{\Sigma(1385)} - m_{\Delta(1232)}.$$

The best values for the latter two differences were

$$m_{\Xi(1530)} - m_{\Sigma(1385)} = 147 \pm 1.5 \text{ MeV}/c^2$$
$$m_{\Sigma(1385)} - m_{\Delta(1232)} = 146 \pm 1.4 \text{ MeV}/c^2.$$

Thus the mass of the Ω^- was predicted to be

$$m_{\Xi(1530)} + 146 = 1675 \text{ MeV}/c^2.$$

With such properties the only $\Delta S = 1$ decays which are kinematically allowed are

$$\Omega^- \to \begin{cases} \Lambda K^- \\ \Xi^- \pi^0 \\ \Xi^0 \pi^-. \end{cases}$$

The discovery of the Ω^-, its decay by the above modes and the measurement of its mass, the best present value of which is

1672.43 ± 0.32 MeV/c^2 have already been described in section 5.10 and represented a triumph for the SU(3) scheme. The Ω^- should have spin $\frac{3}{2}$ and this, too, has been confirmed in measurements in recent experiments. In one of these, some five million photographs of 8.25 GeV/c K$^-$-mesons were taken in the CERN 2 m hydrogen bubble chamber to produce observed Ω^- particles. In another remarkable experiment at CERN a beam of hyperons was created by collisions of very high energy protons from the SPS with a primary target. Even short lived particles like Ξ^- and Ω^- may survive passage down a long beam line to the detection apparatus if their energy is high so that the lifetime is relativistically dilated. In both cases the decay angular distribution of the Ω^- is characteristic of a spin $\frac{3}{2}$ particle.

10.6   Mass formulae and mixing

The masses of the states within a multiplet are (approximately) related by the Gell-Mann–Okubo mass formula

$$m = m_0 + m_1 Y + m_2(I(I+1) - \tfrac{1}{4}Y^2).\qquad\qquad \textbf{10.1}$$

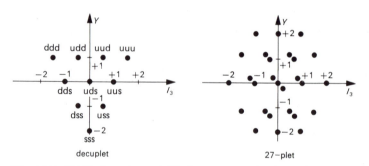

Fig. 10.7. SU(3) decuplet and 27-plet states on the Y–I_3 plane.

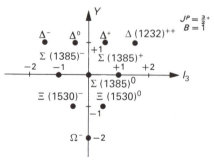

Fig. 10.8. The $J^P = \frac{3}{2}^+$, $B = +1$ decuplet on the Y–I_3 plane.

This formula was originally based on a speculative argument concerning the nature of the forces responsible for the symmetry breaking. We prefer here to treat the formula as an empirical rule which will in due course be accounted for by quantum chromodynamics (see chapter 13). In the baryon decuplet

$$Y = B + S = 2(I - 1)$$

so that **10.1** gives

$$m = (m_0 + 2m_2) + Y(m_1 + \tfrac{3}{2}m_2)$$

giving equal mass spacing. In the $\tfrac{1}{2}^+$ octet **10.1** gives

$$2m_n + 2m_{\Xi^0} = m_{\Sigma^0} + 3m_\Lambda$$

while the measured masses yield 4515 and 4539 MeV/c^2 for the left- and right-hand sides of this equation so that the relation is seen to be moderately well satisfied.

We might expect that the Gell-Mann–Okubo formula could be applied to the meson octets. For mesons the field equations always involve the squares of the masses, so that it is plausible that the formula here should be written with mass-squared instead of mass as in the baryon case. A test of whether mass or mass-squared is the correct quantity is complicated by the phenomenon of mixing, which we shall discuss below.

If we write the Gell-Mann–Okubo formula **10.1** in a general form for mesons where the subscripts $\tfrac{1}{2}$ and 1 refer to the *I*-spin and where m_8 is the mass of the *I*-spin zero member of the octet, then

$$m_{\frac{1}{2}}^2 = \tfrac{1}{4}m_1^2 + \tfrac{3}{4}m_8^2, \qquad\qquad \textbf{10.2}$$

since for the mesons particle and antiparticle appear reflected about the I_3 axis.

We have already noted that the mesons appear to be grouped in nonets rather than octets. In fact, the $I = 0$ member of the SU(3) octet has quantum numbers identical to those of the SU(3) singlet, so that if there is SU(3) breaking we might expect mixing between these particles. We will denote the true SU(3), $I = 0$, $Y = 0$, octet and singlet particles as α_8 and α_0 respectively. However, the observed physical particles may be (specific) mixtures of the α_8 and α_0 and the multiplet containing these mixtures will have nine members. The idea of mixing was proposed by Sakurai (1962), who proposed also a mixing parameter in the form of the 'mixing angle'. This parameterisation makes it easy to preserve the normalisation and has other formally satisfying features. The situation is very similar to that of the $K^0\bar{K}^0$ particles.

Formally both the K^0–$\bar{K}^0$ and octet–singlet mixing can be treated rather similarly by means of the 'mass matrix'. We shall not attempt here to develop the treatment of the mass matrix fully or rigorously, but merely introduce the idea and use it to obtain an expression for the mixing angle.

First note that we expect to obtain the mass m_8 of the pure octet state from the expression 10.2. For instance, in the vector meson nonet we should have

$$m_8^2 = \tfrac{4}{3}m_{K^*(892)}^2 - \tfrac{1}{3}m_\rho^2 = 0.863 \quad (\text{GeV}/c^2)^2.$$

In this nonet the physical singlet particles are the ω and the ϕ for which the mass-squared values are 0.610 and 1.035 $(\text{GeV}/c^2)^2$ respectively.

Now we write the physical particle wave functions β and β', say, in terms of the Sakurai mixing parameter

$$|\beta\rangle = |\alpha_0\rangle \cos\theta + |\alpha_8\rangle \sin\theta$$
$$|\beta'\rangle = -|\alpha_0\rangle \sin\theta + |\alpha_8\rangle \cos\theta. \qquad \textbf{10.3}$$

For a stable stationary state, the solutions of the wave equation contain a factor $e^{im_0 t}$, and the mass of the particle is the eigenvalue of the total Hamiltonian for the interaction. Thus

$$\langle \beta|H_0|\beta\rangle = m_\beta^2 \quad (\text{e.g. } m_\omega^2)$$
$$\langle \beta'|H_0|\beta'\rangle = m_{\beta'}^2. \qquad \textbf{10.4}$$

If there is no mixing then

$$\langle \beta|H_0|\beta'\rangle = \langle \beta'|H_0|\beta\rangle = 0,$$

but if some part of H_0 mixes the basic states this will no longer be true.

For the no-mixing case we could write a two-component time-dependent Schrödinger equation of the form

$$i\frac{\mathrm{d}}{\mathrm{d}t}\begin{bmatrix} \beta \\ \beta' \end{bmatrix} = \begin{bmatrix} m_\beta^2 & 0 \\ 0 & m_{\beta'}^2 \end{bmatrix}\begin{bmatrix} \beta \\ \beta' \end{bmatrix}$$

where in this case the mass matrix is diagonal.

If there is mixing, the off-diagonal terms are no longer zero and we have

$$i\frac{\mathrm{d}}{\mathrm{d}t}\begin{bmatrix} \alpha_0 \\ \alpha_8 \end{bmatrix} = \begin{bmatrix} m_0^2 & m_{0,8}^2 \\ m_{8,0}^2 & m_8^2 \end{bmatrix}\begin{bmatrix} \alpha_0 \\ \alpha_8 \end{bmatrix} \qquad \textbf{10.5}$$

where m_0 and m_8 are the masses of the I-spin zero members of the SU(3) singlet and octet multiplets and $m_{0,8}$ and $m_{8,0}$ are due to mixing. The mass matrices corresponding to the representation of the states in terms of the physical particles and the pure SU(3) states must satisfy the conditions that their traces and determinants are equal:

$$m_0^2 + m_8^2 = m_\beta^2 + m_{\beta'}^2$$
$$m_0^2 m_8^2 - m_{0,8}^4 = m_\beta^2 m_{\beta'}^2$$

$(m_{0,8}$ has been taken equal to $m_{8,0})$.

The two representations must be linked by a rotation which is the operator

$$R = \begin{bmatrix} \cos\theta & -\sin\theta \\ \sin\theta & \cos\theta \end{bmatrix}.$$

Thus θ is the angle of rotation which will diagonalise the mass matrix

$$M = \begin{bmatrix} m_0^2 & m_{0,8}^2 \\ m_{8,0}^2 & m_8^2 \end{bmatrix}$$

of equation **10.5**. Imposing this condition in the form

$$R^{-1}MR = \begin{bmatrix} m_\beta^2 & 0 \\ 0 & m_{\beta'}^2 \end{bmatrix}, \tag{10.6}$$

straightforward algebra shows that the condition that the off-diagnonal elements in RMR^{-1} be zero is

$$\tan 2\theta = \frac{m_{0,8}^2 + m_{8,0}^2}{m_8^2 - m_0^2}. \tag{10.7}$$

10.3 and **10.4** or consideration of the diagonal elements of **10.6** yield

$$m_\beta^2 = m_0^2 \cos^2\theta + m_8^2 \sin^2\theta - (m_{0,8}^2 + m_{8,0}^2)\sin\theta\cos\theta,$$
$$m_{\beta'}^2 = m_0^2 \sin^2\theta + m_8^2 \cos^2\theta + (m_{0,8}^2 + m_{8,0}^2)\sin\theta\cos\theta. \tag{10.8}$$

We may eliminate $(m_{0,8}^2 + m_{8,0}^2)$ and m_0^2 from **10.7** and **10.8** to obtain an expression for the mixing angle in terms of m_8^2 (which can be calculated from **10.2**). This expression can be written

$$\sin^2\theta = \frac{m_\beta^2 - m_8^2}{m_{\beta'}^2 - m_\beta^2}.$$

For the pseudo-scalar mesons we find $\theta = -11.1°$ (mixing η and η'). For the vector mesons $\theta = +38.6°$ (ω, ϕ) and for the tensor mesons $\theta = +28°$ (f_2, f_2'). Thus for the pseudo-scalar mesons the mixing is small and the physical η nearly obeys the Gell-Mann–Okubo formula, but for both the vector and tensor mesons the mixing is large and the physical particle masses do not obey the GMO relation. The vector and tensor nonets exhibit close to 'ideal mixing' ($\theta \approx 35°$) for which ϕ and f_2' are pure s̄s states (see below).

Note that by introducing the new free parameter θ, the mass relation ceases to be a test of the theory or to be a predictive tool. There are other results involving θ which are open to independent test, such as the ratios of certain decay modes of the $I = 0$ mesons within the nonet and, on the basis of a particular quark interaction model, the ratios of production of, say, η and η' in certain reactions. Such results, although in general agreement with the mixing angles obtained from the masses, usually have large errors for θ while the model used in the production reaction analysis is also not well proved.

10.7   Mesons and baryons constructed from quarks

The quarks as building blocks for the mesons and baryons were introduced in section 10.4. From fig. 10.5 and the relations between Y, B, S, Q and I_3, we can deduce the properties of the quarks, which are summarised in table 10.1.

The simplest assumption is that all the quarks have spin $\frac{1}{2}$ (see section 10.11). Thus, for instance, a K^+-meson will be constructed from a $u\bar{s}$ quark combination and a proton from uud.

We have already seen that SU(3) allows singlets, octets, decuplets, antidecuplets, 27-plets and higher multiplets. However, the quark model as described above is more restrictive. For mesons we expect only

$$3 \otimes \bar{3} = 8 \oplus 1,$$

i.e. singlets and octets, while for baryons we expect only

$$3 \otimes 3 \otimes 3 = 1 \oplus 8 \oplus 8 \oplus 10,$$

i.e. no antidecuplet or 27-plet or higher multiplet.

For instance, consider the meson 27-plet shown in fig. 10.9. Some possible decay modes are shown for the 'far-out' states such as a doubly-charged $S = +2$ meson (K^+K^+), a doubly-charged ($I = \frac{3}{2}$) $S = +1$ meson (doubly-charged $K^* \to K^+\pi^+$) and a doubly-charged $S = 0$ meson (decaying to $\pi^+\pi^+$ or $\pi^0\pi^+\pi^+$). In order to construct such an $S = +2$, $I = +1$ meson we require four quarks, $\bar{s}\bar{s}uu$. Similarly, the student may check that a strangeness $S = +1$ and charge -2 baryon must belong to a baryon 27-plet and be constructed from five quarks ($\bar{s}uuud$), while a $B = +1$, $S = +1$ singly-charged particle must belong to an antidecuplet.

Table 10.1. *Properties of the u, d and s quarks*

	B	I	I_3	Q	Y	S
u	$\frac{1}{3}$	$\frac{1}{2}$	$+\frac{1}{2}$	$\frac{2}{3}$	$+\frac{1}{3}$	0
d	$\frac{1}{3}$	$\frac{1}{2}$	$-\frac{1}{2}$	$-\frac{1}{3}$	$+\frac{1}{3}$	0
s	$\frac{1}{3}$	0	0	$-\frac{1}{3}$	$-\frac{2}{3}$	-1
$\bar{u}$	$-\frac{1}{3}$	$\frac{1}{2}$	$-\frac{1}{2}$	$-\frac{2}{3}$	$-\frac{1}{3}$	0
$\bar{d}$	$-\frac{1}{3}$	$\frac{1}{2}$	$+\frac{1}{2}$	$+\frac{1}{3}$	$-\frac{1}{3}$	0
$\bar{s}$	$-\frac{1}{3}$	0	0	$+\frac{1}{3}$	$+\frac{2}{3}$	$+1$

Some evidence of such states has been reported in formation experiments in the form of peaks in the total cross-section for K⁺p scattering (pure $I = 1$) and in the total $I = 0$, K-nucleon scattering cross-section as deduced from the K⁺p and K⁺n(K⁺d) cross-sections. However, the peak in the K⁺p cross-section may probably be explained by the opening up of inelastic pion-production channels, while other possible explanations of the $I = 0$ peak have also been proposed. In addition, a number of double-charged peaks have been found in production experiments. These peaks have, in general, either found explanations as reflections of other resonances or have not been substantiated. Thus, at the present time, all well established meson resonances fit into octets or singlets (nonets) and all the baryon resonances into singlets, octets or decuplets. It seems clear that 'far-out' or 'exotic' resonances are produced only weakly, if at all, and this result must be counted as evidence for the correctness of the quark model.

10.8 The orbital-excitation model for quarks

If the mesons are taken to consist of $\bar{q}q$ pairs, the q and $\bar{q}$ may rotate about each other with relative orbital angular momentum $l\hbar$. We may then have a series of states of increasing values of l. Such a model has been proposed by Dalitz.

Mesons built in this way must have some specific properties. The parity is given by $P = (-1)^{l+1}$ and the charge-conjugation parity in the strangeness-zero case will be $C = (-1)^{l+s}$, where s is the spin of the $\bar{q}q$ pair. Thus, for instance, all natural parity states must have $C = P$.

The lowest $\bar{q}q$ states are then the $^{3}S_{1}$ and $^{1}S_{0}$ states, which have the appropriate J^{PC} for the vector and scalar nonets respectively. From the

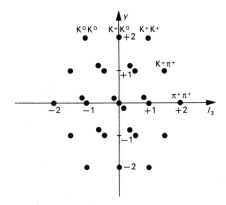

Fig. 10.9. Possible 27-plet meson states.

basic $^3S_1$ and $^1S_0$ nonets there can be generated two series of rotational levels of increasing l. The triplet $^3S_1$ yields P-states $^3P_0$, $^3P_1$ and $^3P_2$, for which $P = +1$, and for the neutral states of which $C = +1$. The singlet P-state is $^1P_1$, with again positive parity, but negative under charge conjugation.

The four P-states may be rendered non-degenerate by spin-orbit SU(3)-breaking forces proportional to $\mathbf{l} \cdot \mathbf{s}$. But since

$$\langle \mathbf{l} \cdot \mathbf{s} \rangle = \frac{j(j+1) - l(l+1) - s(s+1)}{2}$$

we have equal spacing in mass-squared for the P-levels

$$\begin{array}{cccc} ^3P_{L-1} & ^3P_L & ^1P_L & ^3P_{L+1} \\ \leftarrow 2L \rightarrow & \leftarrow 2L \rightarrow & \leftarrow 2L \rightarrow \end{array}$$

For D-states the inner states are twice as close as the other spacings.

It is possible to assign some of the higher-mass mesons to such orbital-excitation multiplets. For the baryon three-quark system the orbital-excitation model is, of course, considerably more complicated.

10.9   Developments of the quark model: magnetic moments

The SU(3) classification has been enlarged to encompass also the spin quantum number. The appropriate group is SU(6), which should be a good symmetry if the interactions are invariant with respect to rotations of the spin and if spin–orbit couplings are not important. This limitation means that SU(6) can be valid only in the non-relativistic region, unlike the SU(3) symmetry which was independent of space-time.

In considering the calculation of magnetic moments we shall need to construct the appropriate states for proton and neutron from the spinning quarks. With certain simple assumptions SU(3) itself can give certain relations, such as, for instance, $\mu(\Sigma^+) = \mu(p)$ (measured values 2.5 ± 0.5 and 2.79 respectively). (This is immediately obvious if it is assumed that the electromagnetic interaction is a U-spin scalar.) The SU(6) quark model, however, while including these relations, yields additional ones and in particular relates the very precisely known proton and neutron magnetic moments.

We give below a naïve calculation of the proton and neutron magnetic moments which yields results in remarkably good agreement with the experimentally measured values. It has lately however become clear from muon scattering experiments that the gluons (see chapter 13) which bind the quarks, and also virtual pairs of s-s̄ quarks, may make important contributions to the spin. Although we should bear in mind that the agreement of the naïve result with experiment may be an accident the calculation is nevertheless a useful exercise of its kind.

We take the proton to be constructed from the quark combination (uud). We now use the Clebsch–Gordan coefficients to construct the spin $\frac{1}{2}$ quark states. We shall take $j_z = +\frac{1}{2}$ so that the $(\frac{1}{2}, +\frac{1}{2})$ physical proton will be built from the three states (A), (B), (C) below, where we make the simplest assumption that the quarks are all in relative S-states. (This implies that the overall wave function is symmetric whereas the generalised Pauli principle (**3.7**) would require (Fermion) spin $\frac{1}{2}$ quarks to have an antisymmetric wave function. The explanation of this paradox lies in the idea of colour – see section 13.1.)

	u	u	d
(A)	$(\frac{1}{2}, +\frac{1}{2})$	$(\frac{1}{2}, +\frac{1}{2})$	$(\frac{1}{2}, -\frac{1}{2})$
(B)	$(\frac{1}{2}, +\frac{1}{2})$	$(\frac{1}{2}, -\frac{1}{2})$	$(\frac{1}{2}, +\frac{1}{2})$
(C)	$(\frac{1}{2}, -\frac{1}{2})$	$(\frac{1}{2}, +\frac{1}{2})$	$(\frac{1}{2}, +\frac{1}{2})$.

States in which the up and down quarks are arranged in a different order, e.g. udu, are all the same, the only point of consequence for this problem being the spin orientations.

Grouping now the two up quarks we see that

$$(A) \equiv (1, +1)(\tfrac{1}{2}, -\tfrac{1}{2}),$$
$$(B) \text{ and } (C) \equiv (1, 0)(\tfrac{1}{2}, +\tfrac{1}{2}).$$

We can write

$$(1, 0)(\tfrac{1}{2}, +\tfrac{1}{2}) = \tfrac{1}{2}(B) + \tfrac{1}{2}(C).$$

Thus, referring to the table of Clebsch–Gordan coefficients, we have

$$\begin{aligned}
p_{\frac{1}{2}} &= \sqrt{\tfrac{2}{3}}(A) - \sqrt{\tfrac{1}{3}}\sqrt{\tfrac{1}{2}}\{(B) + (C)\} \\
&= \sqrt{\tfrac{1}{6}}(2A - B - C) \\
&= \sqrt{\tfrac{1}{6}}(2u\uparrow u\uparrow d\downarrow - u\uparrow u\downarrow d\uparrow - u\downarrow u\uparrow d\uparrow),
\end{aligned} \qquad \textbf{10.9}$$

where the arrows indicate the spin directions.

Similarly, the physical neutron may be written in terms of spinning quarks as

$$n_{\frac{1}{2}} = \sqrt{\tfrac{1}{6}}(2d\uparrow d\uparrow u\downarrow - d\uparrow d\downarrow u\uparrow - d\downarrow d\uparrow u\uparrow), \qquad \textbf{10.10}$$

If we write a basic 'quark magneton' in the usual way as

$$\mu_q = \frac{Q\hbar}{2m_q c}$$

where Q is the charge and m_q the mass of the quark (here assumed approximately equal for the u and d quarks), then the up and down quarks will have magnetic moments expressed in terms of m_q and the z-component of the spin σ_3 as

$$\mu_u = \tfrac{2}{3}\sigma_{3,u}\mu_q$$
$$\mu_d = -\tfrac{1}{3}\sigma_{3,d}\mu_q$$

We now calculate the proton and neutron magnetic moments as the sums of the moments arising from the quark states of **10.9** and **10.10**.

$$\mu_p = \tfrac{1}{6}[4(\tfrac{2}{3}+\tfrac{2}{3}+\tfrac{1}{3})+(\tfrac{2}{3}-\tfrac{2}{3}-\tfrac{1}{3})+(-\tfrac{2}{3}+\tfrac{2}{3}-\tfrac{1}{3})]\mu_q = \mu_q$$
$$\mu_n = -\tfrac{2}{3}\mu_q$$

so that $\mu_n/\mu_p = -\tfrac{2}{3}$. The experimental value is $-1.913/2.792 = -0.68$. The agreement represents a further striking success for the quark model.

10.10   Nucleon structure from scattering experiments

It has long been realised that the nucleons are not point particles but are of finite size $\sim 10^{-13}$ cm while the evidence from symmetries discussed in earlier sections has suggested that they are built from quarks. Probing the nucleons dynamically by means of deep inelastic scattering of leptons (electrons, muons and neutrinos) has led to an understanding of nucleon structure, to the confirmation of the quarks as the elements of the nucleon substructure and to measurement of the quark properties. The leptons, having no strong interaction, probe the nucleon structure by means of the electromagnetic and weak interactions.

10.10.1   *Elements of scattering theory: elastic scattering*

We start from the familiar Rutherford scattering formula which gives the cross-section for elastic scattering of two charged, spinless, point particles as a result of the electromagnetic interaction

$$\left(\frac{d\sigma}{d\Omega}\right)_R = \frac{(Z\alpha)^2 E^2}{4p^4 \sin^4(\theta/2)}.$$

Ze is the charge on the target (assumed infinitely massive so that there is no recoil), e is the incident charge, p and E the incident momentum and energy, θ the scattering angle and α the fine-structure constant.

For scattering of high energy electrons on protons, several factors modify the simple Rutherford formula:

(*a*) the particles are not spinless and the proton magnetic moment contributes to the interaction;

(*b*) the proton is not infinitely massive and recoils;

(*c*) the proton is not a point charge but a charge distribution $e\rho(x)$.

In general, the cross-section for scattering by an extended target can be written in terms of that for a point target by inclusion of a 'form factor' $F(\mathbf{q})$

$$\left(\frac{d\sigma}{d\Omega}\right) = \left(\frac{d\sigma}{d\Omega}\right)_{\text{Point}} |F(\mathbf{q})|^2$$

where $\mathbf{q}$ is the momentum transfer from the projectile to the target, $\mathbf{q} = \mathbf{p}_i - \mathbf{p}_f$, where $\mathbf{p}_i$ and $\mathbf{p}_f$ are the initial and final momenta. The form factor in such a case can be shown to be the Fourier transform of the charge distribution

$$F(\mathbf{q}) = \int \rho(\mathbf{x}) e^{i\mathbf{q}\cdot\mathbf{x}} d^3x.$$

When all the above factors (a), (b), (c) are taken into account, the cross-section formula can be written

$$\left(\frac{d\sigma}{d\Omega}\right)_{\text{lab}} = \frac{\alpha^2}{4E^2 \sin^4(\theta/2)} \frac{E'}{E} \left[\frac{G_E^2 + \tau G_M^2}{1+\tau} \cos^2\left(\frac{\theta}{2}\right) + 2\tau G_M^2 \sin^2\left(\frac{\theta}{2}\right) \right]$$

10.11

where we have written $\tau = -(q^2/4m^2)$, where m is the proton mass, and where E' is the new energy of the projectile after scattering. The G_E and G_M are form factors related to the charge and magnetic moment distributions in the proton (but, due to the complications introduced by the recoil, not simply as the Fourier transforms of these distributions). They are functions of q^2.

10.10.2 *Elements of scattering theory: inelastic scattering*

If we wish to probe to dimensions of order x then the projectile wavelength must, in general, be of this order or less. For scattering as a result of the electromagnetic interaction the process takes place as a result of photon exchange as in fig. 10.10. Thus short-distance structure studies require high energy photon exchange or *large values of* $-q^2$. Such large q^2

Fig. 10.10. Photon exchange in charged-particle scattering.

collisions, where the interaction is with the individual nucleon constituents, will in general result in break up of the nucleon, resulting at high energies in jets including mesons and nucleons, so that the initial particle identities are no longer preserved and the collision becomes *inelastic*. The effective mass W of the final hadronic state is no longer constant as in elastic scattering so that we require *two independent variables rather than one* in order to describe the scattering.

It is convenient at this stage to move to a formalism which is independent of the reference frame (above formulae refer to the laboratory frame) and to this end we take q^2 to be the *four-momentum transfer* (see appendix A1). Since the energy-momentum four-vectors $(\mathbf{p}, i\varepsilon)$ are Lorentz vectors the square of the four-momentum transfer q^2 is an invariant quantity (fig. 10.11):

$$q^2 = (\mathbf{p}_i - \mathbf{p}_f)^2 - (\varepsilon_i - \varepsilon_f)^2$$
$$= -2m^2 - 2p_i p_f \cos\theta + 2\varepsilon_i \varepsilon_f$$
$$\simeq 2p_i p_f(1 - \cos\theta) = 4p_i p_f \sin^2\theta/2,$$

the approximation being good if $m^2 \ll q^2$. In this form q^2 is seen to be positive and (referred to the four-momentum components) is known as 'spacelike'. For real rather than virtual, particle exchange q^2 is negative or 'timelike'. Considering the struck proton vertex

$$q^2 = p'^2 - (E' - M)^2.$$

The energy $(E' - M)$ acquired by the hadronic system is equal to the energy transfer from the virtual photon, i.e. the timelike component of the four-vector $(\mathbf{p}, i\varepsilon)$. In this application ε is normally referred to by the symbol v, so that

$$v = E' - M.$$

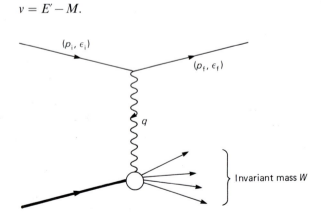

Fig. 10.11. Four-momentum transfer q in scattering processes.

If W is the effective mass of the hadronic system, then

$$W^2 = E'^2 - p'^2$$

with

$$q^2 = M^2 + 2vM - W^2.$$

In *elastic* scattering $W = M$ and $q^2 = 2vM$ so that v and q^2 are not independent variables.

The cross-section in terms of variables q^2 and v can be written

$$\frac{\mathrm{d}^2\sigma}{\mathrm{d}q^2\,\mathrm{d}v} = \frac{4\pi\alpha^2}{q^4}\frac{E'}{E}\frac{1}{M}\left[W_2(q^2, v)\cos^2\left(\frac{\theta}{2}\right) + 2W_1(q^2, v)\sin^2\left(\frac{\theta}{2}\right)\right]. \quad \textbf{10.12}$$

The $W_1(q^2, v)$ and $W_2(q^2, v)$ are form factors related to the transverse and longitudinal polarisation states of the exchanged virtual photon (*virtual photons can have longitudinal polarisation*).

A useful variable in this area is

$$x = q^2/2Mv$$

where it is clear that $0 \leqslant x \leqslant 1$; $x = 1$ corresponding to the elastic, and $x = 0$ to the totally inelastic, situations.

10.10.3 *Bjorken scaling and partons*

If the proton contains point-like scattering centres then, for sufficiently small wavelength photon probes, the cross-section should be expressible in the point-like form. In order to compare more directly the forms of equations **10.12** and **10.11** we write form factors

$$F_1(q^2, v) = W_1(q^2, v)$$

$$F_2(q^2, v) = \frac{v}{M}W_2(q^2, v)$$

so that **10.12** becomes

$$\frac{\mathrm{d}^2\sigma}{\mathrm{d}q^2\,\mathrm{d}v} = \frac{4\pi\alpha^2}{q^4}\frac{E'}{E}\frac{1}{v}\left[F_2(q^2, v)\cos^2\left(\frac{\theta}{2}\right) + \frac{2v}{M}F_1(q^2, v)\sin^2\left(\frac{\theta}{2}\right)\right]. \quad \textbf{10.13}$$

which compares directly with the elastic case **10.11** for which $v/M = 2\tau$. Thus for point-like scattering the functions $F(q^2, v)$ can not depend on two variables such as q^2 and v but should be expressible in terms of a single parameter such as x. This property is known as the Bjorken scaling hypothesis (Bjorken, 1967), and states that in the limit where q and v tend to infinity the form factors depend only on the *ratio* of these quantities determined by the parameter x. The fact that x is dimensionless and has no associated scale leads to the description of the phenomenon as 'scale invariance'. The verification of the independence of form factors with respect to q^2 is illustrated in fig. 10.12. Although scale invariance is not perfect for all values of x the dependence on q^2 is always weak compared with the very strong dependence on q^2 of the elastic form factor for electron–proton scattering where the interaction is with the proton as a whole and where the form factor falls by $\sim 10^6$ as q^2 increases from 1–25 GeV^2. Reasons for some deviation from scale invariance are discussed in chapter 13, section 13.3, but the main features remain unaffected.

The point constituents of the nucleon originally named 'partons' by Feynman (1969) are now naturally identified with the quarks and gluons (see 13.2). Beyond the scaling features discussed here inelastic scattering can be used as a probe to investigate the nature and momentum distributions of the constituents. In addition, the ratio of the form factors in neutrino– and electron–nucleon scattering provides a test of the quark charge assignments which is in good agreement with the $\frac{2}{3}e$ and $-\frac{1}{3}e$ values for u and d quarks discussed in earlier sections. The development of this subject in more detail is beyond the scope of this text and we confine our treatment to just one further important aspect: the parton spins.

10.10.4 *Parton spins*

In our comparison of equations **10.11** and **10.13** we implicitly assumed scattering off a spin $\frac{1}{2}$ target (the proton). We can express this element of the comparison more explicitly in terms of the form factors

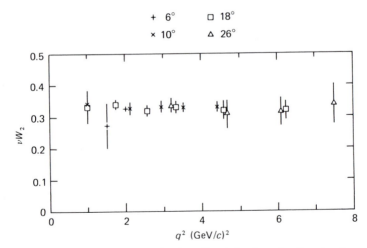

Fig. 10.12. An example of scaling behaviour for the structure function W_2 for electron–proton scattering which is seen to be independent of q^2. (Friedman and Kendall, 1972).

$F_1(x)$ and $F_2(x)$. If we take the point charge version of **10.11** ($G_M = G_E = 1$) and use the relations for q in terms of θ for an elastic collision:

$$q^2 = 2E^2(1 - \cos\theta) \quad \text{and} \quad dq^2 = \frac{E^2\,d\Omega}{\pi} \quad (E \gg m, E = P_i)$$

then **10.11** has the form

$$\frac{d^2\sigma}{dq^2} = \frac{4\pi\alpha^2 Z^2}{q^4}\left(\frac{E'}{E}\right)\left(\cos^2\left(\frac{\theta}{2}\right) + \frac{q^2}{2m^2}\sin^2\left(\frac{\theta}{2}\right)\right) \qquad \textbf{10.14}$$

where we have written Ze for the parton charge and m for the parton mass.

We require to compare this equation with **10.13** or its analogue expressed in terms of the variable x, but we note that a proper comparison requires us to replace M in **10.13** by the parton mass.

We can interpret the physical significance of x and relate it to the quark and proton masses by an argument using the 'infinite momentum' frame in which the proton has a momentum so large that we can neglect its rest mass and write its four-momentum P as $(p, 0, 0, ip)$. The proton thus consists of a bunch of partons moving along the x-axis with momentum such that the parton masses and the transverse components of their momenta can also be neglected. We suppose that each parton carries a fraction f of the total momentum p ($0 < f < 1$) so that a parton momentum is fp. In our model one parton scatters, picking up the momentum transfer q. Thus

$$(fP + q)^2 = -m^2 \simeq 0$$

where m is the parton mass. Then

$$f^2 P^2 + q^2 + 2fP\cdot q \simeq 0.$$

But $|f^2 P^2| = f^2 M^2 \simeq 0$ (cf. q^2) so that

$$f = -\frac{q^2}{2P\cdot q} = \frac{q^2}{2Mv}$$

since we can choose to evaluate the invariant scalar product of the four-vectors in the laboratory where $P = M$ (nucleon at rest) and the time component of $q = v$. Thus we see that, in fact, f is equal to our previously defined parameter x and that x represents the fraction of the proton momentum carried by the struck quark. We may also write the elastic relation $q^2 = 2mv$ to get (in this approximation)

$$x = \frac{m}{M}.$$

We may now substitute this relation in equation **10.13** to obtain

$$\frac{d^2\sigma}{dq^2\,dx} = \frac{4\pi\alpha^2}{q^4}\frac{E'}{E}\frac{1}{x}\left[F_2(x)\cos^2\left(\frac{\theta}{2}\right) + 2xF_1(x)\frac{q^2}{2m^2}\sin^2\left(\frac{\theta}{2}\right)\right]. \qquad \textbf{10.15}$$

Comparing the coefficients in **10.15** and **10.14** we see that for the scattering to behave as that from point-like constituents of spin $\frac{1}{2}$ and normal Dirac magnetic moments (assumed in **10.11**) we must have

$$\frac{2xF_1(x)}{F_2(x)} = 1.$$

This is known at the Callan–Gross relation (1968). It is well satisfied by the experimental measurements establishing the quarks as spin $\frac{1}{2}$ particles.

10.11   **The search for free quarks**

In the preceding sections we have discussed the successes of the quark model in the classification of resonances, in the calculation of some magnetic moments and in accounting for the features of lepton–nucleon scattering.

Although, as we shall see in chapter 13, it may well be the case that the force between quarks is such that free quarks can never be observed, nevertheless this is not presently an inevitable consequence of quantum chromodynamic theory and the search for free quarks is still vigorously pursued.

The most distinctive observable property of the quarks is their fractional charge. The ionisation produced by $\frac{1}{3}e$ and $\frac{2}{3}e$ charged quarks will be $\frac{1}{9}$ and $\frac{4}{9}$ that of singly-charged particles travelling at the same speed. Extensive studies of particles from accelerator targets and in the cosmic radiation have not produced convincing evidence for the existence of fractionally-charged particles. Early results on cloud-chamber tracks of cosmic rays have not been confirmed by subsequent experiments and were probably instrumental artefacts.

The only positive results on the existence of fractionally-charged particles come from a search for fractionally-charged particles in matter (LaRue *et al.*, 1977, 1981). The measurement is a sophisticated version of the classical Millikan oil drop experiment but using very small diamagnetic niobium balls (mass $\sim 9 \times 10^{-5}$ g) levitated between two horizontal plates. The results of the experiment indicated clearly that some of the spheres carried residual charges of $\pm\frac{1}{3}$ of the electron charge.

Such measurements have however been repeated by P. F. Smith *et al.* (1987) using a similar experimental arrangement, and in an extensive set of data these workers find no evidence for fractional charges with a limit of $< 10^{-21}$ quarks/nucleon.

In another long running search for quarks in matter, Marinelli and Morpurgo (1980) levitated small iron cylinders in a magnetic field. An electric field was used to apply a force to the particle and thus to measure the charge. No fractional charge particles have been detected with a limit for free quarks in iron of $\leqslant 3 \times 10^{-21}$ quarks/nucleon.

On balance the evidence to date is against the existence of free quarks in matter; a result which may be accounted for by the nature of the quark–quark interaction discussed in section 13.3.

11

Weak interactions and weak–electromagnetic unification

11.1 'Cabbibo mixing'

We can represent the weak decays of hadrons in terms of transformation of one of the quark constituents while the others act as 'spectators'. A number of examples of this approach are shown in table 11.1.

For the strange particle decays we have $\Delta S = 1$, $\Delta I = \frac{1}{2}$ for the hadronic parts and $\Delta Q = \Delta S = 1$ for the semi-leptonic processes as discussed in section 7.5.

The obvious similarity in structure between the semi-leptonic $\Delta S = 0$ and $|\Delta S| = 1$ processes allows the rate for the $|\Delta S| = 1$ decays to be calculated directly using the couplings from the $\Delta S = 0$ decays. The results of such a calculation give values 20–40 times greater than the measured values, suggesting that the coupling of the strange quark to the u quark is less than the ud coupling.

Cabbibo (1963) proposed that the d and s states should be mixed by a rotation parameter which became known as the Cabbibo angle θ_{C}. The u, d and s quarks were then represented as a doublet in analogy with the lepton doublets (e, ν_e) and (μ, ν_μ) with the mixed d and s state as the second element

$$\begin{pmatrix} \mathrm{u} \\ \mathrm{d}\cos\theta_{\mathrm{C}} + \mathrm{s}\sin\theta_{\mathrm{C}} \end{pmatrix}.$$

Note that it is an arbitrary convention to choose u as the unmixed state. A value of θ_c for which $\cos \theta_c \neq \sin \theta_c$ will then result in a difference in ud and us couplings. The contribution of the initial and final state functions to the rate will then introduce factors such as those given in table 11.1.

In order to eliminate other effects in determining θ_c it is useful to compare processes which are either pure vector (e.g. $p \rightarrow n e^+ v_e$ in ^{14}O with μ-decay) or pure axial vector (e.g. $K \rightarrow \mu v$ and $\pi \rightarrow \mu v$). The results of such a comparison yield consistent values for strangeness-changing and non-changing decays with $\theta_c \sim 0.23$ radians.

The more complex situation resulting from the discovery of additional quarks is briefly discussed in chapter 12.

11.2 *v*-interactions and the unitarity problem

We have already mentioned the use of neutrino beams in probing the quark structure of matter (section 10.10). Neutrino interactions have yielded results of even greater importance for the understanding of the weak interactions.

Table 11.1

Decay	Quark transformation	'Mixing' factor in decay rate	
$n \rightarrow pe^- \bar{v}_e$	$(ud)d \rightarrow ue^- \bar{v}_e(ud)$	$\cos^2 \theta_C$ }	$\Delta S = 0$
$p \rightarrow ne^+ v_e(^{14}O)$	$(ud)u \rightarrow de^+ v_e(ud)$	$\cos^2 \theta_C$ }	
$\Lambda \rightarrow pe^- \bar{v}_e$	$(ud)s \rightarrow ue^- \bar{v}_e(ud)$	$\sin^2 \theta_C$ }	$\Delta S = 1$
$\Sigma^- \rightarrow ne^- \bar{v}_e$	$(dd)s \rightarrow ue^- \bar{v}_e(dd)$	$\sin^2 \theta_C$ }	
$\pi^- \rightarrow \pi^0 e^- \bar{v}_e$	$(\bar{u})d \rightarrow ue^- \bar{v}_e(\bar{u})$	$\cos^2 \theta_C$	$\Delta S = 0$
$K^- \rightarrow \pi^0 e^- \bar{v}_e$	$(\bar{u})s \rightarrow ue^- \bar{v}_e(\bar{u})$	$\sin^2 \theta_C$	$\Delta S = 1$
$\mu^- \rightarrow e^- \bar{v}_e v_\mu$	—	1	
$\Lambda \rightarrow p\pi^-$	$(ud)s \rightarrow u(ud)(\bar{u}d)$	$\sin^2 \theta_C$	$\Delta S = 1$
$\Sigma^- \rightarrow n\pi^-$	$(dd)s \rightarrow u(dd)(\bar{u}d)$	$\sin^2 \theta_C$	$\Delta S = 1$

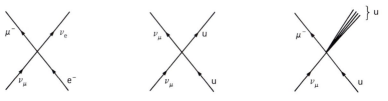

Fig. 11.1. Neutrino–lepton and neutrino–quark interactions according to Fermi theory.

We do not attempt to derive several of the formulae used in this section. For more detail the reader is referred, for instance, to *Quarks and Leptons* (Halzen and Martin, 1984).

We first recall that the four-fermion processes we have discussed in chapter 7 were considered according to Fermi theory as point interactions. Neutrino–quark and neutrino–lepton scattering have the same form (fig. 11.1).

Since an incoming v can be replaced by an outgoing $\bar{v}$ the first diagram is equivalent to muon decay. However, unlike the decay process, in the scattering experiments we have the opportunity to vary the centre-of-mass energy. A detailed analysis of the cross-section for such a scattering process gives total cross-sections which *are proportional to s*, the square of the total *C*-system energy. Thus, although the cross-sections, for instance for v–e scattering, are very small at even moderate energies ($\sim 2 \times 10^{-39}$ cm^2 at 10 GeV), nevertheless the cross-section would eventually increase to infinity and long before that would violate the 'unitarity limit'. We recall (section 9.3) that the maximum cross-section in a given partial wave is given by the expression

$$\frac{2\pi}{p^2}(2j+1)$$

where p is the *C*-system momentum. In the relativistic limit we then have for $\bar{v}e^-$ and ve^- head-on collisions and particles in pure helicity states the situation shown in fig. 11.2 so that $j = 0$ or 1. However, angular momentum conservation does not allow the $\bar{v}$ to scatter through 180° since

Fig. 11.2. Possible pure helicity states in $\bar{v}e^-$ and ve^- head-on collisions.

Fig. 11.3. Lepton–lepton scattering by exchange of an intermediate vector boson W.

this would result in a $j_z = -1$ state, so only one of the three expected states is allowed.

The total ve cross-section is given by

$$\sigma_{ve} = \frac{G^2 s}{\pi}$$

where G is the coupling constant (see, for instance, Halzen and Martin, *op. cit*). Thus for σ_{ve} to remain less than the unitary limit,

$$\frac{G^2 s}{\pi} < \frac{\pi}{p^2}.$$

But if the total energy E is $\gg m$ then $p^2 \sim s/4$ (where $s = $ total cms energy squared) and

$$\frac{4G^2 p^2}{\pi} < \frac{\pi}{p^2} \quad \text{or} \quad s^2 < \frac{4\pi^2}{G^2}.$$

Taking G from low energy data we find that the unitarity limit is reached at a cms momentum $\sim 300 \text{ GeV}/c$. This simple argument neglects 'higher-order' effects due to more complex couplings involving higher powers of the coupling constant. Such higher-order processes only make the problem more severe. In fact, the underlying difficulty lies in the dimensionality of the coupling constant G ([energy]$^{-2}$, see section 4.2), since, from a purely dimensional argument, we see that with such a coupling the cross-section *must* be proportional to $E^2(s)$:

$$\sigma \propto G^2 \ E^2$$
$$[L^2] \quad [L^4][L^{-2}].$$

This feature also renders the theory 'unrenormalisable', by which is meant that the ultra-violet divergences can not be isolated in a limited number of parameters, such as masses and coupling constants. With a coupling with dimensions of a negative power of the energy an unlimited number of arbitrary constants is needed to absorb the infinities.

These difficulties led to the introduction of an exchange quantum into the theory. The force is taken to be propagated as shown in fig. 11.3 by exchange of the so-called intermediate vector boson $W^{\pm}$. The effect of such an exchange is to introduce into the amplitude a 'propagator' term

$$\frac{g_W^2}{M_W^2 + q^2} \qquad \qquad \textbf{11.1}$$

where q is the momentum transfer in the process and g_W is the coupling constant at both the W-lepton vertices. In the low energy limit with $q^2 \ll$

M_W^2 we get an expression of the same form as for the point interaction but at high energies the cross-section tends to the constant value

$$\frac{G^2 M_W^2}{\pi}$$

where G is the coupling constant as defined in the Fermi theory. Comparing the square of the matrix element **11.1** in the case of low energies where $q \to 0$, with the corresponding point interaction quantity $G^2/2$, we see that

$$g_W^2 = \frac{G M_W^2}{\sqrt{2}}. \qquad\qquad \textbf{11.2}$$

Anticipating the discussion of electromagnetic-weak unification in section 11.4 we estimate the value of M_W by writing g_W equal to the electromagnetic coupling constant e. From the low energy data we know that $G = 10^{-5}/M_p^2$ ($\hbar = c = 1$) and inserting the values we get $M_w \sim 100$ GeV.

Even apart from this estimate it is clear that the W must be heavy so that its effects are negligible at low energies. For such a heavy quantum the range of the interaction will be correspondingly short (cf. section 2.2). Even with this modification the problem of the unitarity limit is not solved but simply postponed to higher energies. Neither is the theory rendered renormalisable by this development.

The remaining problems are associated with diagrams like fig. 11.4(a) which still diverge at high energies. We also note that the W introduces diagrams in quantum electrodynamics such as fig. 11.4(b). A way out of these difficulties is to cancel offending diagrams such as fig. 11.4(b) by introducing diagrams involving the exchange of new vector particles. Thus

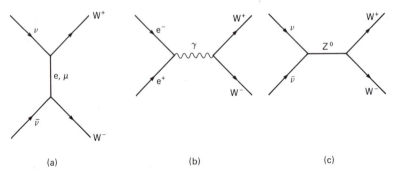

(a)   (b)   (c)

Fig. 11.4. (a) and (b): examples of divergent diagrams in weak and electromagnetic processes. The process (c) can cancel the divergence in process (a).

it is possible to cancel the amplitude of fig. 11.4(a) with the process shown in fig. 11.4(c). This new process can also cure the problem in the electrodynamic process of fig. 11.4(b) providing an important indication of the close connection between the weak and electromagnetic processes.

11.3 Neutral currents

The process of fig. 11.4(c) involves *neutral currents*, in contrast to the *charged-current* processes involved in W-exchange.

We shall see in subsequent sections that neutral currents are a key element in the unification of weak and electromagnetic interactions. Here, we turn to look at the experimental implications and observations of neutral currents in experiments with neutrinos.

We have already discussed the production of neutrino beams in the experiments which demonstrated the existence of distinct electron and muon types of neutrino (section 3.11). With the advent of accelerators to

Fig. 11.5. The WA1 neutrino detector, 20 m long and weighing 1500 tons, used in experiments at the CERN SPS 400 GeV proton synchrotron by a CERN, Dortmund, Heidelberg and Saclay collaboration. The detector measures the energy produced in the neutrino interaction and the momentum of the muons which emerge (courtesy Photo CERN).

produce proton beams of several hundred GeV, major programmes of neutrino physics have been a feature of the CERN and FERMILAB laboratories.

The detectors have been large bubble chambers (heavy liquid and hydrogen) and massive calorimetric-type detectors containing thousands of tons of material (fig. 11.5).

In a neutral current event the neutrino maintains its identity and there is no charged lepton in the final state as in fig. 11.6(b). Such events predicted by theory were first observed in the heavy liquid chamber, Gargamelle, at CERN in 1973 in an experiment originally designed for other neutrino studies. The rate for the neutral current events is less than, but of the same order as, the rate for charged current processes.

A particularly simple reaction to be expected only if neutral currents exist is muon neutrino–electron scattering

$$\bar{v}_\mu + e^- \to \bar{v}_\mu + e^- \quad v_\mu + e^- \to v_\mu + e^-$$

and, although the cross-section is small, an exhaustive search of pictures taken in the Gargamelle chamber filled with freon (CF_3Br) exposed to an antineutrino beam at the CERN proton synchrotron revealed three such events (fig. 11.7) in 1.4×10^6 pictures each with 10^9 $\bar{v}_\mu$ per pulse. Subsequent experiments have confirmed this result.

11.4   Gauge theories and the Weinberg–Salam model

It is beyond the scope of this book to give a proper account of this topic, but its importance is such that we include a qualitative discussion of the key elements even though this will involve a number of assertions without proof. The reader who wishes to pursue this subject in more detail is referred to one of the several texts where this subject is treated more thoroughly (e.g. Halzen and Martin, *op. cit.*, Close, *An Introduction to Quarks and Partons* (1979)).

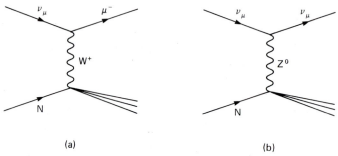

(a)   (b)

Fig. 11.6. Charged and neutral–current neutrino interaction diagrams.

11.4.1 *Global gauge transformations*

We have already seen (sections 3.8 and 8.2) that the transformations corresponding to internal symmetries are of the form

$$\psi' = e^{i\alpha}\psi$$

where α is a scalar and is the same for all values of the coordinates x. Invariance under such transformations implies that the phase of the wave function is arbitrary and unobservable.

In considering invariance under such a transformation we must look not only at the effect on the wave function but also at the space and time derivatives of the wave function which are involved in the equation of motion (the wave equation) for the system. These derivatives are conventionally written as $\partial_\mu\psi$ ($\mu = 1, 2, 3, 4$) and are also invariant under global gauge transformations. If we discuss interactions in terms of the appropriate fields, we will also be concerned with the invariance of the field amplitudes under the same transformations. Global gauge invariance is sometimes referred to as gauge invariance 'of the first kind'.

11.4.2 *Local gauge transformations*

The gauge transformation can be generalised by allowing α to be a function of space–time ($\alpha = \alpha(x)$). The transformation may then differ at different points in space–time and is known as a *local* gauge transformation. Invariance under such transformations (sometimes known also as gauge invariance of the second kind) imposes conditions different from global gauge invariance. It may well seem surprising that any theory should be invariant under such transformations but as we shall see below the familiar theory of electromagnetism in fact has this property.

Again we require that $\partial_\mu\psi$ must transform in the same way as ψ if the Lagrangian is to be invariant, so that if

$$\psi' = e^{ie\alpha(x)}\psi$$

then

$$\partial_\mu\psi' = e^{ie\alpha(x)}\partial_\mu\psi$$

(we have explicitly extracted a constant e, the electron charge, from α for convenience, as will be clear below). It is clear, on the other hand, that when α is a function of x then this last equation cannot be true, since

$$\partial_\mu\psi' = e^{ie\alpha(x)}(\partial_\mu\psi + ie\psi\,\partial_\mu\alpha(x))$$

However, we see that if, instead of the derivative ∂_μ, we use the covariant derivative

$$D_\mu = \partial_\mu - ieA_\mu$$

(where A_μ is the four-vector field potential), then the system acted upon by this operator is, in fact, invariant under local gauge transformations

$$\mathrm{D}_\mu \psi' = e^{ie\alpha(x)} \mathrm{D}_\mu \psi.$$

We recall that we have already come across this phenomenon for the electromagnetic field (section 8.2). The local gauge invariance is thus preserved only if we include a massless vector field with which the particles interact. In electromagnetic interactions the charged particles interact with the four-vector field A_μ.

Inclusion of the appropriate field means that the local gauge transformation applies to the A_μ as well as to the wave function. This is a result of profound significance in that it implies the existence of a field A_μ of massless particles where the form of the particle–field interaction is specified by the requirement to satisfy the invariance.

The Maxwell equations are invariant with respect to the addition of the derivative of any scalar $\alpha(x)$ to the vector potential and if we take the local gauge transformation on the field itself as

$$A'_\mu(x) \rightarrow A_\mu(x) + \frac{\partial \alpha(x)}{\partial x}$$

Fig. 11.7. Examples of neutral current interactions in the Gargamelle bubble chamber at CERN. (*Left*) shows a hadronic interaction in which a neutrino collides with a nucleon in the chamber to yield only hadrons and no muons. (*Above*) shows an interaction with an electron (which radiates a photon and loses most of its energy) and where again no muon is produced.

then it is straightforward to show that

$$D_\mu \psi'(x) \rightarrow e^{ie\alpha(x)} D_\mu \psi(x).$$

The theory is renormalisable with cancellation of higher order divergent terms. There can be no terms of the kind $A_\mu A_\mu$ in the Lagrangian, since such a term would imply a self-coupling of the photon to generate a non-zero mass and would destroy the gauge invariance.

Although most students of electromagnetism never meet the gauge property in this form the effects of it are familiar. If we consider a number of electric charges moving in an electric field having any kind of spatial variation then the global gauge transformation corresponds to the same change of electric potential throughout the space. It is clear that such a change will make no difference to the motion of the charges. A local transformation corresponds to changes of potential which differ as a function of position. The electric field does not itself satisfy local gauge invariance but the moving charges generate magnetic fields with associated potentials which exactly compensate for the changes so that the overall electromagnetic field is indeed invariant under the local gauge transformation.

11.4.3   *Yang–Mills gauge theory and weak I-spin*
The infinite set of phase transformations

$$\psi'(x) \rightarrow e^{ie\alpha(x)} \psi(x)$$

constitute a unitary group U(1) (cf. section 10.3), $\alpha(x)$ is a *scalar* and the group is said to be *Abelian*.

These ideas can, however, be generalised and more complex phase transformations are also possible where $\alpha(x)$ is not a scalar – non-Abelian transformations. Yang and Mills (1954) extended the idea to apply to *I*-spin space and constructed a gauge theory for rotations in such a three-dimensional isotopic spin space. Local gauge invariance in such a space requires three gauge fields of massless particles, which we label W_μ^i ($i = 1$, 2, 3). SU(2) is the appropriate operator group based on the non-commuting Pauli matrices τ (cf. section 10.3). We recall that conservation of *I*-spin involves invariance under rotations in *I*-spin space (cf. section 3.6).

$$\psi' = e^{ig\tau \cdot \Lambda} \psi$$

where g is a constant and Λ is an arbitrary reference vector in *I*-spin space and for *local* invariance $\Lambda = \Lambda(x)$.

We have already seen that isotopic spin invariance in its strong interaction sense does not apply to the weak processes. However, the *idea* may be taken over and the W_μ^i fields are then associated with massless intermediate vector boson fields. The covariant derivative in this '*weak isospin*' formalism is

$$D_\mu = \partial_\mu - ig\boldsymbol{\tau} \cdot \mathbf{W}_\mu.$$

The *vector* nature of the fields leads to an additional vector product term in the field transformation

$$\mathbf{W}_\mu' \rightarrow \mathbf{W}_\mu + \partial_\mu \boldsymbol{\Lambda} - g\boldsymbol{\Lambda} \times \mathbf{W}.$$

This implies an interaction between W and all particles which carry weak isospin including itself, so that the Ws are both the sources and the carriers of the weak isospin field.

11.4.4 *Spontaneous symmetry breaking*

The theory as outlined above involves only *massless* gauge bosons. In fact the symmetry described by local gauge invariance requires all particles to be massless including not only the gauge bosons which are the quanta of the field but also the quarks and leptons. We know very well on the other hand that particles such as the electron are massive and also that the short range of the weak force and the asymptotic behaviour of the theory require massive gauge bosons. At the same time the quantum of the electromagnetic field, the photon, *is* massless clearly demonstrating that the gauge symmetry must be broken in a unified theory.

We have already met another piece of evidence of symmetry breaking between the electromagnetic and weak aspects of the unified electro-weak theory in the phenomenon of parity violation in weak but not in electromagnetic processes (chapter 6 and section 8.8). An indication of the connection between the symmetry breaking and particle mass is evident in this case. As we have seen (sections 6.5 and 7.4) the parity violation implies a definite helicity for massless neutrinos, which have *only* weak interactions and which always travel at the velocity of light. For a particle of finite mass however we can always choose a reference frame moving with constant velocity *greater* than that of the particle. In such a frame the particle helicity will be reversed so that for massive particles both helicities must be present at some level. Thus whereas there is no fundamental reason why we should not have different symmetries for right- and left-handed weak processes and an interaction which prefers left-handed fermions, such a symmetry must be broken in order to include the massive particles which may be right- as well as left-handed.

A way of generating massive particles by a mechanism known as 'spontaneous symmetry breaking' has been proposed by a number of authors (Englert and Brout, 1964; Higgs 1964, 1966; Guralnick, Hagen and Kibble, 1964; Kibble, 1967) and is now generally known as the 'Higgs mechanism'. Spontaneous symmetry breaking can arise when the Lagrangian and the equations of motion of the system possess symmetries which do *not* however hold for the *ground state* of the system. There are in fact many familiar examples of such situations. One of the simplest and most graphic is a flexible rod under longitudinal compression. The Lagrangian is symmetrical with respect to rotations about the axis of the rod. When the compressive force exceeds a critical value however the rod buckles sideways into an arc because the energy in such a state is less than in the metastable compressed linear state. The buckled state does not have the symmetry of the Lagrangian although whatever the plane of buckling the energy is the same. Another often quoted example is that of a ferromagnet. An infinitely extended set of iron atoms has a Lagrangian which is rotationally invariant. The ground state of the ferromagnet on the other hand is one in which the spins of all the atoms are aligned in some particular direction and the symmetry is spontaneously broken. The Higgs mechanism is a gauge theoretic realisation of such spontaneous symmetry breaking.

In constructing the unified electro-weak theory Weinberg and Salam introduced a pair of 'Higgs fields' which transform as a doublet under SU(2). The fields are such that the lowest energy state is no longer empty space but a space permeated by the neutral Higgs field. The W and Z couple to this field generating terms in the Lagrangian corresponding to massive particles. The effect is analagous to situations where the photon 'acquires mass' by interaction with an existing electromagnetic field as in a plasma or in a superconductor. The zero mass of the photon is preserved in the Weinberg–Salam theory because it does not couple to the Higgs field. The quanta of the Higgs field should be manifest as 'Higgs bosons'. We shall return to the search for Higgs particles in a later section (section 11.7).

We have already seen that any plausible theory must be 'renormalisable' so that it does not give rise to infinities with the inclusion of higher order processes. The proof that the spontaneous symmetry breaking did not spoil the renormalisability of the Weinberg–Salam theory was achieved in 1971 by G. t'Hooft, by B. W. Lee and by M. Veltmann.

The reader will notice that in all the analogies which we have mentioned the symmetry breaking and the associated ground state are due to relatively complicated manifestations of well understood elementary interactions whereas in the case of the Higgs mechanism a new, extra, field is postulated to account for the observed phenomena. The Higgs mechanism also apparently gives rise to a serious problem in cosmology which we shall touch on in chapter 15. These features lead many physicists to feel that this mechanism can at best provide only a limited insight to the problem of the symmetry breaking and the origin of mass. It seems likely that new experimental data will be needed to make progress with this fundamental problem. In the meantime the 'Standard Model' incorporating the Weinberg–Salam theory allows the development of a rich field in theory and experiment independent of the correctness or otherwise of the Higgs mechanism as an explanation of the symmetry breaking and the origin of mass.

11.4.5   *The 'Standard Model'*

The theory of weak and electromagnetic interactions developed over the period 1961–68 by Glashow, Salam and Weinberg has come to be known as the 'Standard Model'.

We start with the following constituents:

> Leptons (spin $= \frac{1}{2}$)
> Quarks (spin $= \frac{1}{2}$)
> Vector bosons (spin $= 1$)
> A scalar boson (spin $= 0$) – the Higgs particle.

We have seen that the leptons have very specific helicity properties, so that electrons, for instance, are described in weak isospin space by a left-handed doublet

$$\begin{pmatrix} v_e \\ e^- \end{pmatrix}_L \quad \left. \begin{array}{l} I = \frac{1}{2}, I_3 = +\frac{1}{2}, Q = 0 \\ I = \frac{1}{2}, I_3 = -\frac{1}{2}, Q = -1 \end{array} \right\} Y = -\frac{1}{2}$$

and a right-handed singlet

$$e_R \quad I = 0, \quad I_3 = 0, \quad Q = -1, \quad Y = -1$$

which must be suitably normalised to give unity for the unpolarised electron state. The fact that the weak interaction does not conserve parity demands that the interaction differs between the right- and left-handed particles.

We have already discussed the invariance with respect to the weak isospin SU(2) but we now note that we must have a Lagrangian which is also invariant with respect to rotations in weak hypercharge space. This additional one-dimensional requirement implies invariance under the operator group U(1) so that the full symmetry is described by SU(2) × U(1). This latter symmetry requires an additional isoscalar field conventionally labelled as B_μ, so that before the symmetry is broken we have four massless quanta, $W_\mu^{1,2,3}$ and B_μ.

The isovector field W_μ couples gauge invariantly with a coupling constant g to all particles carrying weak isospin – the leptons, quarks and Higgs particle.

The isoscalar field B_μ couples gauge invariantly with a coupling constant g' (which may be different from g) to all particles carrying hypercharge – again leptons, quarks and Higgs.

The Higgs effect breaks the 'hidden' symmetry and the gauge bosons observable in nature need no longer be massless. The observable particles are well defined mixtures of the symmetric fields and there result two charged quanta of equal mass $W^\pm$, a massive neutral particle Z^0 and a massless particle, the photon. In particular the photon is the superposition

$$A = \cos\theta_W\, B_\mu + \sin\theta_W\, W_\mu^3$$

and the Z^0

$$Z^0 = -\sin\theta_W\, B_\mu + \cos\theta_W\, W_\mu^3$$

where θ_W, the Weinberg angle, specifies the mixing. The ratio of the isoscalar and isovector couplings, g'/g, is then given by

$$g'/g = \tan\theta_W$$

θ_W is then the *only free parameter* in the standard model predictions concerning neutral currents.

The leptons and quarks also acquire mass as a result of their couplings to the Higgs field but the neutrinos remain massless.

The Lagrangian can be written in terms of θ_W and the currents (see section 7.3) J_μ for weak isospin, $J_\mu^{(Y)}$ for hypercharge. Since $Y = Q - I_3$

$$J_\mu^{(Y)} = J_\mu^{(e)} - J_\mu^{(3)}$$

where $J_\mu^{(e)}$ is the electromagnetic current. Then defining

$$J_\mu^\pm = J_\mu^1 \pm J_\mu^2$$

the Lagrangian becomes

$$\mathscr{L} = \frac{g}{\sqrt{2}}[J_\mu^- W_\mu^+ + J_\mu^+ W_\mu^-] + \frac{g}{\cos\theta_W}[J_\mu^{(3)} - J_\mu^{(e)}\sin^2\theta_W]Z_\mu^0$$

$$+ gJ_\mu^{(e)}A_\mu \sin\theta_W. \qquad\qquad \textbf{11.3}$$

The first term relates to the *weak charged current*, the last to the *electromagnetic (neutral) current* and the second describes a *weak neutral current.* The observation of the predicted effects of weak neutral currents (section 11.3) provided the first major triumph for the theory.

Some interesting consequences follow immediately from the Lagrangian **11.3**. The electromagnetic coupling constant is e, so comparing with the third term of **11.3** we see that

$$e = g \sin \theta_W = g' \cos \theta_W. \qquad \textbf{11.4}$$

This relation encapsulates the unification condition and the fact that the coupling to $W^{\pm}$ and Z^0 is the same as in the purely electromagnetic case, the apparent difference between weak and electromagnetic interactions being associated with the heavy W and Z propagators. A detailed analysis yields $g_W = g/2\sqrt{2}$ (cf. first term of **11.3** – except for factor $\frac{1}{2}$!) so that using **11.2** and **11.4** we get

$$M_W = \left(\frac{\sqrt{2}\,e^2}{8G \sin^2 \theta_W} \right)^{\frac{1}{2}}$$

and inserting values

$$M_W = \frac{37.4}{\sin \theta_W} \quad \text{GeV}/c^2. \qquad \textbf{11.5}$$

The mixing of the massless fields to give the physical particles yields

$$M_W^2 = \tfrac{1}{4}g^2\eta^2 \quad \text{and} \quad M_{Z^0}^2 = \tfrac{1}{4}\eta^2(g^2 + g'^2)$$

(η is the vacuum expectation value associated with the self-interaction potential Higgs-effect field). Thus

$$M_W^2 = M_{Z^0}^2 \frac{g^2}{g^2 + g'^2} = M_{Z^0}^2 \cos^2 \theta_W$$

and

$$M_{Z^0} = \frac{75}{\sin 2\theta_W} \quad \text{GeV}/c^2.$$

The Weinberg angle, θ_W, the only free parameter in the theory, is *a parameter* in a large number of measurable processes such as neutrino–hadron and neutrino–electron interactions, electron and muon processes, e^+e^- reactions and in atomic processes.

These measurements span a range of four-momentum transfer from 10^{-6} to 10^4 GeV^2 and their agreement within errors represents very strong support for the Standard Model. A global fit to these data gives a value

$$\sin^2 \theta_W = 0.230 \pm 0.0048.$$

As we shall see later in this chapter (section 11.6.4) direct measurements of the Z^0 mass yield a value for $\sin \theta_W$ consistent with the above and with a much smaller error.

11.5   Weak–electromagnetic interference measurements

The most striking verification of the standard model has been the discovery of the W and Z particles. Before describing this triumph of the model, however, we look at two different examples of experiments which preceded the W and Z discoveries and which illustrate in different ways the phenomena of electro–weak interference expected from the unified model.

11.5.1   *Electro–weak interference in* $e^+e^- \to \mu^+\mu^-$

The process

$$e^+e^- \to \mu^+\mu^-$$

has been studied at the PETRA storage ring at DESY using a number of large detector systems. The muons are identified by their ability to penetrate substantial amounts of material (usually the iron of the analysing magnet yoke) without interaction (muons suffer only weak and electromagnetic interactions) and without producing electromagnetic showers as do electrons (too heavy).

The process may be described by weak and electromagnetic amplitudes A_{em} and A_{wk} corresponding to the diagrams of fig. 11.8.

The cross-section is proportional to the square of the sum of the amplitudes

$$\sigma \propto |A_{em} + A_{wk}|^2 = |A_{em}|^2 + |A_{wk}|^2 + 2Re\,(A_{em}\,A_{wk}^*).$$

At a total energy $= s^{\frac{1}{2}} = 34$ GeV the approximate values of the three terms are 0.1 nb, 1.5×10^{-4} nb and 8×10^{-3} nb respectively, so that the interference term is significant. As is often the case, the interference term shows up most readily as a forward/backward asymmetry for the μ^+/μ^- measured by a parameter (which is a function of s)

$$A(s) = \frac{N_F - N_B}{N_F + N_B}$$

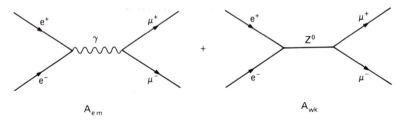

Fig. 11.8. Weak and electromagnetic amplitudes in $e^+e^- \to \mu^+\mu^-$.

where N_F and N_B are the numbers of events with a positive lepton in the forward and backward (with respect to the incident positron) hemispheres respectively.

As shown in fig. 11.9, the effect is greatest for energies near (but not too near) the Z^0-mass actually passing through zero and changing sign *at* the Z^0-mass. Some data from the PETRA detectors are shown in fig. 11.10 and the asymmetry is quite clear. Similar effects are observed for $\tau^+\tau^-$ production.

11.5.2 *Parity non-conservation in inelastic electron scattering*

Electro–weak theory predicts that the interference between weak and electromagnetic amplitudes should give rise to very small parity violating effects in electromagnetic processes. Such an effect was first observed in a beautiful experiment carried out using a polarised electron beam at SLAC to study inelastic scattering off deuterium

$$\vec{e} + d \rightarrow e' + x.$$

Polarised electrons were produced by a source consisting of a gallium

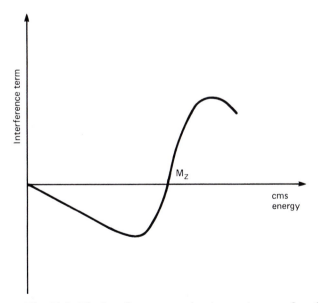

Fig. 11.9. The interference term in $e^+e^- \rightarrow \mu^+\mu^-$ as a function of centre-of-mass energy.

arsenide crystal pumped by circularly polarised photons from a pulsed dye laser at a wavelength of 710 nm. The linearly polarised light from the laser was converted to circularly polarised light by a crystal with birefringence proportional to the applied electric field (Pockels cell). The helicity of the polarised photons and thus of the produced electrons was readily reversed by reversing the field on the Pockels cell.

As the longitudinally polarised electron beam passes through the magnets of the beam-transport system the electron spin precesses by an amount which depends on the electron anomalous magnetic moment, the

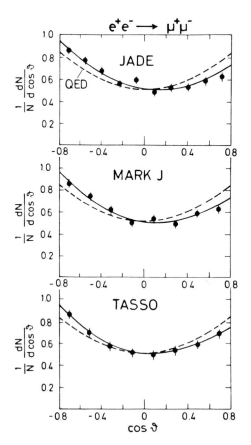

Fig. 11.10. Asymmetries in $e^+e^- \rightarrow \mu^+\mu^-$ from detectors at the PETRA storage ring at the DESY laboratory at Hamburg. The data is for total energy at or close to 34.5 GeV. The dashed lines give the QED (lowest order) prediction of $1+\cos^2\theta$. The solid lines are fits using the Standard Model (A. Bohm, 1983).

bending angle and the energy. The value of the angle of precession was given by

$$\theta_{\text{Precession}} = \frac{E_0(\text{GeV})}{3.237} \cdot \pi \text{ radians}$$

directly proportional to the electron energy E_0.

The layout of the experiment is shown in fig. 11.11. The Møller scattering (elastic electron–electron scattering) polarimeter served to

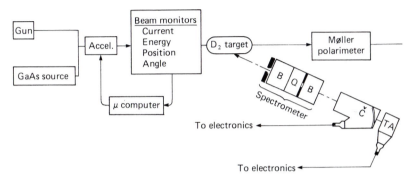

Fig. 11.11. Layout of the polarised-electron scattering experiment at SLAC which demonstrated parity non-conservation in an 'electromagnetic' interaction due to electro-weak interference.

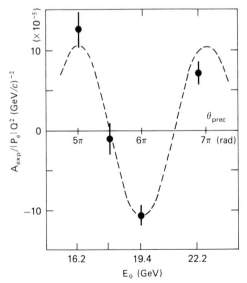

Fig. 11.12. The asymmetry in the SLAC e–d scattering experiment (note scale) as a function of the energy E_0 or, equivalently, the angle of precession of the electron spin.

measure the polarisation of the beam which averaged 0.37. Electrons scattered from the deuterium target were analysed by a spectrometer consisting of bending and focusing magnets (BQB) and detected first by a nitrogen-filled Cerenkov counter and then by a lead-glass shower counter (TA in figure). Extensive tests established that asymmetries could be measured to a precision of about 10^{-5}.

The asymmetry measured was

$$A_{exp} = \frac{N_+ - N_-}{N_+ + N_-}$$

where N_+ and N_- are the (normalised to equal beam intensity) scattered-electron intensities for opposite voltages on the Pockels cell (opposite helicities). If for a fully polarised beam A is the true asymmetry arising from the weak–electromagnetic interference we should then have

$$A_{exp} = |P_e| A \cos\left[(E_0(\text{GeV})/3.237)\pi\right]$$

where P_e is the actual electron polarisation.

A is expected to be proportional to the momentum-transfer squared, Q^2, to the recoiling hadronic system. The behaviour of $(A_{exp}/Q^2|P_e|)$ as a function of E_0 is shown in fig. 11.12. Although the maximum asymmetry is tiny ($\sim 1 \times 10^{-4}$) the experiment is sufficiently sensitive to show quite unambiguously the $\cos(E_0)$-dependence arising from the precession. The best value obtained was $A/Q^2 = (-9.5 \pm 1.6) \times 10^{-5}$ $(\text{GeV}/c)^{-2}$.

The result thus for the first time demonstrated clearly a non-parity conserving effect in electromagnetic interactions as expected from the Weinberg–Salam theory and was fitted by $\sin^2\theta_W = 0.224 \pm 0.020$, in complete agreement with the value of $\sin^2\theta_W$ obtained in the apparently very different neutrino scattering experiments.

11.6   Experimental discovery of the $W^\pm$ and Z^0

11.6.1   *Masses and branching fractions*

Although data from many earlier experiments provided strong evidence in support of the Weinberg–Salam theory, the definitive test remained the actual observation of the $W^\pm$ and Z^0 particles.

Fig. 11.13. Possible leptonic and hadronic decays for W.

We have seen that the theory predicts the masses of the $W^\pm$ and Z^0 in terms of the Weinberg angle θ_W. Taking into account radiative corrections and using an appropriately renormalised value of θ_W the predicted values were

$$M_{W^\pm} = 83 {}^{+3.0}_{-2.8} \text{ GeV}/c^2$$

$$M_{Z^0} = 93.8 {}^{+2.5}_{-2.4} \text{ GeV}/c^2.$$

The $W^\pm$-bosons can decay into lepton or quark pairs so that (fig. 11.13)

$$W^- \rightarrow l^- \bar{\nu}_l \quad (e^- \bar{\nu}_e, \mu^- \bar{\nu}_\mu, \tau^- \bar{\nu}_\tau)$$
$$\rightarrow q\bar{q}' \quad (d\bar{u}, s\bar{u} \dots \text{cf. table 11.1}),$$

with corresponding decays for W^+ with particles replaced by antiparticles. The first term in **11.3** then gives the widths for these decay modes. For instance (cf. also **11.2**), for each of the leptonic modes

$$\Gamma(W \rightarrow l\bar{\nu}_l) = \frac{GM_W^2}{6\pi\sqrt{2}} = 250 \text{ MeV}.$$

For the quark–antiquark pairs a multiplying factor of 3 arises from the 'colour' (see later, section 13.1) and some effects due to non-negligible quark masses may also enter. The leptonic branching ratio for the Ws turns out to be

$$\frac{\Gamma(W \rightarrow l\bar{\nu}_l)}{\Gamma(W \rightarrow \text{all})} = 8.5\%.$$

in the case of three generations of quarks and leptons.

For the Z^0 we expect decays

$$Z^0 \rightarrow l\bar{l} \quad (e^+e^-, \mu^+\mu^-, \tau^+\tau^-, \nu_x\bar{\nu}_x)$$
$$\rightarrow q\bar{q} \quad (d\bar{d}, \dots).$$

For these neutral-current processes the second term in **11.3** gives the partial-decay widths. The partial decay width for all leptons is

$$\frac{\Gamma(Z^0 \rightarrow l^+l^-)}{\Gamma(Z^0 \rightarrow \text{all})} = 3.1\%.$$

Fig. 11.14. Z^0 formation and decay in e^+e^- collisions.

A particularly interesting feature of the Z^0-width is that it depends on the number of different kinds of neutrinos (and thus leptons). Each different neutrino adds a width

$$\Gamma(Z^0 \to \nu\bar{\nu}) = 166.5 \pm 1.0 \, \text{MeV}.$$

Thus a measurement of the width will indicate whether there exist as yet unobserved lepton families, a question of fundamental interest in both particle physics and cosmology.

Equipped with the expected masses and leptonic branching ratios, we are in a position to understand the experiments in which the Z^0 and Ws were discovered. In order to make such very heavy objects it is necessary to have a correspondingly large amount of energy available in the centre of mass. This points to the use of colliding-beam machines and the ideal and cleanest way to form Z^0 is certainly in an e^+e^- colliding-beam machine via the process of fig. 11.14.

This clearly requires a minimum electron or positron energy of $M_Z/2 = 46.5 \, \text{GeV}$. In 1982 such e^+-e^- colliders were still some seven years in the future but a proton–antiproton collider had been developed at CERN using the existing Super Proton Synchrotron, in order to look for the Ws and Z^0 in another (though less clean) way.

Proton–antiproton collisions are at the constituent level, $q-\bar{q}$ collisions. The $q-\bar{q}$ can annihilate to yield $W^{\pm}$ or Z^0 (the inverse of the decay processes we discussed above) as illustrated in fig. 11.15. The difficulties of achieving very high energy colliding electrons and positrons (synchrotron radiation problems) are exchanged for the problem of achieving a high intensity, well focused antiproton beam. In addition, the hadron collisions are much more complicated, since three quarks and three antiquarks are present and interactions produce many secondaries (see fig. 11.16). It is for this reason, and because the leptons from the $W^{\pm}$ or Z-decay emerge unchanged while quarks must transform into complicated jets of particles, that the search for W and Z concentrated on the minority leptonic decay modes.

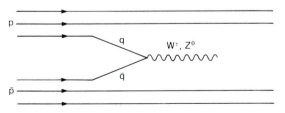

Fig. 11.15. W or Z formation in $\bar{p}p$ collisions.

The jets due to the non-interacting quarks will, in general, tend to result in particles with relatively low transverse momentum. For leptons from $W^{\pm}$ or Z^0-decay the kinematics yields a momentum distribution peaked around $p_T = M_W/2$. Since, if we neglect the sea quarks and gluons the quarks on average will each carry $\sim \frac{1}{3}$ of the total energy of the particle of which they form a part, we might expect a *minimum* energy for each of the proton and antiproton to be ~ 150 GeV in order to produce the new particles. The cross-section for the interaction is, however, expected to increase with energy. A plausible model for the $q\bar{q}$ process yields cross-sections for $\sqrt{s} = 540$ GeV (CERN collider) of

$$\sigma_{W^{\pm}} \sim 4 \text{ nb} \quad \text{and} \quad \sigma_{Z^0} \sim 2 \text{ nb}.$$

11.6.2 *The UA1 detector*

The W and Z were first observed at CERN in 1983 using the SPS in its $\bar{p}p$ collider mode. Essential to the success of the work was the use of the technique of stochastic cooling to achieve a useful flux of antiprotons (see section 1.2.5). Carlo Rubbia, leader of the UA1 collaboration, and Simon Van der Meer were awarded the Nobel prize for this achievement.

The W and Z bosons were detected in two independent elaborate detectors known as UA1 (Underground Area 1) and UA2. Here we shall describe briefly only the UA1 system, which is shown in fig. 11.17.

The UA1 detector aims to cover a very large fraction of the full 4π solid angle with track detection and calorimetry. The *central track detector* consists of a large segmented drift chamber with many wire planes (22 800 wires in total) 5.8 m in length and 2.3 m in diameter providing an overall spatial resolution ~ 290 μm in the drift-time coordinate. The chamber lies

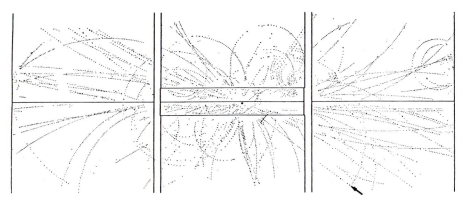

Fig. 11.16. A $\bar{p}p$ event in the UA1 detector at CERN showing the typical high multiplicity (Photo CERN).

between the poles of a dipole magnet which produces a field of 0.7 T perpendicular to the beam direction.

Surrounding the central detector is an *electromagnetic* (e–m) *calorimeter* (which is also within the magnet) built from layers of scintillator and lead.

(a)

(b)

Fig. 11.17. The UA1 apparatus: (a) perspective view; (b) side view.

The full thickness is 26.4 radiation lengths which is split into four segments which are read out independently. The calorimeter is segmented but additional spatial localisation of the signals is achieved by comparing signals from multipliers at opposite ends of the scintillator sheets.

The *hadron calorimeter* is built from iron–scintillator sandwich layers with the iron layers in the central region forming the return yoke of the magnet.

A *muon detector* consisting of orthogonal layers of drift tubes surrounds the whole detector.

11.6.3 *Event selection and data*

As discussed above (section 11.6.1), the W- and Z-bosons are likely to be most readily detected in their leptonic decays. During 1982 and 1983 an integrated luminosity of 136 nb^{-1} was accumulated for $W \rightarrow e\nu$ and $Z^0 \rightarrow l^+l^-$ (e *or* μ) and 108 nb^{-1} for $W \rightarrow \mu\nu$. Several trigger conditions were operated simultaneously, of which we mention only those directly relevant here:

'Electron trigger': at least 12 GeV of energy in two adjacent elements of the e–m calorimeter.

'Muon trigger': at least one penetrating track in the muon chambers and pointing in both projections to the interaction vertex (within a specified limit).

Additional criteria were applied to events satisfying these triggers in order to tighten the lepton identification. For instance, the e–m calorimeter signal must be matched by an appropriate track in the central detector while the shower profile in the e–m calorimeter had to be consistent with that expected for an electron.

In the case of the $W^{\pm} \rightarrow l\nu$ search there had to be a missing energy associated with the neutrino. Longitudinal energy balance is not particularly useful since inevitably it is substantially affected by the loss of particles which go undetected down the beam pipes, while the spectator quarks also produce predominantly longitudinal particles. Attention is therefore concentrated on the *transverse* momentum and energy balance.

The electron trigger yielded 150 000 reconstructed events of which only 346 remained after application of all the cuts: 10^6 events had a 'muon-trigger' flag. A series of cuts on the matching and momentum of the track in the central detector reduced this number to 689 with $p_T > 15$ GeV/c or $p > 30$ GeV/c. These reduced samples of events were studied in detail by physicists using reconstructions of the events on interactive visual displays.

For the electron events a selection was made of events with no particle jet opposite to the electron. In fact, most of this group of 55 events have no jet at all. This selection excludes events with no missing energy and jet–jet events where one jet can fake the electron signature. The sample was reduced further to 43 events from the region of the detector where the accuracy in electron energy determination is satisfactory and where there is good agreement between electron energy determination in the central detector and the calorimeter. A typical $W \to e\nu$ candidate from among these events is shown in fig. 11.18. The vector momentum of the electron and that calculated for the ν are found to be essentially back to back. Figure 11.19 shows an excellent balance between the transverse energies of electron and neutrino. These features are exactly as expected from the decay of a very heavy slow particle decaying to $e\nu$.

Since the unseen W does have *some* momentum it is not possible to measure the exact mass of the W for each individual event. One method to extract this quantity is to use the 'transverse mass', m_T^W, where

$$m_T^W = [2E_T^e E_T^\nu (1 - \cos\theta_{e\nu})]^{\frac{1}{2}}$$

where E_T^e and E_T^ν are the transverse energies of the electron and neutrino respectively and $\theta_{e\nu}$ is the angle between the electron and neutrino directions. For a W with zero longitudinal momentum $m_T^W = m_W$ (exercise in kinematics) otherwise $m_T^W < m_W$. It is then possible to calculate (with some plausible assumptions to which the result is rather insensitive) the expected distribution of m_T^W as a function of m_W. The result yielded $m_W = (80.3^{+0.4}_{-1.3})$ GeV/c^2. A better value is obtained by using a reduced sample for which E_T^ν and E_T^e are both > 30 GeV. The fit to this distribution yielded $m_W = 80.9 \pm 1.5$ GeV/c^2.

Extensive simulations of a variety of possible sources of background yielded no events satisfying the cuts.

Fig. 11.18. A W candidate event from the UA1 detector. The arrow shows the isolated high-p_T, identified electron.

The totally-independent UA2 experiment found a value of $m_W = 83.1 \pm 1.9 \pm 1.3 \, \text{GeV}/c^2$ (first error statistical, second systematic) from a sample of 37 events.

The UA1 experiment had also the sample of muon-trigger events where the selection problems are different. A final sample of 14 events yielded a mass of $81^{+6}_{-7} \, \text{GeV}/c^2$.

The (cross-section × branching ratio) measured for the electron events was

$$(\sigma B) = 0.53 \pm 0.08(\text{stat}) \pm 0.09(\text{syst}) \quad \text{nb} \qquad \text{UA1}$$
$$(\sigma B) = 0.53 \pm 0.1(\text{stat}) \pm 0.1(\text{syst}) \quad \text{nb} \qquad \text{UA2}$$

which is in excellent agreement with the theoretical expectation.

Turning now to the search for $Z^0 \to e^+e^-$ in the same samples of events cuts were imposed to demand two well-defined, isolated showers in the electromagnetic calorimeter with $E_T > 25 \, \text{GeV}$ and with matching tracks in the central detector. Only four events survived this set of cuts in UA1. One of these events is shown in fig. 11.20. These events are fully consistent

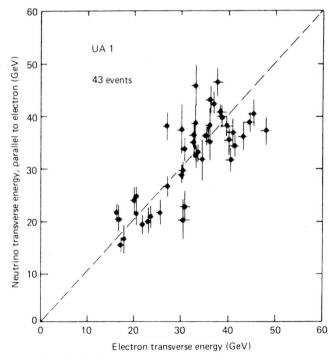

Fig. 11.19. The component of the missing (neutrino) energy parallel to the electron momentum plotted against the electron energy for the $W \to e\nu$ events from UA1.

with Z^0-decay and on fitting yielded a mass $m_{Z^0} = 95.6 \pm 1.4(\text{stat}) \pm 2.9(\text{syst})\,\text{GeV}/c^2$ and width $\Gamma_{Z^0} < 8.5\,\text{GeV}$. Taking into account the acceptance, the cross-section times branching ratio is $(\sigma B)Z^0 \rightarrow e^+e^- = 0.05 \pm 0.02(\text{stat}) \pm 0.009(\text{syst})\,\text{nb}$.

Event 7433. 1001.

(a)

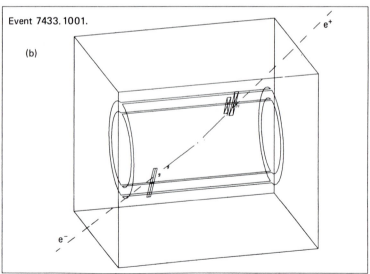

Event 7433. 1001.

(b)

e^+

e^-

Fig. 11.20. An e^+e^-, Z^0 event from UA1. (a) shows all reconstructed vertex associated tracks while (b) shows the same event but including only tracks with $p_T > 2\,\text{GeV}/c$ and calorimeter hits with $E_T > 2\,\text{GeV}$. Only the e^+e^- pair survives the cuts.

The UA2 experiment found eight $Z^0 \to e^+e^-$ candidates of which four had measurements good enough to be used for a mass determination yielding $m_{Z^0} = 92.7 \pm 1.7(\text{stat}) \pm 1.4(\text{syst})\,\text{GeV}/c^2$ with $\Gamma_{Z^0} < 6.5\,\text{GeV}$. UA1 also has 12 $\mu^+\mu^-$ events which satisfy the cuts.

The distribution of these e^+e^- and $\mu^+\mu^-$ events is shown in fig. 11.21, where the Z^0-peak is clearly evident.

11.6.4 Z^0 *production in* e^+–e^- *colliders*

As mentioned in section 11.6.1 the cleanest way to produce Z^0 is in e^+–e^- collisions via the process illustrated in fig. 11.14. The e^+–e^- annihilate forming the Z^0 which then decays into lepton–antilepton or quark–antiquark pairs. No spectator particles are present to confuse the issue and the cross-section should peak as the total energy varies through the Z^0 mass, with a peak width dependent on the number of lepton generations.

e^+–e^- colliders with energy high enough to study these processes have been built at CERN (LEP) and at SLAC (SLC). The LEP machine has already been described in section 1.2.5. The SLAC Linear Collider uses the SLAC electron linac to accelerate bunches of electrons and positrons which are then guided round opposing arcs to collide in a straight section around which is situated the detection system. The SLC is thus based on an existing machine which although modified and enhanced by the addition of cooling rings and collision arcs is limited in flux by features of the original accelerator. Also in such a 'linear collider' the bunches collide once only and are then dumped, in contrast to a circular machine such as LEP where the bunches collide many times. The luminosity in the SLC is optimised by focusing the beams down to very small dimensions (less than $3\,\mu$ diameter) at the crossing point but as a result of the limitations mentioned above the luminosity of the SLC is several orders of magnitude less than that of LEP.

The SLC started operation in the Spring of 1989 and slowly accumulated 106 Z^0 decays which showed a clear Z^0 peak at energies spanning the Z^0 mass region allowing a new determination of the Z^0 mass and width. LEP started operation for physics in September 1989 and by October each of the four detectors at LEP had recorded several thousand Z^0 decays. Both laboratories have since increased their data samples and current determinations of the Z^0 parameters by the CERN experiments are based on tens of thousands of Z^0 decays.

The e^+–e^- collider detectors are intended to study a wide range of physics. They differ in design depending upon the emphasis placed upon different aspects such as particle identification (for instance using ring image Cherenkov detectors), precision in energy measurement using calorimeters, fine granularity to separate close tracks, etc., but tend to have many characteristics in common. One of the most important of such characteristics is 'hermiticity' or coverage of practically 4π solid angle (with the exception of the beam pipe) in order to detect all particles produced except neutrinos which can then be inferred by transverse momentum balance. The detectors turn out to have cylindrical geometry around the beam pipe with solenoidal magnetic fields for track momentum measurement.

As an example fig. 11.22 illustrates the ALEPH detector at LEP. The e^+ and e^- bunches collide in the beam pipe at the centre of the detector. In layers around the beam pipe are situated:

(*a*) a silicon strip microvertex detector which gives high precision position information on charged tracks near the interaction point; particularly important for study of decays of short lived particles.

(*b*) an eight-layer high-precision drift chamber (the 'Inner Tracking Chamber') to give track coordinates near the origin and track information to be used in the trigger.

(*c*) a large time projection chamber (see figure 1.12) giving charged track information including momenta in the solenoidal magnetic field.

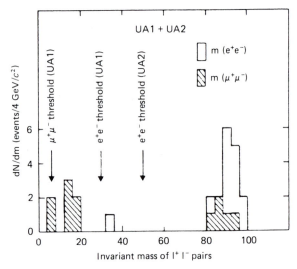

Fig. 11.21. Mass distribution for e^+e^- and $\mu^+\mu^-$ events from the UA1 and UA2 experiments at the CERN pp collider showing the Z^0 peak.

Fig. 11.22. Cut-away schematic of the ALEPH detector at LEP indicating the main components.

Vertex
Detector

Inner Track
Chamber

Time Projection
Chamber

Electromagnetic
Calorimeter

Superconducting
Magnet Coil

Hadron
Calorimeter

Muon Detection
Chamber

e^-

e^+

(*d*) an electromagnetic calorimeter consisting of layers of lead interleaved with wire chambers working in proportional mode and backed by cathode pads arranged in a tower geometry pointing at the interaction point. The 73 728 towers subtend small angles ($1° \times 1°$) at the interaction point providing a 'fine grain' for good track separation. This calorimeter measures energy for electrons and γ-rays (and thus also for $\pi°$ mesons).

(*e*) a superconducting solenoid to provide a uniform axial magnetic field of 1.5T for momentum measurement in the time projection chamber.

(*f*) a hadron calorimeter consisting of the iron of the magnetic field return path interleaved with streamer tubes and also providing 4608 projective towers. This calorimeter measures hadron energies.

(*g*) two layers of wire chambers to locate the tracks of muons which pass practically unaffected through all the inner detector layers.

The whole detector is backed by an array of electronics and computers to provide trigger decisions and to collect and record the data. Each event will typically produce some 40 particles generating up to 500 000 pieces of data. Bunch collisions take place every 22 μs, which sets a limit on the time available to collect data and take trigger decisions.

Figure 11.23 shows the cross section plotted as a function of total energy at 11 energies, for events from the ALEPH detector.

Values of fitted mass and width of the Z^0 from the first three weeks of LEP running and the initial SLC paper are shown in table 11.2. The values from all these experiments based on about 11 400 events, are in good agreement. Combining these data yields values of

$$M_Z = 91.094 \pm 0.032 \pm 0.045 \, \text{GeV}/c^2$$
$$\Gamma_Z = 2.57 \pm 0.07 \, \text{GeV}/c^2$$

11.6.5 *Comparison with the Standard Model and the number of neutrino generations*

We have seen in section 11.4.5 that the Z^0 mass is directly related to the value of $\sin^2\theta_W$. In fact the relationship is more complicated than first presented there in that, at a detailed level, account must be taken of certain radiative effects. The purely electromagnetic effects are well understood but there are also effects due to the exchange of virtual particles including the top quark (see section 12.7) and the Higgs boson. The masses of the top quark and the Higgs boson are not known although experiments which have sought the top quark now place a lower limit on this mass of about 89 GeV/c^2. Upper bounds on m_{top} can be set from neutrino–nucleon scattering and from low energy neutral current interactions between neutrinos and electrons. Taking a plausible estimate

Table 11.2. *Z^0 mass and width and number of neutrino generations from initial runs at LEP and SLC. The first error in the mass is statistical, the second arises from the uncertainty in the beam energy.*

	M_z (GeV/c^2)	Γ_z (GeV/c^2)	N_ν
ALEPH	$91.174 \pm 0.055 \pm 0.045$	2.68 ± 0.15	3.27 ± 0.30
DELPHI	$91.06 \pm 0.09 \pm 0.045$	2.42 ± 0.21	$2.4 \pm 0.4 \pm 0.5$
L3	$91.132 \pm 0.057 \pm 0.045$	2.588 ± 0.137	3.42 ± 0.48
OPAL	$91.01 \pm 0.05 \pm 0.045$	2.60 ± 0.13	3.12 ± 0.42
SLAC Mark II	$91.11 \pm 0.23 \pm 0.030$	$1.61 + 0.60 \\ -0.43$	3.8 ± 1.4

Fig. 11.23. Cross-section for $e^+e^- \rightarrow$ hadrons as a function of LEP energy. The standard model predictions for $N_\nu = 2$, 3 and 4 are shown (ALEPH detector).

for the top mass of 130 GeV/c^2 and a Higgs boson mass equal to M_Z and using $M_Z = 91.09$ GeV/c^2 as above gives

$$\sin^2\theta_W = 0.2276 \pm 0.0004$$

Uncertainty in m_{top} would only contribute an error $\sim \pm 0.001$ to this value while it is even less sensitive to the Higgs boson mass. We note that this value is perfectly consistent with the value of 0.230 ± 0.0048 obtained from the global fit to other data discussed in section 11.4.5, once more vindicating the Standard Model.

The best values for M_W now available are

$$M_W = 80.0 \pm 0.4 \pm 0.4 \pm 1.2 \, \text{GeV}/c^2 \quad \text{from UA2 at CERN}$$
$$M_W = 80.0 \pm 0.2 \pm 0.6 \quad \quad \text{GeV}/c^2 \quad \text{from CDF at FNAL}$$

The first error quoted is statistical, the second systematic and the third error quoted for the UA2 experiment at CERN is due to uncertainty in the overall energy scale. These values are fully consistent with the values we have quoted above for M_Z and $\sin^2\theta_W$, while the difference M_Z–M_W provides a constraint on m_{top} consistent with the value we have used above and which will become increasingly important as the uncertainty on M_W is further reduced.

We have already seen (11.6.1) that the Z^0 width depends on the number of open channels to which it can decay and that each different neutrino species adds 166.5 MeV/c^2 to the decay width. The determination of this parameter has thus been a prime objective of all Z^0 studies and has been effectively settled by the recent e^+e^- measurements.

If we write the number of $-\frac{1}{3}e$ charge quarks as $N_{-\frac{1}{3}}$, of $+\frac{2}{3}e$ charge quarks as $N_{\frac{2}{3}}$, the number of charged lepton families as $N_{l^\pm}$ and the number of neutrino families as N_ν the Standard Model gives

$$\Gamma_Z = 0.1665 N_\nu + 0.0835 N_{l^\pm} + 0.38 N_{-\frac{1}{3}} + 0.30 N_{\frac{2}{3}} \, \text{GeV}/c^2$$

We are only concerned with particles into pairs of which the Z^0 can decay i.e. having mass $< \frac{1}{2}M_Z$ so we set $N_{\frac{2}{3}} = 2$, $N_{-\frac{1}{3}} = 3$ (see section 12.7), $N_{l^\pm} = 3$ and then calculate Γ_Z for different values of N_ν. In fact the peak cross section for Z^0 production is also a function of N_ν and proves statistically more powerful in distinguishing between different values of N_ν. Values of N_ν calculated from the data of different experiments are given in table 11.2 and the expected shapes for $N_\nu = 2$, 3, 4 are shown compared with the data from $\sim 20\,000$ events in ALEPH in figure 11.23. The fitted value for this data is

$$N_\nu = 3.01 \pm 0.15(\text{exp.}) \pm 0.05(\text{theo.})$$

There thus seems to be no doubt that the number of light (mass less than half the Z^0 mass) neutrino families in the universe is three, corresponding to the three known neutrinos associated with the electron, muon and tau leptons.

11.7   Search for the Higgs boson

The existence or otherwise of the Higgs boson (and its properties if found) is presently the single most fundamental outstanding question in particle physics.

The Standard Model does not however predict the mass of the Higgs boson, a fact which makes searches for this particle especially difficult.

The Higgs must be a scalar particle and electrically neutral as otherwise its coupling to the photon would generate a photon mass. A most important property is that the dimensionless coupling of the Higgs boson to a pair of fermions or bosons is *proportional to the mass* of the fermions or bosons involved. Thus the coupling is greatest to the heaviest decay products which have mass $< m_H/2$ and such decays will dominate. For instance if the Higgs mass is greater than $2M_Z$ then the ratios of couplings for decay to

$$Z^0Z^0, W^+W^-, \tau^+\tau^-, p\bar{p}, \mu^+\mu^-, e^+e^-$$

are in the ratio

$$1.00:0.88:0.02:0.01:0.001:5.5 \times 10^{-6}$$

The Higgs boson will not couple to massless quanta such as $\gamma\gamma$ (or gluons – see section 13.2).

The width of the Higgs decay peak will depend on the number of kinematically open channels, being relatively small for $m_H < M_Z$. For very high Higgs mass such as $1\ \text{TeV}/c^2$, the lifetime would be $\lesssim 10^{-28}$ s and the Higgs width would be comparable to its mass so that in this case the Higgs boson would not be detectable as a particle or resonance but only through less direct effects.

Since the coupling to e^+e^- and $p\bar{p}$ is small the cross-section for Higgs production in e^+e^- or $p\bar{p}$ colliders will be small but if the mass is not too high we may expect to see it at the Fermi laboratory (FNAL) Tevatron $p\bar{p}$ collider or at LEP at CERN.

Very low masses for the Higgs are excluded by the fact that it has not been observed in nuclear decays, in kaon decays, or in the decays of B or Υ particles (see chapter 12), although in the latter cases there are some theoretical reservations regarding this conclusion. Mass less than $14\ \text{MeV}/c^2$ seems to be definitely ruled out and mass less than $\sim 5\ \text{GeV}/c^2$ to be unlikely on these grounds.

Production processes which should be readily distinguishable against background are

$$e^+e^- \to Z^0 \to H^0\gamma$$
$$Z^0 \to H^0Z^*$$

where Z^* is a virtual Z^0 decaying to fermion pairs – $e^+e^-, \mu^+\mu^-, \tau^+\tau^-, \nu\bar\nu, q\bar q$. The selection of such candidate events by appropriate triggers and cuts involves a level of technical detail inappropriate to this text. Recent studies at LEP exclude Higgs masses between 3 and 19.3 GeV/c^2 (OPAL collaboration, 1989) and between 32 MeV/c^2 and 24 GeV/c^2 at 95% confidence level (ALEPH collaboration, 1990).

We may note that if m_H is less than $2\,m_\mu$ so that the only open channel is decay to e^+e^- then the Higgs would have a lifetime (0.2–several) ns allowing it to travel observable distances in detectors. Searches for such events have found no candidates.

Searches to higher mass will continue at FNAL and at CERN with availability of more data but it may well be that we will need to wait for accelerators of higher energy to answer the question of the existence or otherwise of the Higgs boson.

12

New flavours

12.1 Prediction of 'charm'

No neutral current processes (see section 11.3) are observed which involve change of strangeness. Indeed, this fact militated against early theories of weak interactions which involved neutral currents.

For example, decays such as $K^+ \to \pi^+ \nu \bar{\nu}$, $K^+ \to \pi^+ e^+ e^-$, $K^0 \to \mu^+ \mu^-$ are absent or highly suppressed relative to corresponding charged current processes. The ratio $K^+ \to \pi^+ \nu \bar{\nu} / K^+ \to \pi^0 \mu^+ \nu < 10^{-5}$ and $K_L^0 \to \mu^+ \mu^- / K_L^0 \to$ all $= (9.1 \pm 2).\ 10^{-9}$.

We see that these observations require a mechanism which will suppress the $\Delta S = 1$, $\Delta Q = 0$ weak quark processes $s \leftrightarrow d$.

We can write the general neutral current coupling term as (see 11.1)

$$\underbrace{u\bar{u} + (d\bar{d}\cos^2\theta_C + s\bar{s}\sin^2\theta_C)}_{\Delta S = 0} + \underbrace{(s\bar{d} + \bar{s}d)\sin\theta_C\cos\theta_C}_{\Delta S = 1}$$

where θ_C is the Cabbibo mixing angle. The proposal of Glashow, Iliopoulos and Maiani (1970) was to *cancel* the $\Delta S = 1$ neutral current term with a new term arising from the existence of a new quark not known at that time and carrying a new 'flavour' which they called 'charm'. The four observable quarks with flavours up, down, strangeness and charm correspond to two doublets in each of which one of the elements is a mixture of the d and s states and which are arranged to be mutually orthogonal. The doublets are then postulated to be

$$\begin{pmatrix} u \\ d' \end{pmatrix} = \begin{bmatrix} u \\ d\cos\theta_C + s\sin\theta_C \end{bmatrix} \quad \text{and} \quad \begin{pmatrix} c \\ s' \end{pmatrix} = \begin{bmatrix} c \\ s\cos\theta_C - d\sin\theta_C \end{bmatrix}.$$

This proposal adds new terms to the neutral current term which becomes

$$u\bar{u} + c\bar{c} + (d\bar{d} + s\bar{s})\cos^2\theta_C + (s\bar{s} + d\bar{d})\sin^2\theta_C$$
$$+ (s\bar{d} + \bar{s}d - s\bar{d} - \bar{s}d)\sin\theta_C\cos\theta_C$$

so that the last, strangeness changing, term cancels to zero. This scheme allows predictions concerning the relative strengths of decays involving the charmed quark c in the same way as we constructed table 11.1. Thus we expect

$c \rightarrow s$ decays proportional to $\cos^2\theta_C$

$c \rightarrow d$ decays proportional to $\sin^2\theta_C$

so that with $\theta_C = 0.23$ radians the $c \rightarrow s$ decay is 'Cabbibo favoured' ($\cos^2\theta_C \sim 0.95$) and will dominate over the $c \rightarrow d$ decay.

We shall see shortly (section 12.6) that one and possibly two or more further quark flavours exist. The scheme of doublets can be extended to cover this situation (section 12.8).

12.2   Discovery of the J/ψ

Glashow had predicted the existence of the charmed quark and in 1974 Gaillard and Lee estimated its mass to be $\approx 1.5\,\text{GeV}/c^2$ from an argument based on the $K_L^0 - K_S^0$ mass difference. It made its debut in dramatic fashion in independent experiments at SLAC (Stanford) and at Brookhaven National Laboratory (Long Island, N.Y.).

In the SLAC experiment a detailed study was being carried out in the e^+e^- storage ring, SPEAR, of the total cross-section for e^+e^- interactions as a function of total energy which was increased progressively in small steps. At a total energy of about 3.1 GeV the cross-section showed a narrow peak rising above the background *by more than two orders of magnitude*. The peak in the cross-section was seen in the partial cross-sections for $e^+e^- \rightarrow$ hadrons, $\rightarrow \mu^+\mu^-$ and to e^+e^- (fig. 12.1) with a measured width of a few MeV. Most of the observed width, however, is due to the spread in the momentum of the circulating beams and the true (unfolded) width $\sim 0.07\,\text{MeV}$. The SLAC group named this resonance the ψ.

While the SLAC work was being done a group at Brookhaven National Laboratory was using the Brookhaven alternating gradient synchrotron

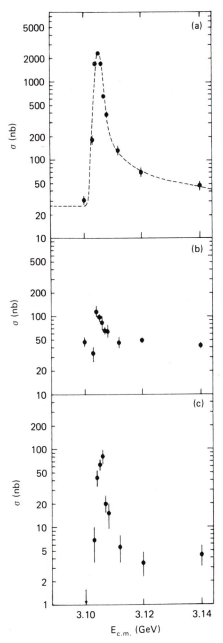

Fig. 12.1. Data from the e⁺e⁻ storage ring SPEAR at SLAC which provided the first evidence for the ψ in e⁺e⁻ collisons. Cross-section for (a) multihadron final states, (b) e⁺e⁻ final states, (c) $\mu^+\mu^-, \pi^+\pi^-$ and K⁺K⁻ final states (Augustin *et al.*, 1974).

(AGS) to study the production of direct e^+e^- pairs from a Be target bombarded with protons. The electrons were identified and their energies measured in magnetic spectrometer arms downstream of the target. The effective mass of the e^+e^- could thus be reconstructed and showed a narrow peak at a total mass of $3.1\ GeV/c^2$. In this case the width is dominated by the resolution of the spectrometers. The data from the AGS experiment is shown in fig. 12.2. The Brookhaven group named the resonance the J and the resonance is now normally referred to as the J/ψ or $\psi(1S)$ or $\psi(3097)$.

J. B. Richter and S. Ting, the leaders of the SLAC and Brookhaven groups respectively, were later awarded the Nobel Prize for the discovery of the J/ψ.

12.2.1 *Properties of the J/ψ*

The narrow width of the J/ψ distinguishes it sharply from all the strongly decaying resonances with widths ranging from a few MeV to over 100 MeV. The branching ratio of 7% for decay to electrons, which is a purely electromagnetic process, is high compared with the electromagnetic width of resonances such as ρ and ω. These facts, coupled with the high mass, suggest a system built from *heavy* quarks (compared with the u, d, s) and subject to a conservation law which inhibits normal hadronic decay processes. It was immediately proposed that the J/ψ was the first example of a state built from the postulated new charmed quarks. The simplest and successful interpretation of the J/ψ was as a bound 'charmonium' state consisting of a charmed quark and its antiparticle ($c\bar{c}$) similar to positronium (e^+e^-) and to the $s\bar{s}$ ϕ-meson. The resonance production and decay are exactly analogous to other 'formation' processes (see, for example, section 9.2) as in fig. 12.3.

For the other quantum numbers we can gain some information from the shape of the resonance peak in, for instance, $\mu^+\mu^-$ production (fig. 12.4). This cross-section is well fitted by two interfering amplitudes corresponding to direct production and production of the ψ via an intermediate virtual γ-ray. In such a case the ψ must have quantum numbers identical to the photon, $J^P = 1^-$. The I-spin may be obtained, as we have seen in other cases (e.g. $\Lambda(1520)$, section 9.7), from the ratio of hadronic modes involving different charge combinations, in particular decay to $\rho^+\pi^-, \rho^0\pi^0, \rho^-\pi^+$. The three $\rho\pi$ charge combinations are found to be equally likely, so that (see appendix B for the C–G coefficients) the I-spin must be 0. Since it decays to three pions the G-parity should be -1, in agreement with $G = (-1)^{I+J}$. The ψ is thus an isosinglet $^3S_1$ state.

12.3 **Charmonium**

Very shortly after the discovery of the ψ at SLAC a second, heavier, vector particle was discovered, the $\psi(3685)$ or $\psi(2S)$. About half of these $\psi(2S)$ decay via the lower mass states in modes such as $\psi(2S) \to \psi(1S)\pi\pi, \psi(1S)\eta$. The width is again narrow and the $\psi(2S)$ is most readily interpreted as a bound $(c\bar{c})$ state which is a 'radial' excitation of the ψ, i.e. we assume principal quantum number 1 for ψ and 2 for $\psi(2S)$, hence the name.

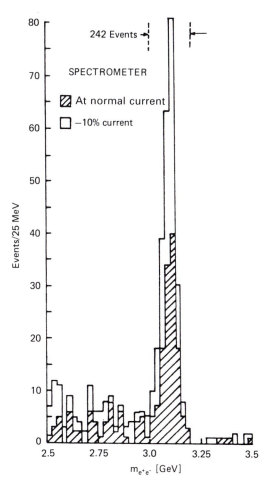

Fig. 12.2. Mass spectrum showing the first evidence for the production of the J in hadron collisons. Results from two spectrometer current settings are plotted showing that the peak is independent of spectrometer current. The run at reduced current was taken two months after the original run (Aubert *et al.*, 1974).

The $\psi(1S)$ and $\psi(2S)$ $^3S_1$ states, however, turn out to be only the beginning of the charmonium story. *A priori* we might expect that there might exist also 0^- (singlet S states), triplet P states and possibly also D states in the $c\bar{c}$ system. Neither the 0^- nor the 0^+, 1^+, 2^+ ^3P states will

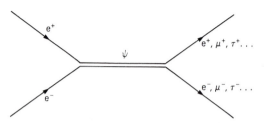

Fig. 12.3. Formation of ψ in e^+e^- collisions and subsequent decay. In fact, the formation also takes place via a process in which an intermediate virtual γ-ray links the e^+e^- to the ψ.

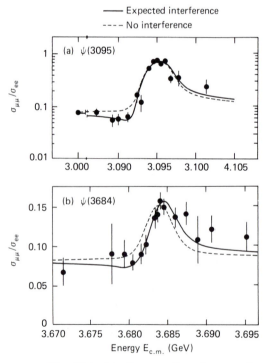

Fig. 12.4. Interference effects between direct and virtual production of $\psi(1S)$ and $\psi(2S)$ in e^+e^- interactions (SLAC–LBL group).

be accessible via $e^+e^- \to \gamma \to c\bar{c}$ since the photon has quantum numbers $J^P = 1^-$.

Depending upon the masses, however, they may be formed by photon decay from the $\psi(2S)$ and $\psi(1S)$ states. Such transitions will give rise to γ-rays of energy of a few hundred MeV (cf. $M_{\psi(2S)} - M_{\psi(1S)} \sim 600$ MeV).

The spectroscopy of the charmonium states has been intensively studied, with the results shown in fig. 12.5. One of the best instruments for such studies was the 'Crystal Ball' spectrometer in which the interaction region at SLAC was surrounded by an array of NaI crystals covering nearly the whole solid angle. As discussed in section 1.3.5, such crystals give a high resolution and high efficiency for gamma detection. In the transitions where the $\psi(2S)$ decays to a $\chi(^3P)$ state which subsequently decays to J/ψ by emission of a second gamma, one can observe the two gammas in coincidence $(E_{\gamma_1} + E_{\gamma_2} = M_{\psi(2S)} - M_{\psi(1S)})$ followed by decays characteristic of the $\psi(1S)$. We shall return to the charmonium spectrum in chapter 13.

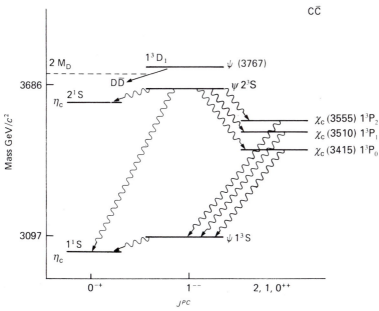

Fig. 12.5. Charmonium spectrum showing γ-ray transitions.

12.4   Charmed hadrons

The charmonium states consist of a bound (c$\bar{\text{c}}$) pair. The charm quantum numbers characterising the new 'flavour' carried by c and $\bar{\text{c}}$ add to zero in the (c$\bar{\text{c}}$) states, which are sometimes said to have 'hidden' charm.

We note that for such states all the quantum numbers could be conserved by, for instance, a decay

$$\psi \to 3\pi.$$

If such a decay could proceed freely by strong interaction it would be fast and the ψ-width would be tens or hundreds of MeV. The situation is similar to that for the ϕ-decay where the 3π-decay is inhibited relative to K$\bar{\text{K}}$ (despite the fact that the phase-space factor favours the 3π-mode):

$$\frac{\phi \to 3\pi}{\phi \to K\bar{K}} = \frac{1}{5.6}.$$

This phenomenon is an example of the Okubo–Zweig–Iizuka (OZI) rule illustrated in fig. 12.6.

In $\phi \to K\bar{K}$ the s, $\bar{\text{s}}$ quark lines link initial and final states, while in $\phi \to \pi^+\pi^-\pi^0$ the initial and final states appear to have no connection (though we shall return to this question in chapter 13). The OZI rule suggests that such 'unlinked' diagrams are inhibited.

We may then ask where are the ψ-decays of the type (a) in which the decay products will be charmed mesons ('naked charm') with charm quantum number not equal to zero? The answer turns out to be that the (c$\bar{\text{d}}$) and (c̄d) states, the D-mesons, have mass $> M_\psi/2$ so that ψ (and also $\psi(2S)$ decay to these states is energetically impossible. In fact, there were found in e$^+$e$^-$ collisions at energies above the $\psi(2S)$ state a series of even more massive c$\bar{\text{c}}$ states: $\psi(3770)$, $\psi(4040)$, $\psi(4160)$, $\psi(4415)$. These are all

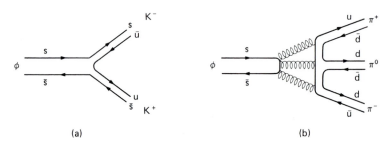

(a)   (b)

Fig. 12.6. The OZI rule inhibits the process $\phi \to \pi^+\pi^-\pi^0$ (b) where no quark lines link the initial and final states in contrast to $\phi \to K^+K^-$ (a).

1^- states and *are much wider* than the $\psi(1S)$ and $\psi(2S)$ having widths of from 25–78 MeV, indicating that there exist decay channels open to these states which are not accessible to the $\psi(1S)$ and $\psi(2S)$.

Particles with explicit charm were sought by setting the e^+e^- energy on one of these states such as $\psi(3770)$ and then looking for the Ds as peaks in the mass spectrum of the produced particles. Some results from SPEAR are shown in fig. 12.7.

Peaks are obvious in the $K\pi$ and $K\pi\pi$ effective mass distribution, allowing the identification of the $D^\pm$ and D^0-mesons with masses $1869.4 \pm 0.6 \, \text{MeV}/c^2$ ($D^\pm$) and $1864.7 \pm 0.6 \, \text{MeV}/c^2$ ($D^0, \bar{D}^0$) as well as an excited state D^* with mass 2010 MeV/c^2 decaying to D by emission of a π-meson or a γ-ray. For the $D^\pm \sim 60\%$ of decay modes are to channels including a K-meson and $\sim 20\%$ to electron channels, while for the D^0, $\bar{D}^0 \sim 85\%$ decay to channels with K-mesons. The natural widths are so small as to be swamped by the experimental resolution.

These properties of the Ds are in excellent agreement with the expectations of the charm hypothesis:

(a) $2 \times M_D \sim 3730 \, \text{MeV}/c^2$, which is greater than $M_{\psi(1S)}(3097 \, \text{MeV}/c^2)$ or $M_{\psi(2S)}(3685 \, \text{MeV}/c^2)$ so that $\psi(1S)$ and $\psi(2S)$ can not, therefore, decay to $D\bar{D}$ and have widths which are correspondingly small;

(b) the narrow D-widths are consistent with weak decay, as is the parity violation clearly involved in decay to both two and three pseudo-scalars (cf. '$\tau - \theta$', chapter 6). The dominance of decays involving strange particles is exactly to be expected as the Cabbibo-favoured channel ($c \rightarrow s$ quark transition so that $-\Delta C = \Delta S$) as discussed in section 12.1.

The decay

$$D^+ \rightarrow K^-\pi^+\pi^+$$

(c$\bar{\text{d}}$)$\qquad$(s$\bar{\text{u}}$)

is clearly observed corresponding to a transition $c \rightarrow s$, but the decay

$$D^+ \nrightarrow K^+\pi^+\pi^-$$

for which we would have to have $c \rightarrow \bar{s}$ is not observed.

The weak decay of the D-mesons raised the possibility that the lifetime could be sufficiently long to allow the meson to travel a measurable distance before decay. This proves to be indeed so, as shown for example in fig. 12.8. This example is from an experiment at SLAC which has provided a substantial number of visible and measurable charmed particle decays. Interactions of mono-energetic photons (laser light backscattered from a 30 GeV electron beam) in a liquid hydrogen bubble chamber were photographed by a special high resolution camera. Such photographs were taken only when a trigger system of counters and MWPCs outside the chamber indicated the presence of a hadronic interaction. Such measurements give D-meson lifetimes $\sim 10^{-12} - 10^{-13}$ s.

The advent of the charmed quark opens up a whole new spectroscopy of both mesons and baryons which include a c or $\bar{c}$ quark as one of the constituents. Others so far clearly identified are:

$D_s^{\pm}$ – charmed, strange mesons ($c\bar{s}$ and $\bar{c}s$); mass 1969 MeV/c^2; decay modes $\phi\pi, \phi3\pi, K^*K$

Λ_c^+ – non-strange, charmed baryon (cud); mass 2284.9 MeV/c^2; many decay modes, e.g. $p\bar{K}^0, pK^-\pi^+, \Lambda\pi^+ \cdots$.

$\Sigma_c^{++,+,0}(2455)$ – non-strange, charmed baryon (uuc, udc, ddc); mass 2453 MeV/c^2; decay to $\Lambda_c^+\pi$.

Ξ_c – strange, charmed, baryon (cus); mass 2.46 GeV/c^2; decay to $\Lambda K^-\pi^+\pi^+$.

Also to be expected but not yet confirmed, is the Ω_c^0(ssc).

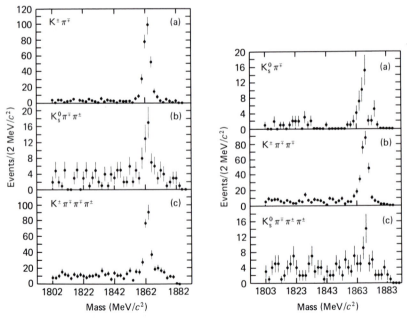

Fig. 12.7. Evidence for charmed D-meson decays from the Mark II detector at SLAC.

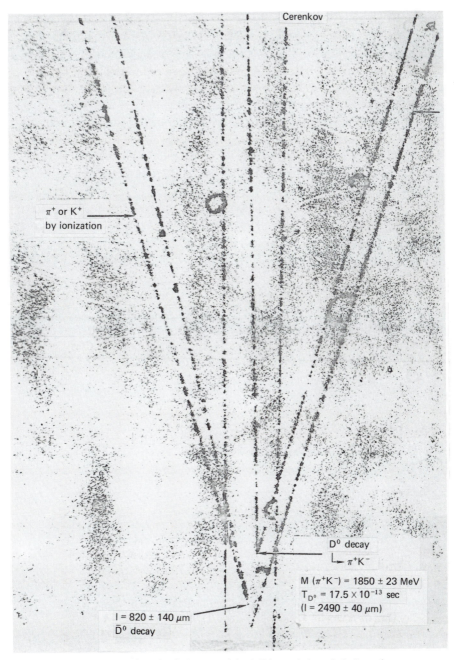

Fig. 12.8. D^0 and $\bar{D}^0$ decays with visible path length before decay. The photograph is from a high-resolution bubble chamber at SLAC using a mono-energetic γ-ray beam.

12.5   Symmetry including charm – SU(4)

We have seen that the u, d, s quark flavours formed the fundamental triplet for SU(3) with a basic weight diagram in $I_3 - Y$ space which is a triangle (section 10.4). We now have to include a fourth basic state corresponding to the c-quark with properties charge $+\frac{2}{3}e, I = 0, S = 0, B = \frac{1}{3}, C = 1$.

A new dimension is required to describe the four-flavour system and the triangle becomes a pyramid (fig. 12.9). The sides of the triangle show the three SU(2) subgroups which make up SU(3) and the faces of the pyramid show the four SU(3) subgroups (uds), (dsc), (duc), (usc) which constitute SU(4).

The extension of the *basic* multiplet also extends the *observed* multiplets of particles. For mesons the SU(3)

$$3 \otimes \bar{3} = 1 \oplus 8$$

octet and singlet structure becomes

$$4 \otimes \bar{4} = 1 \oplus 15$$

implying the existence of seven new states which we expect to see in *each of* the multiplets. The data for D and D* states are consistent with assignment to the pseudo-scalar and vector multiplets respectively, so we readily identify the seven pseudo-scalar states as:

$$\mathbf{D^0} \quad \mathbf{D^+} \quad \mathbf{D_s^+} \qquad \mathbf{\bar{D}^0} \quad \mathbf{D^-} \quad \mathbf{D_s^-} \qquad \eta_c$$

$$c\bar{u} \qquad c\bar{d} \qquad c\bar{s} \qquad \bar{c}u \qquad \bar{c}d \qquad \bar{c}s \qquad c\bar{c}$$

Charm $+1$ ($\bar{3}$ of SU(3)) Charm -1 (3 of SU(3)) Charm 0 (singlet of SU(3)).

The 1^- states are then the D*, $\mathbf{D_s^*}$ and ψ states.

For the baryons the SU(3) situation

$$3 \otimes 3 \otimes 3 = 1 \oplus 8 \oplus 8 \oplus 10$$

becomes

$$4 \otimes 4 \otimes 4 = 4 \oplus 20 \oplus 20 \oplus 20$$

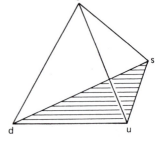

Fig. 12.9. Basic SU(4) multiplet.

The old octets of SU(3) are augmented by 12 and the decuplet by 10 charmed states. The extensions of the $J^P = \frac{1}{2}^+$ octet and the $J^P = \frac{3}{2}^+$ decuplet are shown in fig. 12.10.

12.6 Upsilon and the b-quark

The discovery of the new flavour charm stimulated the search for further new quarks. A further encouragement to such a search was the analogy between the three lepton doublets $(\nu_e, e), (\nu_\mu, \mu), (\nu_\tau, \tau)$ and the quark doublets $(u, d), (c, s)$ and $(?, ?)$ (see chapter 14). As in the case of the charmed quark, the first evidence for the b – beauty or bottom – quark was in the observation of a bound $b\bar{b}$ state. Figure 12.11 shows the effective mass spectrum of muon pairs produced in proton–nucleus interactions as observed in a double-arm spectrometer using 400 GeV incident protons at Fermi Laboratory by Herb *et al.* (1977) and Innes *et al.* (1977). As well as the $\psi(1S)$ and $\psi(2S)$ enhancements, there is evidence for peaks at much higher mass in the 9–10 GeV/c^2 region. These states, known as the upsilon, $\varUpsilon$, have since been studied in detail at the Cornell electron accelerator in an electron–positron storage ring of energy ideally matched to this mass range. As with the J/ψ states, $e^+ - e^-$ collisons have proved the best method of generating such states without the complications of the spectator particles present in hadron collisions. Results from the Cornell experiments are shown in fig. 12.12. Four $\varUpsilon$ states are clearly visible with masses:

$$M_{\varUpsilon(1\mathrm{S})} = 9.4603 \pm 0.0002\,\mathrm{GeV}/c^2, \qquad M_{\varUpsilon(2\mathrm{S})} = 10.0233 \pm 0.0003\,\mathrm{GeV}/c^2$$
$$M_{\varUpsilon(3\mathrm{S})} = 10.3553 \pm 0.0005\,\mathrm{GeV}/c^2, \qquad M_{\varUpsilon(4\mathrm{S})} = 10.5800 \pm 0.0035\,\mathrm{GeV}/c^2.$$

The first three states are all narrow (0.03–0.04 MeV/c^2) but the $\varUpsilon(4S)$ is significantly wider (~ 20 MeV), just as in the case of the ψs where the $\psi(1S)$ and $\psi(2S)$ are narrow but the $\psi(3770)$ has a width of 25 MeV. As in that case, we may assume that the first three states have masses too low to allow decay to mesons with explicit beauty (corresponding to the D-mesons for charm) but that such a decay channel *is* open for the $\varUpsilon(4S)$. This would place the mass of the lightest mesons containing a b-quark in the interval

$$5.170 \lesssim M \lesssim 5.280\,\mathrm{GeV}/c^2.$$

The b-flavoured mesons have been found with exactly the expected properties (Behrends *et al.*, 1983). The Cornell e^+e^- storage ring was set to produce the $\varUpsilon(4S)$. The b-quark is expected to decay predominantly to the c-quark (see also section 12.8) so, in order to sort out a small number of b-meson decays in a complicated situation, attention was concentrated on events with D-mesons known to carry charm. In particular, the chain

$$B^- \to D^0 \pi^-$$
$$ \hookrightarrow K^- \pi^+$$

and its charge conjugate, were sought where each flavour change is accompanied by a charge change (cf. $-\Delta Q = \Delta C$), where the decays do not involve unseen neutrals which would make mass determination difficult or impossible and where low multiplicity decays are selected to cut down combinatorial backgrounds.

Particle combinations were studied using the CLEO detector, which consists of a cylindrical central drift chamber within a solenoidal magnetic field plus an electromagnetic shower counter, time of flight systems and muon detectors. Combinations contributing to possible $K^- \pi^+$ and $K^- \pi^+ \pi^+$ peaks from D^0- or D^{*+}-decay respectively were in turn combined with additional pion tracks to seek for the decays

$$B^- \to D^0 \pi^-$$
$$\bar{B}^0 \to D^0 \pi^+ \pi^-$$
$$\bar{B}^0 \to D^{*+} \pi^-$$
$$B^- \to D^{*+} \pi^- \pi^-.$$

The resulting spectrum is shown in fig. 12.13. Background estimates do not show a peak at 5275 MeV/c^2. The fitted mass value (average of charged and neutral) is $5272.3 \pm 1.5(\text{stat}) \pm 2.0(\text{syst})$ MeV/c^2.

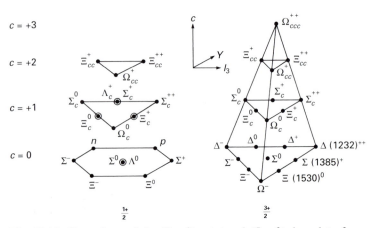

Fig. 12.10. Extensions of the $J^P = \frac{1}{2}^+$ octet and $J^P = \frac{3}{2}^+$ decuplet of SU(3) to include the charm coordinate.

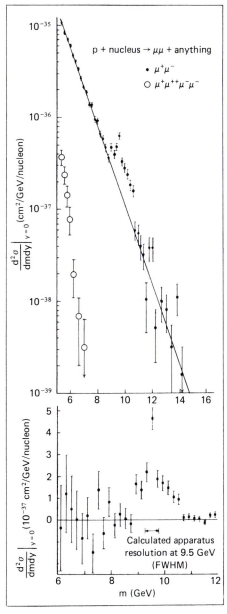

Fig. 12.11. First evidence for the existence of the Upsilon (Υ) at Fermilab. The effective mass of $\mu^+\mu^-$ pairs from proton–nucleus collisions at 400 GeV shows a peak above background at $\sim 9.5\,\mathrm{GeV}/c^2$.

12.7   Search for the t-quark

The discovery of the b-quark states increases confidence that the quarks, like the leptons, fall into three doublets and that there will exist a sixth quark flavour known as 'top' (or 'truth') to give the pattern:

$$\begin{pmatrix} u \\ d' \end{pmatrix} \begin{pmatrix} c \\ s' \end{pmatrix} \begin{pmatrix} t \\ b' \end{pmatrix}.$$

The properties of the t-quark are easily forecast with the exception of its mass, for which there is no reliable prediction. The relationship

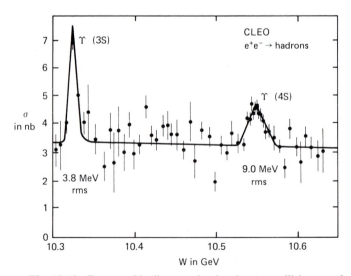

Fig. 12.12. Data on Upsilon production in e^+e^- collisions at the Cornell storage ring CESR showing the existence of four states.

between the masses of different quark flavours and between the lepton masses remains one of the outstanding problems of particle physics.

The relationship of the flavour quantum numbers to the charge applied to quarks (recall section 5.13) now becomes

$$Q = I_3 + \tfrac{1}{2}(B + S + C + B^* + T).$$

The search for 'toponium', the bound state $(t\bar{t})$ has been pursued at the PETRA e^+e^- storage ring at DESY to the highest energy available from the machine $(22.5 + 22.5)\,\text{GeV}$. A toponium state should appear, like the ψ and $\varUpsilon$, as a narrow peak in the total and partial cross-sections and the cross-sections have been measured at close intervals up to the maximum energy of the machine with no sign of any peaks above the masses of the upsilon states.

In addition to the sharp peaks corresponding to the 'onium' masses we also expect changes in the level of the continuum cross-section for annihilation to hadrons. For electromagnetic interactions between point-like leptons and point-like quarks the annihilation and elastic scattering processes are equivalent (fig. 12.14).

Thus the annihilation cross-section to a given $q\bar{q}$ pair will, as in the Rutherford scattering formula, be proportional to e^2 where e is the charge of q and the total cross-section for annihilation to quarks (and thus to hadrons) will be proportional to $\sum_i e_i^2$ where the sum is over all the

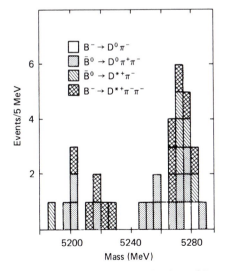

Fig. 12.13. Mass distribution of B-meson candidates (Behrends *et al.*, 1983).

accessible quark charges. If we use the cross-section for annihilation to muons as a normalisation, we can calculate the expected values for the ratio R where

$$R = \frac{\sigma(\text{hadrons})}{\sigma_{\mu\mu}} = \frac{\sum\limits_i e_i^2}{1}.$$

For e^+e^- total energies below the J/ψ threshold we have accessible only the u, d, s quarks and we might expect a ratio

$$R = (\tfrac{2}{3})^2 + (\tfrac{1}{3})^2 + (\tfrac{1}{3})^2 = \tfrac{2}{3}.$$

Above the ψ-threshold we expect a step up to $R = \tfrac{10}{9}$ (c-quark added), above the Υ to $R = \tfrac{11}{9}$ (b-quark added) and above the toponium threshold to $R = \tfrac{5}{3}$ (t-quark added). The measured behaviour of R is shown in fig. 12.15. Sharp peaks are observed for the J/ψ states, the upsilon states and, at lower masses, for the $J^P = 1^-$ resonances. Steps in level are also seen as

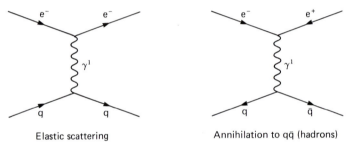

Elastic scattering   Annihilation to q$\bar{\text{q}}$ (hadrons)

Fig. 12.14. Equivalence of scattering and annihilation processes.

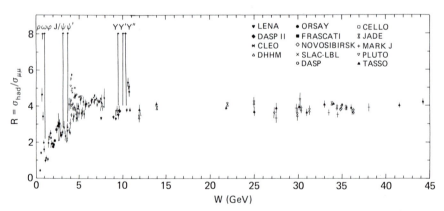

Fig. 12.15. R, the ratio of hadron production to $\mu^+\mu^-$ production in e^+e^- collisions as a function of total energy (compilation by Söding and Wolf, 1981).

the high-mass quarks become accessible. The constant value of R above the upsilon indicates the point-like nature of the quarks and leptons at least at this energy scale and the absence of top quarks at mass less than $45 \, \text{GeV}/c^2$. The absolute values of R, however, are about three times greater than we might have expected from the argument presented above. This fact has its explanation in the existence of another degree of freedom known as 'colour' which multiplies the number of states by three and to which we shall return in section 13.1.

The top quark may also be formed in other very high energy processes such as $\bar{p}p$ interactions or as a decay product of $W^\pm$ or Z^0. Intensive searches have been carried out at the CERN $p\bar{p}$ collider and at the Fermi Laboratory Tevatron where p and $\bar{p}$ collide at a total energy of 2TeV. These searches have so far failed to find evidence for the top quark. The best data from the Tevatron as of 1990 place a *lower limit* $\sim 89 \, \text{GeV}/c^2$ on the mass of the top quark.

Since effects due to the top quark enter into the calculation of radiative corrections (see section 11.6.5) for other processes some estimates of the top mass can be derived from the W and Z masses and widths, their production cross sections and the value of $\sin^2\theta_w$ as well as from some other particle decays. Present data suggest a top mass $\sim 130 \, \text{GeV}/c^2$ with an uncertainty (1 s.d.) $\pm 30 \, \text{GeV}/c^2$. These figures should be taken as indicative rather than conclusive.

On the basis of the $p\bar{p}$ collider data present e^+e^- colliders apparently have too low an energy to produce $t\bar{t}$ pairs. When the LEP energy is increased to its full design value of 90–100 GeV per beam $t\bar{t}$ production *may* be possible but the indirect theoretical estimates suggest that top may still be tantalisingly out of reach.

12.8 Six flavours and weak decays

We recall (section 12.1) that the absence of strangeness-changing neutral current processes led to a picture in which the mass eigenstates d, s are mixed and related to the weak interaction eigenstates d′, s′ by

$$\begin{pmatrix} d' \\ s' \end{pmatrix} = \begin{pmatrix} \cos\theta_C & \sin\theta_C \\ -\sin\theta_C & \cos\theta_C \end{pmatrix} \begin{pmatrix} d \\ s \end{pmatrix} \qquad \textbf{12.1}$$

where θ_C is the Cabbibo angle.

Writing the unitary matrix which performs the Cabbibo rotation as U

$$U = \begin{pmatrix} \cos\theta_C & \sin\theta_C \\ -\sin\theta_C & \cos\theta_C \end{pmatrix}$$

we can rewrite **12.1** above as

$$d'_i = \sum_j U_{ij} d_j$$

where $d_1 = d$ and $d_2 = s$ (both left-handed quark states). We note that the unitarity of U forbids the $s \leftrightarrow d$ flavour changing transitions. With three quark doublets we generalise to a column vector and U becomes a 3×3 matrix. An orthogonal $N \times N$ matrix has $N(N-1)/2$ real parameters yielding one (the Cabbibo angle) for 2×2 and three for the 3×3 case. In addition, consideration of the relative phases of the quark states reveals that it is not possible in general to define the phases to make U real and that for the $N \times N$ matrix there will remain

$$(N-1)(N-2)/2$$

residual phase factors, i.e. one such factor $e^{i\delta}$ in the 3×3 case. It can be shown (see, for instance, Halzen and Martin (1984)) that such a residual phase will in general give rise to *CP violation* so that there is a fascinating connection between this apparently gratuitous symmetry-breaking effect and the existence of the (b, t) quark doublet.

Returning to the 3×3 U-matrix (Kobayashi and Maskawa, 1972) we may write it in a form which illustrates the significance of the modulus of each element in terms of the quark transitions involved (compare also table 11.1).

$$U = \begin{bmatrix} U_{ud} & U_{us} & U_{ub} \\ U_{cd} & U_{cs} & U_{cb} \\ U_{td} & U_{ts} & U_{tb} \end{bmatrix}.$$

We can already obtain values for many of these terms from experimental measurements:

$|U_{ud}|$ (analogous to $\cos\theta_C$ in 2×2) from β-decay $= 0.9747 \pm 0.0011$;

$|U_{cd}|$ from dimuon production by neutrinos $= 0.21 \pm 0.03$

$$\nu_\mu d \rightarrow \mu^- c$$
$$\qquad\qquad \hookrightarrow s\mu^+ \nu_\mu$$

$|U_{us}|$ (analogous to $\sin\theta_C$ in 2×2) from kaon and hyperon semi-leptonic decays $= 0.2197 \pm 0.0019$

$|U_{cs}|$ from D-meson decay, e.g. $D^+ \rightarrow \bar{K}^0 e^+ \nu_e$, and from dimuon production on strange quarks (involving knowledge of the strange quark density in the 'sea' in nucleons) $= 0.65$–0.98;

$$\nu_\mu s \rightarrow \mu^- c$$
$$\qquad\qquad \hookrightarrow s\mu^+ \nu_\mu$$

$|U_{ub}|$ from B-meson decays directly to non-strange particles

$(b \to ue^-\bar{v}_e)$. In practice very difficult to measure (single lepton spectrum shape in B decays, particularly near and beyond the $B \to Dlv$ kinematic limit). What is generally measured is the ratio $|U_{ub}/U_{cb}|$ for which the best value is 0.10 ± 0.03;

$|U_{cb}|$ in principle from B-meson decays to charmed particles (b $\to$ clv) $= 0.046 \pm 0.009$.

For the third row of the matrix applying to top transitions we have, of course, no data, though the unitarity condition will force U_{tb} to be $\sim$ unity. The dominance of the diagonal elements implies that the preferred decay for mesons containing a top quark will be by cascades such as

We have discussed Cabbibo mixing for the quarks but have assumed no mixing for the lepton doublets. In fact, for massless neutrinos any such mixing is unobservable, since any rotation would still leave the neutrino-mass eigenstates unchanged. We discuss neutrino mixing in detail in section 15.3.2.

13
Quark and gluon interactions

13.1 Colour

The property of 'colour' is most obviously manifest in two very different ways. The idea was first introduced by Greenberg (1964) to account for the apparent breaking of the spin-statistics theorem for certain members of the $J^P = \frac{3}{2}^+$ decuplet.

The members of this decuplet are apparently the lowest lying states with orbital angular momentum zero and all three quark spins parallel to achieve the quantum numbers $J^P = \frac{3}{2}^+$. However, for the Δ^{++}(uuu) and the Ω^-(sss) the three quarks are identical and the wave function is totally symmetrical. This is in sharp contradiction to the spin-symmetry relationship, according to which particles with odd half-integral spin should have asymmetrical wave functions (section 3.7). Greenberg's suggestion was that there existed a hidden degree of freedom which differentiated the quarks and allowed the overall wave function to be asymmetrical, thus restoring agreement with the spin-statistics theorem. He called the proposed degree of freedom colour and postulated:

(a) that there should exist three colours;

(b) that real particles are always 'colour singlets'.

All particles must then contain all three colours or colour-anticolour combinations such that they are overall 'white' or colourless.

The second obvious manifestation of colour has already been mentioned in section 12.7. The value of the ratio R

$$R = \frac{\sigma(e^+e^- \to \text{hadrons})}{\sigma(e^+e^- \to \mu^+\mu^-)} = \sum_i e_i^2$$

where e_i are the charges of the quark pairs accessible at any energy. This simple expression is subject to small higher-order corrections but yields values of $\frac{2}{3}$, $\frac{10}{9}$, $\frac{11}{9}$ in the regions below the $\psi, \psi \to \Upsilon$ and above Υ respectively. The corresponding experimentally measured values, however, are *three times* these values below the ψ and above the upsilon, i.e. $R = 2$ and $\frac{11}{3}$ in these regions. In the intermediate u, d, s, c region the situation is complicated by the existence of the ψ states and R is apparently 4.5. This phenomenon has a simple explanation if the quarks can be of three colours, immediately multiplying R by 3

$$R = 3 \sum_i e_i^2$$

Although the first evidence of colour appeared in these rather indirect ways, this property has turned out to be fundamental to an understanding of the interactions between quarks.

13.2   Quantum chromodynamics

The idea of colour first introduced for quite different reasons is now believed to play a fundamental role in the interaction between quarks. The great success of the gauge theory of quantum electrodynamics in accounting for the interactions between charges to an extraordinary degree of precision has encouraged physicists to seek a similar theory for the strong interactions. Quantum chromodynamics (QCD) is such a gauge theory which closely parallels quantum electrodynamics (QED) in the following ways

QED	QCD
Electric charge	— Colour
Force between charges due to photon exchange	— Force between coloured quarks due to exchange of quanta called 'gluons'
(Massless bosons, spin 1)	— (Massless bosons, spin 1)

But note:

Two charges	Three colours and three anticolours
Quantum is *neutral*	Gluons are *coloured*
Gauge group U(1)	Gauge group SU(3)

We first consider the fact that the gluons are coloured. We shall give our colours the (arbitrary) names red, green and blue and consider, for instance, an interaction between a red and green quark.

The interaction is taken to proceed, as shown in fig. 13.1, by exchange of a $r\bar{g}$ (or $\bar{r}g$) gluon so that the gluon itself is coloured (neither rgb nor colour-anticolour). For a three-colour system there are nine possible independent states built from colour–anticolour combinations (cf. meson octet + singlet states built from u, d and s quarks). These are

$$r\bar{b}, b\bar{r}, r\bar{g}, b\bar{g}, \bar{r}g, \bar{b}g$$

plus two orthogonal, antisymmetric octet states from colour–anticolour pairs which are (normalised to unity)

$$(b\bar{b} - r\bar{r})/\sqrt{2} \quad \text{and} \quad (r\bar{r} + b\bar{b} - 2g\bar{g})/\sqrt{6}$$

and the singlet state

$$(r\bar{r} + b\bar{b} + g\bar{g})/\sqrt{3}$$

Since the singlet carries no net colour charge it will not contribute to the colour force and we are left with only eight coloured gluons.

An analysis of the couplings for the three quark and quark–antiquark colour singlets (baryons and mesons) shows that the binding is attractive in these cases, while for other combinations which are not colour singlets

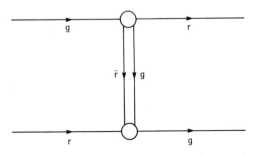

Fig. 13.1. Interaction of quarks via gluon exchange.

the force is either less attractive or even repulsive. This result is in itself encouraging in that it accounts for the fact that only these two varieties of the many possible colour combinations exist in nature.

The existence of three colours compared with only one kind of charge naturally enriches the possible structures compared with the electromagnetic case. In addition to the analogues of the familiar rules – like colours repel and colour–anticolour attracts – we have the additional principle that different colours attract if the quantum state is antisymmetrical and repel if the state is symmetric, under exchange of quarks. This means that if we take the three possible pairs red–green, green–blue and blue–red then a third quark is attracted only if its colour is different and if the quantum state is antisymmetric under pair exchange, thus leading to red–blue–green baryons. A fourth quark is then repelled by one quark and attracted by two, but only in the antisymmetric combinations, thus introducing a factor of $\frac{1}{2}$ in the attractive component and making the overall force zero. Although the hadrons are overall colourless, they feel a residual strong force due to their coloured constituents.

13.3   Quark–quark interactions

The fact that the gluons are coloured has a particularly important effect on the way in which 'charge screening' operates in QCD.

We first consider the corresponding effect in QED. An electron in quantum field theory can emit photons which annihilate into electron–positron pairs so that at all times the electron is effectively surrounded by a cloud of such pairs. Due to the electrostatic attraction the positrons of the pairs will tend to be closer to the parent electron than will the electrons. Such an effect is familiar for an electric charge in a dielectric and the situation for a charge in a vacuum is analogous though for a somewhat different reason. In terms of Feynman diagrams, this picture of the electron involves higher-order diagrams with additional loops and branches (fig. 13.2). If we attempt to measure the electron charge by scattering another charge upon it then the result we get will depend on the extent to which the incident charge penetrates the screening cloud. For a very small impact parameter and very large four-momentum transfer the incident charge will see an electron with little screening and the measured charge will increase with increasing momentum transfer. In the limit the scattering charge goes to infinity (corresponding to summation of an infinite number of higher-order diagrams) but the calculation can be handled by working in terms of a renormalised charge e defined at a selected (arbitrary) value of $Q^2 = \mu^2$ ($Q^2 = -q^2$ where $q^2 = $ four-momentum squared transfer from the incident to the target particle.) The charge variation is then calculable as a function of Q^2. A convenient way to express this result is in terms of a variable or 'running' coupling constant $\alpha(Q^2)$ which is a function of Q^2. The result of such a calculation yields

$$\alpha(Q^2) \simeq \frac{\alpha(\mu^2)}{1 - \dfrac{\alpha(\mu^2)}{3\pi} \log\left(\dfrac{Q^2}{\mu^2}\right)}$$

the approximation being best at large Q^2. In fact, α increases from $\frac{1}{137}$ only very slowly with Q^2.

In QCD there are also screening effects but the higher-order diagrams contain additional loops since gluons can interact with gluons and also because the coloured gluons *carry away colour* from the quark into the surrounding cloud. The overall effect is that as the incident probe penetrates the cloud it sees less and less colour and the measured colour charge *decreases*. Again the situation can be described in terms of a running coupling constant $\alpha_s(Q^2)$;

$$\alpha_s(Q^2) = \frac{\alpha_s(\mu^2)}{1 + \dfrac{\alpha_s(\mu^2)}{12\pi} (33 - 2n_f) \log\left(\dfrac{Q^2}{\mu^2}\right)}$$

where n_f is the number of quark flavours. The positive sign in the denominator (as long as the number of quark flavours is not more than 16) produces a screening effect opposite to that in QED. It has become customary to use a parameter Λ where

$$\Lambda^2 = \mu^2 \exp\left[\frac{-12\pi}{(33 - 2n_f)\alpha_s(\mu^2)}\right]$$

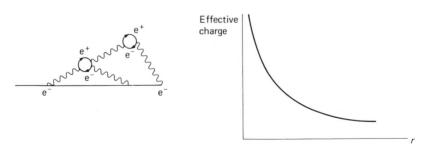

Fig. 13.2. Screening of the electron charge as a result of vacuum polarisation: Feynman diagram and effective charge as a function of distance.

so that

$$\alpha_s(Q^2) = \frac{12\pi}{(33 - 2n_f)\log\left(\dfrac{Q^2}{\Lambda^2}\right)} = \frac{\text{const}}{\log\left(\dfrac{Q^2}{\Lambda^2}\right)}$$

Λ is not predicted by the theory but can be measured experimentally. The situation is complicated by the fact that in the above expression Λ depends on (a) the particular renormalisation scheme and (b) the energy relative to the various flavour thresholds. It is conventional to refer as standard to a renormalisation known as the 'modified minimal subtraction' scheme, $\overline{\text{MS}}$, with Λ written as $\Lambda_{\overline{\text{MS}}}$. The number of effective flavours is often written as a superscript $\Lambda_{\overline{\text{MS}}}^{(2)}$, $\Lambda_{\overline{\text{MS}}}^{(3)}$, $\Lambda_{\overline{\text{MS}}}^{(4)}$, etc. It is straightforward to convert from one value of Λ to another and it is conventional to use $\Lambda_{\overline{\text{MS}}}^{(4)}$ as standard. Available data gives

$$\Lambda_{\overline{\text{MS}}}^{(4)} = 230 \pm 20(\text{stat.}) \pm 60(\text{syst.})\ \text{MeV}$$

with a corresponding value of $\alpha_s(5\ \text{GeV}) = 0.191$.

At small $Q^2 \sim \Lambda^2$ the coupling is large, but for $Q^2 \gg \Lambda^2$ the coupling decreases towards a situation of 'asymptotic freedom' where very close quarks will scarcely interact at all. Asymptotic freedom has a particularly important consequence for calculations based on QCD. In situations where the freedom holds, such as in 'hard' (very close) quark–quark collisions which produce for instance high transverse-momentum scattering (cf. Rutherford scattering through large angles), then the coupling is weak and a perturbation theory approach can yield reliable results. Where the coupling is very strong, i.e. for the large-range interactions, a perturbation theory treatment is not applicable.

At the opposite extreme from the close-encounter situation we can draw conclusions about the long distance interaction from the fact that free quarks have not so far been found in experiments at accelerators. Since the energies available are certainly adequate to provide the quark masses it is usually assumed that the interaction between the quarks is responsible for '*confinement*' of the quarks. This implies that the force between the quarks increases as the separation increases. Such forces are familiar in the case of two bodies connected by an elastic string. As the string stretches, the force increases. The quark 'string' can, however, break when the energy stored is sufficient to produce a new $q\bar{q}$ pair at the break (fig. 13.3).

It is notable that confinement is not presently understood as an intrinsic or fundamental feature of QCD so that the possibility of free quarks (section 10.11) is not absolutely excluded.

With the above features in mind, and resting on the analogy with electric charge for the short-distance interaction, a quark–antiquark potential of the form

$$V_{q\bar{q}} = -\frac{k_1}{r} + k_2 r \tag{13.1}$$

(and minor modifications thereof) has been frequently employed. Such a potential can be used to account for the charmonium spectrum of fig. 12.5. If we take the P(χ) levels to be split by a 'spin orbit' effect then the ordering of the 1S, 'mean' 1P and 2S levels is consistent with such a potential.

It is interesting to compare the mass difference between the triplet S ground state and the first S-state radial excitation $(2^3S - 1^3S)$ for $\psi(1S)$, $\psi(2S)$ and $\Upsilon(1S)$, $\Upsilon(2S)$

$$\Delta m(\psi(2S) - \psi(1S)) = 589.1 \pm 0.1 \, \text{MeV}/c^2$$
$$\Delta m(\Upsilon(2S) - \Upsilon(1S)) = 563.0 \pm 0.4 \, \text{MeV}/c^2$$

The remarkable similarity in spacing may, however, be accidental, since the two terms in the potential **13.1** have different dependences on the constituent quark masses, the first term being proportional to the mass (cf. positronium) and the second proportional to $(\text{mass})^{-\frac{1}{3}}$. If the similarity in spacing is a coincidence arising from the values of the constants and masses, then for the much heavier top quarks we may expect the spacing in toponium, if it is found, to be significantly different.

13.4   Jets and quark fragmentation

We have seen above that quarks apparently do not appear as free particles but rather generate additional q$\bar{q}$ pairs as they fly apart from other quarks and gluons in the interaction. We may expect a similar behaviour for gluons.

Experiments at high energies exhibit features which are strikingly consistent with this picture. Figure 13.4 shows an example of particle *jets* produced in the JADE detector at the PETRA storage ring at DESY. The jets find a natural explanation as the materialisation of a q$\bar{q}$ pair produced in the e$^+$e$^-$ interaction e$^+$e$^-$ → γ → q$\bar{q}$.

Fig. 13.3. String model for the force between members of a $\bar{q}$q pair.

The intermediate virtual γ is at rest in the overall e^+e^- (laboratory) system so that the q$\bar{\text{q}}$ should come out from the interaction in a '*back-to-back*' configuration as the jets are indeed seen to do. Back-to-back jets are in themselves an important piece of evidence in the QCD theory of quarks.

Thus the quarks 'fragment' or 'dress' as they emerge from the collision, being manifest only as jets of hadrons including nucleons and anti-nucleons, pions, kaons and even heavier baryons and mesons.

If the jet axes in fact represent the directions of parent quarks, then the jet angular distribution with respect to the beam direction may be shown to be simply related to the quark spin. For spin $\frac{1}{2}$ quarks the distribution should be

$$I(\theta) \propto 1 + \cos^2 \theta.$$

Such a distribution was indeed measured by Hanson *et al.* (1975) at the SPEAR e^+e^- ring at SLAC at a centre of mass energy W = 7.4 GeV. Data from the TASSO detector at DESY at 34 GeV is shown in fig. 13.5.

The idea of *bremsstrahlung* leads to one of the most important pieces of evidence in support of the quark–gluon picture. An electrically charged particle will radiate photons if subject to acceleration arising, for instance, by interaction with another electric charge. In the same way we might expect a coloured quark to emit QCD bremsstrahlung, i.e. gluons, in

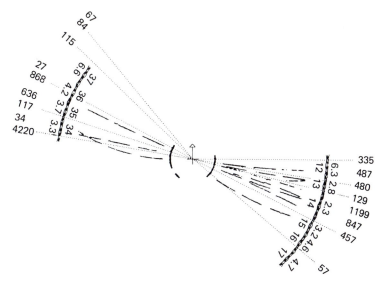

Fig. 13.4. An example of back-to-back hadron jets from the JADE detector at the e^+e^- storage ring PETRA.

interaction with other coloured particles. In an e^+e^- interaction producing a $q\bar{q}$ pair, one of which radiates a gluon in escaping from the interaction, we would then expect to see *three* jets (a quark jet, an antiquark jet and a gluon jet). Striking examples of such events have been seen at PETRA (fig. 13.6).

Another vivid piece of evidence for quarks as jets comes from the SPS collider experiments studying $\bar{p}p$ collisions. Quark–antiquark collisions may be expected to be closely similar to charged particle collisions as, for instance in α-particle scattering on nuclei where the angular distribution of the scattered αs obeys the classical Rutherford formula and the cross-section is proportional to the inverse fourth power of the sine of half the scattering angle. The UA1 experiment sees many examples of jet–jet events (fig. 13.7). The distribution in scattering angle is shown in fig. 13.8 and is well fitted by QCD which, as might be expected from the potential discussed in 13.3, yields a shape quite close to $\sin^{-4}(\theta/2)$.

One might expect that the ratio of the number of 3-jet and 2-jet events would give a direct measurement of the coupling constant α_s since the 3-jet process involves one additional vertex. Unfortunately the jet fragmentation process complicates the analysis required to sort out the 2- and 3-jet samples. The results of such analysis are somewhat dependent on

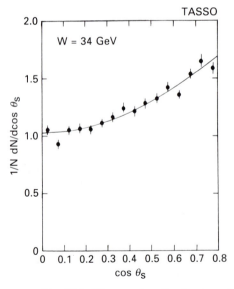

Fig. 13.5. The angular distribution of the jet axis in two-jet events from e^+e^- collisions at total energy of 34 GeV in the TASSO detector at PETRA laboratory. The curve is proportional to $1 + \cos^2\theta$.

the QCD fragmentation model and give $\alpha_s \simeq 0.15 \pm 0.02$ consistent with values obtained by other methods.

13.5 Gluons and gluon couplings

Lepton scattering experiments (section 10.10) showed that the nucleon mass was not all accounted for by the quarks themselves, and we may ask whether there is evidence for the gluons from deep inelastic scattering of leptons.

We first consider the gluon 'Compton scattering' process

$$\gamma' q \rightarrow q g \qquad \qquad \textbf{13.2}$$

where γ' is a *virtual* photon emitted by a charged lepton such as an electron

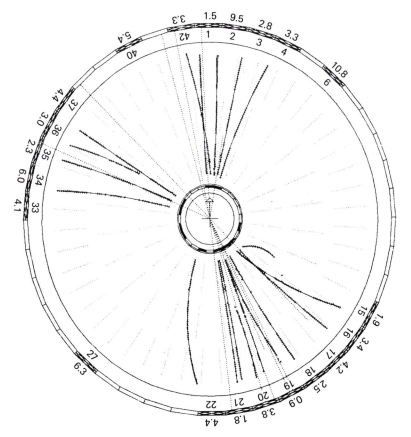

Fig. 13.6. Three-jet event as expected from 'gluon bremsstrahlung', in the JADE detector at the e^+e^- storage ring PETRA.

or muon (fig. 13.9(a)). Turned around, the same diagram (fig. 13.9(b)) also represents the process

$$\gamma' g \rightarrow q\bar{q} \qquad\qquad \textbf{13.3}$$

The calculation of the cross-sections is closely analogous to the calculation of the Compton cross-section in QED with appropriate replacement of coupling constants at the vertices and proper averaging of the quark colours. The 'Compton' process of fig. 13.9(a) results in a quark jet and a gluon jet, neither of which lies along the direction of the virtual photon. The inclusion of the gluon diagram gives an increase in the cross-section for hadron production, particularly at high transverse momentum.

The effect is clearly seen in the cross-section as a function of hadron transverse momentum for lepton–nucleon scattering (in fact, for high-energy incident muons, though the situation is the same for electrons) as shown in fig. 13.10.

The gluon diagrams must clearly also affect the structure functions (see section 10.10). We recall that in the simple quark (or parton) model the structure functions depended only on the ratio q^2/Mv where q is the four-momentum transfer between projectile and target (γ' and quark), M is the quark mass and $v = E - E'$ where E and E' are the initial and final projectile energies. This is the property known as scaling. The cross-section for the gluon Compton diagram, however, introduces an additional term proportional to $\log (q^2/\mu^2)$ (where μ is a lower limit on the transverse momentum of the gluon introduced to avoid a divergence as

$E_T = 70.1$ GeV

$E_T = 66.8$ GeV

Fig. 13.7. Two-jet event in $\bar{p}p$ collisions from the UA1 detector at the CERN SPS illustrating $\bar{q}q$ scattering.

$p_T^2 \to 0$) which does not scale so that the structure function will no longer scale. The violation of scaling is small, depending only on the logarithm of q^2 but is evident for high q^2 (fig. 13.11) and is a clear indication of the effects of the gluon.

A particularly intriguing property of the gluons arises from the fact that, being themselves coloured, they should be able mutually to interact to form pure gluon states known as gluonium or 'glueballs'. The existence of bound colour singlet two-gluon and three-gluon states would appear to be an important test of QCD theory. However, it may be very difficult to distinguish such states from ordinary $q\bar{q}$ (meson) states unless the gluonium happens to have a spin-parity not possible for $q\bar{q}$. More detailed analysis suggests that gluonium states may couple strongly to decay modes not favoured for mesons, such as $\phi\phi$ for gluonium of sufficiently

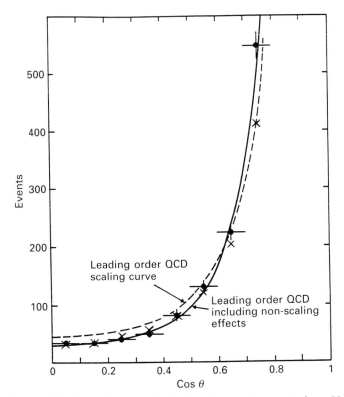

Fig. 13.8. Angular distribution for two-jet events from UA1 at CERN. The two-jet effective mass is selected to be within the interval 150–250 GeV/c^2 and the plot is based on 1142 events. The crosses X, correspond to a distribution proportional to $\sin^{-4}(\theta/2)$ for comparison.

high mass. Several experiments have studied the $\phi\phi$ and other likely systems for evidence of gluonium but, as of early 1990, no compelling evidence has been found for its existence.

Turning back to the problem of the OZI rule discussed in section 12.4,

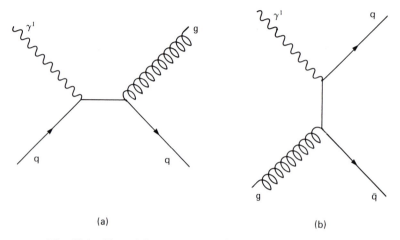

(a)   (b)

Fig. 13.9. Gluon 'Compton scattering'.

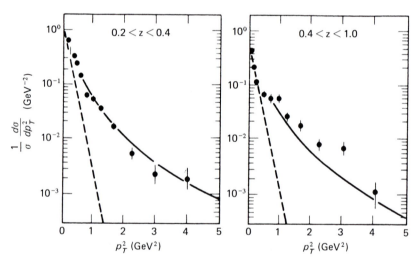

Fig. 13.10. Effect of gluons on the p_T^2 distribution of hadrons in μ-nucleon collisions. The dashed line gives the expected distribution in the absence of gluon emission, while the continuous line gives the distribution expected including gluons and is seen to give a good account of the measured points. The kinematic factor z is related to the gluon momentum. The data are from the European Muon Collaboration at the CERN SPS.

we recall that diagrams with no quark links such as those representing $\phi \rightarrow$ pions, $\psi \rightarrow$ anything (since no charm–anticharm states are energetically accessible), $\Upsilon \rightarrow$ anything, are quite strongly inhibited relative to decays with quark links (e.g. $\phi \rightarrow K^+K^-$). Without quark links we may ask how such decays are possible at all? Gluon links provide the likely answer. Since initial and final states are colour singlets at least two gluons are needed to build a singlet to link the states. In addition, however, we must conserve the charge conjugation parity C which must be -1 for J/ψ and for Υ, since they couple directly to single photons for which $C = -1$ (e–m fields produced by charges which have $C = -1$). We know that the ψ and Υ are $^3S_1$ states and thus, since the gluon state responsible for the decay must have the same quantum numbers (as the ψ or Υ and as a single photon), a two-gluon state is not possible and three gluons represent the

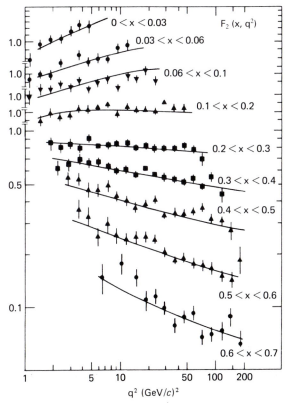

Fig. 13.11. The structure function $F_2(x, q^2)$ as determined in lepton–nucleon scattering. Some variation with q^2 is evident, indicating breaking of the scaling property.

simplest possibility $(C = (-1)^n$ where $n =$ number of gluons). The situation is identical to that for positronium decay with gluons substituted for photons. The process is illustrated in fig. 13.12. An interesting piece of evidence in support of this argument is that for the Υ, where the energy of the decay gluons averages 3.4 GeV, there is an increase in the frequency of three-jet events in e^+e^- collisions as the total e^+e^- energy is varied through the Υ mass.

Finally, we may note that information on the gluon spin can be obtained from an analysis of the angular distribution of the most energetic

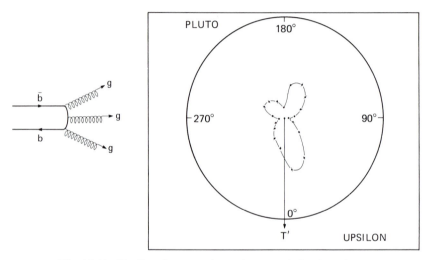

Fig. 13.12. Upsilon decay to three gluons and the three-jet structure of Upsilon events from e^+e^- collisions at the PETRA storage rings.

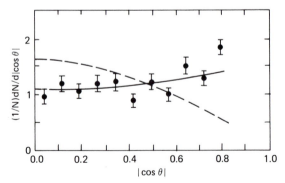

Fig. 13.13. The jet-axis angular distribution as measured by the PLUTO detector at DESY, and the theoretical (normalised) curves for vector gluons (full line) and scalar gluons (dotted line). The data clearly favour vector gluons (Koller and Krasemann, 1979).

jet in three-jet events with respect to the beam direction (Koller & Krasemann (1979)). The data exclude spin zero and are fitted by spin one (fig. 13.13).

13.6 **Status of QCD**

In summary, we see that quantum chromodynamics is a gauge theory which accounts qualitatively in a compelling way for many of the principal features of quark interactions.

Confinement, however, does not spring naturally from the theory, nor does the theory give any clue to the number of flavour generations or the relationship between them.

Detailed quantitative comparison between experiment and theory is difficult, due partly to the great difficulty in making calculations involving strong couplings except in the high Q^2 limit where the quarks are asymptotically free. Nevertheless wherever the theory has been tested it has proved successful and must surely represent a large part, though not the whole, of the truth.

14

Higher symmetries

14.1 Grand unification

We first recall some facts which emerge from the studies and analysis described in the preceding chapters concerning the properties of *quarks* and *leptons*. Both quarks and leptons

(*i*) have spin $\frac{1}{2}$;

(*ii*) obey the Pauli principle;

(*iii*) exhibit no internal structure (at the present limit $\sim 10^{-18}$ m), i.e. they act as point particles;

(*iv*) are left-handed in respect of weak processes;

(*v*) fall into the same kind of doublets for weak processes;

(*vi*) obey similar gauge theories and interact by exchange of spin 1 bosons.

The only difference we recognise between quarks and leptons is that the quarks carry colour while the leptons do not.

These similarities and the successful unification of the weak and electromagnetic interactions naturally lead to intensive efforts to unify the electro–weak with the strong interactions. Such unification implies the existence of a larger group encompassing

$$SU(2) \times U(1) \times SU(3)$$

In such a group, the couplings for the different processes would not be independent, so that, for instance, the Weinberg angle θ_W linking the weak and electromagnetic couplings would be predicted by the theory. All the interactions are then described in terms of a single coupling to which the other couplings can be related once the grand unified group is recognised. Such grand unified theories are frequently referred to by the abbreviation 'GUTS'.

We recall that the strong coupling decreases as a result of the logarithmic screening term (section 13.3) while the electromagnetic coupling increases due to the screening effect, as a function of momentum transfer. If we write α_3, α_2 and α_1 for the couplings corresponding to the SU(3) of colour, the SU(2) and U(1) subgroups, then we can calculate their variation as a function of momentum Q (or impact parameter $1/Q$) and, using the measured values at accessible energies, we can determine the momentum at which the couplings become equal. The corresponding energy is known as the unification mass M_X and its value is found to be $\sim 10^{15}$ GeV/c^2 ($\sim 10^{-10}$ gm). The behaviour of the couplings is illustrated in fig. 14.1. The unification mass is so large that there is no chance that we can approach it with current accelerator techniques. Nevertheless, grand unification can still have small effects even at presently attainable energies.

In order to illustrate such possible effects we return to consider the unification group required to encompass $SU(2) \times U(1) \times SU(3)$. Georgi and Glashow (1974) have shown that the smallest such group is SU(5). An SU(N) gauge theory has $(N^2 - 1)$ gauge bosons (cf. SU(3) with eight coloured gluons), so that for SU(5) we expect 24. The 24 break down into multiplets

$$24 = (8, 1) \oplus (1, 3) \oplus (1, 1) \oplus (3, 2) \oplus (\bar{3}, 2)$$

where the brackets represent an $(SU(3)_{colour}, SU(2)_L)$ decomposition. The $(8, 1)$ corresponds to the gluons, the $(1, 3) \oplus (1, 1)$ to the W, Z and γ while the $(3, 2) \oplus (\bar{3}, 2)$ describes a new *weak doublet* of 12 *coloured* bosons responsible for the unification. These superheavy bosons, usually called X and Y, will have mass $\sim 10^{15}$ GeV/c^2 and are responsible for transforming quarks to leptons by processes such as

$$u + d \rightarrow Y \rightarrow e^+ + \bar{u}$$

or

$$u \rightarrow Y + \bar{d}.$$

We may compare this process with the corresponding electro-weak and strong processes

$$d \rightarrow u + W^-$$

and

$$u_R \rightarrow u_B + g(r\bar{b}).$$

Such processes fix the charges of the X and Y as $\frac{4}{3}e$ and $\frac{1}{3}e$ or $-\frac{4}{3}e$ and $-\frac{1}{3}e$; for example,

$$u \rightarrow Y + \bar{d}$$
$$+\tfrac{2}{3} \quad +\tfrac{1}{3} \quad +\tfrac{1}{3}$$
$$u \rightarrow X + \bar{u}$$
$$+\tfrac{2}{3} \quad +\tfrac{4}{3} \quad -\tfrac{2}{3}.$$

We return to the transformation of quarks to leptons in the next section.

A fascinating aspect of the grand unification group is that it accounts for the relationship between the quark charges and the fact that the electron and proton charges are *identically equal*. The argument depends on the fact that since the photon is one of the gauge bosons of SU(5) it follows (not proved here) that the charge operator is a generator of the group. In addition, it is a property of simple groups such as SU(3) and SU(5) that the trace of each generator vanishes. In order to exploit this

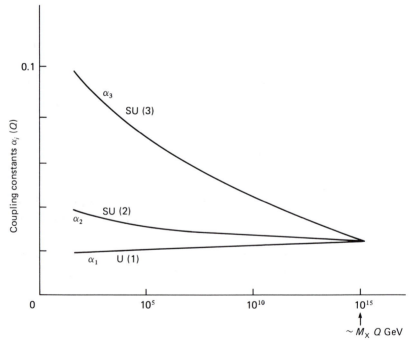

Fig. 14.1. Variation of the running coupling constants for the electro-weak and strong interactions as a function of Q to the (speculative) 'grand unification' mass at $\sim 10^{15}$ GeV.

property we have to identify multiplets of the overall gauge group. For the SU(5) group the multiplets comprise a $\bar{5}$ and a 10 (for *each* generation) where the $\bar{5}$ is usually taken to consist of

$$\bar{5} \equiv \begin{Bmatrix} v_e \\ e^- \\ \bar{d}_r \\ \bar{d}_b \\ \bar{d}_g \end{Bmatrix}_{LH}.$$

On the basis of the argument just presented the sum of the charges in the $\bar{5}$ should equal zero. Thus

$$3Q_d + Q_v + Q_e = 0$$

so that $Q_d = \frac{1}{3}Q_e$.

For the 10 the same argument yields

$$Q_u = -2Q_d.$$

It follows that

$$Q_p = -Q_e.$$

Thus

(a) charge *must* be quantised;

(b) because quarks and leptons appear in the same multiplet their charges are related;

(c) because there are three colours while the electron is colourless the quarks must have $\frac{1}{3}e$ and $\frac{2}{3}e$ fractional charges;

(d) the electron and proton charges are equal and opposite.

The $\overline{10}$ is the antisymmetric combination of two $\bar{5}$s (and the 10 of two fives) and we can see that such a $\overline{10}$ plus the $\bar{5}$ yield all the required states: the u and d quarks in three colours and two helicities, the electron in two helicities and the neutrino in one helicity only (table 14.1).

A satisfying feature of the SU(5) group is that the predicted value of $\sin^2\theta_w$ is ~ 0.20 to be compared with the measured value of 0.2276 ± 0.0004.

SU(5) is the *simplest* satisfactory group required to account for the structure of the electro-weak and QCD subgroups but it is not the only possible group.

14.2 Proton decay

14.2.1 *Lifetime and decay modes*

We have mentioned that grand unification can have small effects at low energies. The most dramatic such effect is the decay of the proton.

Such a possibility is immediately obvious from the discussion of section 14.1 above. Figure 14.2 gives diagrams corresponding to processes which we have already discussed but with the addition of a 'spectator' quark. The diagrams correspond to the process

$$p \to \pi^0 + e^+.$$

From a slightly different point of view it is clear that such transitions must follow from inclusion of quarks and leptons in the same multiplet.

The decay rate for a diagram of the kind shown in fig. 14.2 can be estimated in the same way, as for instance, for muon decay by the weak interaction via a W-propagator. The result has the form

$$\tau_p = \frac{4\pi A M_X^4}{g^2 m_p^5}$$

where g is the grand unified SU(5) coupling, the factor M_X^4 is introduced

Table 14.1.

$\bar{5}$ state combination Left-handed	$\overline{10}$ Right-handed
$\nu_e e^-$	e^-
$\nu_e \bar{d}_r$	$\bar{d}_r$
$\nu_e \bar{d}_b$	$\bar{d}_b$
$\nu_e \bar{d}_g$	$\bar{d}_g$
$e^- \bar{d}_r$	$\bar{u}_r$
$e^- \bar{d}_b$	$\bar{u}_b$
$e^- \bar{d}_g$	$\bar{u}_g$
$\bar{d}_r \bar{d}_b$	u_g
$\bar{d}_r \bar{d}_g$	u_b
$\bar{d}_b \bar{d}_g$	u_r

Fig. 14.2. Some mechanisms for proton decay to $e\pi^0$.

by the X-propagator, m_p is the proton mass and the factor A depends on the hadronic matrix elements. We have seen (fig. 14.1) that $g/4\pi \sim \frac{1}{40}$ with $M_X \sim 10^{15}$ GeV while $A \sim 1$, so that we get $\tau_p \sim 5 \times 10^{31}$ years (cf. lifetime of the Universe $\sim 10^{10}$ years). The lifetime is clearly highly sensitive to the unification mass M_X. More sophisticated calculations based on SU(5) (see Langacker, 1981) yield values $\tau_p = 3 \times 10^{29 \pm 1.3}$ years.

The principal decay channels predicted under SU(5) are:

$$
\begin{array}{llll}
p \to e^+\pi^0 & 30\% & n \to e^+\pi^- & 54\% \\
 e^+\rho^0 & 14\% & e^+\rho^- & 23\%. \\
 e^+\omega & 30\% & & \\
 \bar{\nu}_e\pi^+ & 11\% & &
\end{array}
$$

We may note that most GUTS other than SU(5) also require proton decay though the unification mass and the lifetimes are, in general, different. Another potentially important difference for the nature of detectors designed to study this phenomenon is that other decay modes such as $p \to \bar{\nu}_\tau K^+$ are more favoured in some models.

We shall return to consider the cosmological relevance of GUTs in chapter 15.

14.2.2. *Measurements*

For studies aiming to detect proton decay with lifetimes as long as, say, 10^{32} years, a very large quantity of detector material is required. At such a lifetime, for instance, we require ~ 1000 tons of iron to yield one $e^+\pi^0$-decay per year.

A critical factor in the design of such experiments is the background which could simulate proton-decay events. The dominant background is due to atmospheric neutrinos, the interaction rate for which is *also* proportional to the mass of the detector material, so that the decay and neutrino interaction rates become equal for a lifetime $\sim 10^{31}$ years. Distinction between background and real events and identification of decay modes other than $e^+\pi^0$ demand sophisticated detector systems.

Two main varieties of detector have been developed to attack the problem. The simpler system consists of a very large tank of water monitored by photomultipliers surrounding or even within the tank. The multipliers detect Cerenkov light from relativistic particles, in particular from e^+ and e^- generated in decays such as $p \to e^+\pi^0$, $\pi^0 \to \gamma\gamma$, $\gamma \to e^+e^-$. Pure water is transparent to the Cerenkov light over a large proportion of the spectrum, allowing the use of very large water volumes. The electro-magnetic showers from e^+ and the π^0-decay γs will appear back-to-back so that the Cerenkov light detected by opposite groups of multipliers can be used to measure the decay product momenta. In addition, muons can be detected via their electron decay as pulses with a delay $\sim 2\,\mu s$.

One of the most ambitious detectors of this kind to date is the IMB detector (Bionta *et al.*, 1983) which consists of an approximately cubical mass of 7000 tons of water ($18 \times 17 \times 22.5$ m) monitored by 2408 5-in photomultiplier tubes distributed over the surface of the tank on a 1 m grid (fig. 14.3). The detector is situated at a depth equivalent to 1570 m of water underground in the Morton Salt Mine in Ohio in order to minimise the cosmic-ray background. Nevertheless some 2×10^5 muons per day cross the detector to give triggers (> 12 PMs firing within 50 ns or > 3 PMs in any two of 32 groups of 64 firing within 150 ns). The muons serve to calibrate the detector and are distinguished from decay events by an energy selection plus a cut which requires the reconstructed vertex to lie within a fiducial region everywhere > 2 m within the detector surface. This cut reduces the effective mass of material to 3300 tons. After elimination of the muon events there remained from a run of 132 days 112 events attributable to neutrino interactions. The absolute number, the energy distribution and the up–down ratio were all consistent with the expectation for neutrino interactions. In addition, the topology of the energy distribution for 109 of the events was such that most of the energy was concentrated in one hemisphere ('one-track hypothesis') inconsistent with $p \rightarrow e^+\pi^0$ or $\mu^+\pi^0$. The remaining three events are also found to be inconsistent with $p \rightarrow e^+\pi^0$ on detailed examination and are attributed to inelastic neutrino absorptions $\nu N \rightarrow \mu\pi N$ or $\nu N \rightarrow e\pi N$.

Although detectors like the IMB experiment have given the best limits to date on the decay lifetime, it is of interest to look briefly at the other main type of detector which can yield better track information. Although such detectors to date have not approached the large mass of the IMB system, some very massive devices of this kind are presently under construction.

Examples of such detectors are the KGF calorimeter at a depth of 7600 m water-equivalent in the Kolar Gold Mines in India (which claimed several examples of proton decay not, however, generally accepted as convincing decay events) which was the first such operating detector, and the NUSEX detector, operating at a depth of 5000 m water-equivalent in the Mont Blanc tunnel. This latter detector consists of a sandwich construction of 1 cm-thick iron plates interleaved with 1 cm × 1 cm streamer tubes. The tubes are of plastic with a central high-voltage anode wire. The cathodes consist of two orthogonal sets of pick-up strips above and below the tubes, so that each tube signal gives a three-dimensional coordinate. The mass of the detector is 150 tonnes. Compared with the water Cerenkov, detector devices like NUSEX have a high vertex-reconstruction precision (~ 1 cm compared with ~ 1 m), much better track definition and better containment of events due to the high stopping power of the material. Thus a larger fraction of the detector is useful as a fiducial volume.

Limits on decay have been measured for about 30 modes. In particular the large detectors yield

$$\tau(p \to e^+\pi^0) > 6.8 \times 10^{32}\,y$$
$$\tau(p \to \nu K^+) > 9 \times 10^{31}\,y$$

In summary, we can say that there is at present no totally convincing evidence for nucleon decay. In addition, the large detector results appear

Fig. 14.3. Schematic of the IMB proton-decay detector.

to set a limit for the $p \rightarrow \pi^0 e^+$ decay which is inconsistent with the SU(5) prediction.

Although the great proton decay detectors have so far not found any events we shall however see in the next chapter that they are proving to be very powerful detectors of neutrinos from extraterrestrial sources.

14.3   Magnetic monopoles

Another prediction of grand unified theories is that *magnetic monopoles* should exist as stable particles appearing at the breaking of the grand unified group such as SU(5) into the subgroups which include U(1) (t'Hooft, 1974, and Polyakov, 1974). The magnetic charge is given by the Dirac (1931) relation between magnetic and electric charge

$$g_{\text{Dirac}} = \frac{\hbar c}{2e} \cdot n = \frac{ne}{2\alpha}$$

where α is the fine-structure constant ($\alpha = e^2/\hbar c = \frac{1}{137}$) and the integer n in the Dirac theory is unity. Thus

$$g_{\text{D}} = 68.5 e = 3.29 \times 10^{-8} \text{ cgs units.}$$

The monopole mass M_{M} is related to the mass of the X-boson mass M_{X} by

$$M_{\text{M}} \gtrsim \frac{M_{\text{X}}}{g} \sim 10^{16} \text{ GeV} = 0.02 \ \mu\text{g}$$

(g = strong-interaction coupling strength). This mass is so enormous that there is no prospect of creating monopoles at any accelerator and the only possible source of such particles is as a residue from the Big Bang where grand unification held until the temperature dropped below about 10^{15} GeV at around 10^{-35} s after time zero (see chapter 15). Simple cosmological models, in fact, predict a flux of monopoles at the Earth 10^7 times larger than the upper bound from existing measurements while modifications involving an inflationary Universe scenario yield a vanishingly small flux. The cosmological question must be regarded as open, so that experimental searches for monopoles are of high interest both from the GUT and cosmological points of view. We return to this topic in section 15.1.7.

The dimensionless coupling constant is

$$g_{\text{D}}^2/\hbar c = 34.25$$

to be compared with $e^2/\hbar c = \frac{1}{137}$ so that perturbative calculation techniques are inapplicable.

A number of experimental techniques have been used to search for monopoles: track-etch detectors; ionisation detectors (ionisation is $(g_D/e)^2 \sim 4700$ times the ionisation for an electron charge with the same velocity); bulk matter searches; acoustic wave detection; superconducting induction devices and, indirectly, by searching for monopoles acting as a catalyst for proton decay. We discuss further only the last two of these in that there is *one* observation of an event in an induction device and in view of the interesting cross connection with the proton decay. If a monopole passes through a superconducting coil it produces a step change in the magnetic flux through the coil which generates a step in the electric current in the coil. The method, although technically sophisticated in that it is important to provide very effective magnetic shielding of the coil, is nevertheless particularly attractive in that the signal will be independent of the monopole velocity, mass and electric charge. A group at Stanford (Cabrera, 1982) has used this technique and in 1982 recorded a single current jump the size of which corresponds to a magnetic charge g_D passing through their 5 cm-diameter coil. Although the sensitivity of the Stanford experiment and other independent searches is now at least three orders of magnitude greater than that at the time of the original observation no further events have been observed. Many other experiments using a variety of techniques have set limits on the cosmic ray monopole flux as $\leqslant 10^{-14} \, \mathrm{cm}^{-2} \, \mathrm{s}^{-1} \, \mathrm{sr}^{-1}$.

Finally on this topic we mention the possibility that monopoles may catalyse proton decay with a significant cross-section (Rubakov, 1981, 1982, and Callan, 1982, 1983) via processes of the kind

$$M + p \rightarrow M + e^+ + \pi^0.$$

The cross-section σ_c is expected to be proportional to $1/\beta$ as in exothermic capture processes in nuclei at low velocities where β is the relative velocity of monopole and proton. Thus

$$\sigma_c = \sigma_0/\beta$$

and σ_0 is expected to be ~ 0.1 mb while for slow monopoles $\beta \sim 0.1$ due to the Fermi motion of the protons in the nucleus. The mean free path is then

$$\lambda(\mathrm{cm}) = \frac{1}{N\sigma_c} = \frac{\beta}{N\sigma_0} \sim \begin{cases} 2\text{--}3\,\mathrm{m} \text{ for iron} \\ 16\,\mathrm{m} \text{ for water} \end{cases} \quad (N = \text{nucleons cm}^{-3})$$

so that in the larger proton-decay detectors multiple proton decays could be catalysed along the track of a single monopole. Detection of such multiple decays will depend on appropriately short dead time following the first event or a sufficiently long sensitive time to catch all such events in the same time window. For single detected events it may be impossible to distinguish such catalysed decays from neutrino interactions. No multiple events have so far been observed.

14.4   Supersymmetry

SU(5), as already mentioned, is not the only candidate theory for grand unification and at a stage where GUTs are still far from established as a correct description of nature it may seem over ambitious to attempt to formulate even more general theories. There are however good reasons why theoreticians have devoted immense effort to such studies over the past two decades. First we recognise that the Standard Model and even GUTs are incomplete, leaving a large number of apparently arbitrary features unaccounted for such as the number of generations of quarks and leptons and many of their properties which have to be 'put in by hand'. The number of such constants differs in different models but is always ~ 20. A feature of the Standard Model which is, in the broad picture, disturbing is the distinction between the particles (fermions) and the quanta of the forces which act between them (bosons).

There are also more technical problems with ordinary GUTs. For instance higher order corrections involving X and Y exchange give rise to corrections to the W and Z masses $\sim g^2 M_X^2$ ($\sim 10^{14}$ GeV/c^2) which must somehow be got rid of to achieve the real measured masses ~ 90 GeV/c^2.

On the other hand quantum gravity theory is recognised to be an *unrenormalisable* field theory. A theory which offers an escape from this difficulty at the same time as achieving grander unification is a synthesis of general relativity with *supersymmetry* (SUSY) in what has become known as *supergravity*.

14.4.1.   *General features of supersymmetry.*

In supersymmetry every 'normal' fermion (quarks and leptons) has a supersymmetric spin zero partner (squarks and sleptons) while every normal boson (photon, gluons, W, Z, Higgs) has a spin $\frac{1}{2}$ partner. The quantum of the gravitational force, the graviton, which has spin 2, has a supersymmetric partner, the gravitino, which has spin $\frac{3}{2}$. The list of supersymmetric particles with the usual nomenclature is given in table 14.2. Two doublets of Higgs fields are needed to give masses to $+\frac{2}{3}$ and $-\frac{1}{3}$ charge quarks and to cancel certain anomalies. In the minimal version of supersymmetry (see below) these give rise to five physical boson states: two charged scalar bosons heavier than the W; two neutral scalar bosons, one heavier than the Z^0 and one lighter than the Z^0, and one neutral pseudoscalar boson heavier than the lightest scalar boson.

Thus we have

$$Q^i|F\rangle = B \quad \text{and} \quad Q^i|B\rangle = F$$

where Q^i carries no internal quantum numbers so all particles and their supersymmetric partners have identical quantum numbers other than the spin. We see that no presently known particle can be the SUSY partner of another. The index $i = 1 \dots N$ designates the complexity of the particle multiplets in different versions of supersymmetry and turns out to be equal to the number of gravitinos. For instance in 'minimal supergravity' $N = 1$ and there is one gravitino while in 'extended supergravity' theories N can range up to 8 with 8 gravitinos.

In this picture the 'superspin' direction in the appropriate superspace defines the particle as a boson or a fermion. The theory satisfies local gauge invariance and remarkably it turns out that transformation of bosons to fermions also shifts the particles slightly in ordinary space-time so that the Pauli principle which otherwise would be violated, is preserved intact.

The infinites in fermion exchange have the opposite sign from those in boson exchange so in a supersymmetric theory cancellations are possible which render supergravity finite up to two loops of exchanged particles. Unfortunately this is not true for all higher orders so the problem is only partly solved. (Compare for instance the situation in weak interactions where the introduction of the W was the *first stage* in solving the infinity problem). One important gain is the cancellation of the higher order

Table 14.2

Particle	Spin	Supersymmetric partner		Spin
quark	$\frac{1}{2}$	squark	$\tilde{q}$	0
lepton	$\frac{1}{2}$	slepton	$\tilde{l}$	0
photon	1	photino	$\tilde{\gamma}$	$\frac{1}{2}$
gluon	1	gluino	$\tilde{g}$	$\frac{1}{2}$
W	1	wino	$\tilde{W}$	$\frac{1}{2}$
Z	1	zino	$\tilde{Z}$	$\frac{1}{2}$
$H^\pm$	0	Higgsino	$\tilde{H}^\pm$	$\frac{1}{2}$
H^0	0		$\tilde{H}^0$	$\frac{1}{2}$
graviton	2	gravitino		$\frac{3}{2}$

corrections mentioned above, to the masses M_W and M_Z. Cancellation of infinities also goes some way to explaining the existence of the *two* mass scales of 100 GeV and 10^{15} GeV.

It is of course apparent that the supersymmetry is a *broken* symmetry (like I-spin symmetry, electro-weak symmetry...) since it is clear from experiment that since the SUSY particles have not been observed they can not have the same masses as their ordinary partners and are probably heavy. The SUSY couplings always involve pairs of supersymmetric particles so these particles should be produced in associated pairs and each sparticle must have an odd number of sparticles among its decay products. Thus there must be at least one stable sparticle, being the lightest such particle. Arguments mainly based on the failure to observe stable sparticles in ordinary matter suggest that the lightest is probably the photino, though there is no reason why the photino should be massless. The photino is expected to be weakly interacting with ordinary matter so will not be easy to detect. The $\tilde{W}^{\pm}$ and $\tilde{H}^{\pm}$ are expected to mix to give mass eigenstates different from pure winos or Higgsinos usually referred to as charginos and designated χ.

14.4.2. *The search for SUSY particles*

A priori it is possible that the masses of some or all of the sparticles are sufficiently low that they could be produced by existing accelerators and searches for such particles have been vigorously pursued. Bearing in mind the features of sparticles discussed in the previous section we can see how such particles might be expected to be produced and decay.

As an example let us consider the production of sleptons at an e^+e^- collider of sufficient energy. We may expect two effective diagrams for production: (a) via the same process involving an intermediate virtual Z^0 or γ as is effective for normal leptons or quarks and (b) via photino exchange (fig. 14.4).

The sleptons should commonly decay to a photino and an ordinary lepton, e.g. the selectron should decay to an electron and a photino. The photinos being weakly interacting will generally escape without detection carrying off substantial energy and transverse momentum. Such interactions will thus appear as dileptonic events which are acoplanar i.e. where the beam line is not in the plane of the two secondary particles. Background to such events will come from QED processes and in particular from $l^+l^-\gamma$ and $l^+l^-l'^+l'^-$ final states. Such background processes are however well understood and calculable so that it is possible to determine whether these backgrounds account for the data or whether there are events which could arise from slepton production.

There is presently no convincing evidence for the production of SUSY particles in the processes described above or in any of the other processes in which we might expect to find them. Lower mass limits have been obtained for the sparticles which generally depend on the assumption that the photino is the lightest sparticle. Recent data from ALEPH at LEP for example, give lower limits for the masses of sleptons $\tilde{e}$, $\tilde{\mu}$, $\tilde{\tau}$ as 43.5, 42.6, and 40.4 GeV/c^2 respectively while a pure wino of less than 45.5 GeV/c^2 is excluded as is a pure Higgsino of mass less than 44.5 GeV/c^2. These limits are roughly independent of the photino mass (assumed lightest) up to $m_{\tilde{\gamma}} \sim 28$ GeV/c^2.

Supersymmetry must thus for the present be considered as an attractive *hypothesis* awaiting more data and possibly higher energy accelerators for its establishment or its rejection into the limbo of interesting ideas which do not apply to our universe as it is.

14.5 Superstrings

During the past decade many theorists have devoted themselves to the development of an alternative and very ambitious 'theory of everything' known as *superstring* theory. These ideas originated in work with a somewhat different orientation by Veneziano in the late 1960s but owe their development and present popularity primarily to the work of John Schwarz and Michael Green over the past two decades.

The basic idea is that *point* particles are replaced by *strings*, that is

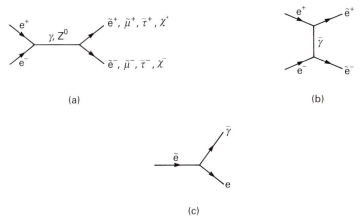

Fig. 14.4. Production (a), (b), and decay (c), processes for supersymmetric particles in e^+e^- collisions.

objects having a linear dimension. Originally both open and closed (loop) strings figured in such theories but latterly the closed loop theories have appeared to be the most promising. Thus instead of particles generating 'world lines' in space-time the loops as they move generate tube-like 'world sheets'. Different types of particle correspond to different oscillations of the loop although the oscillations may be in a space-time of 10 dimensions!

One of the early successes of string theory was the way in which a massless spin 2 particle, identified with the graviton, fell out of the mathematics, encouraging Schwarz and his then collaborator Scherk to see the theory as including gravity (and indeed also the other fundamental forces) in addition to the strong nuclear forces which they had set out to study. In fact the superstring theory has impressive success in eliminating infinities in quantum gravity – possibly to all orders – a unique theoretical achievement.

The natural scale to use in a theory including gravity is the Planck scale where the Planck mass is given by (G_N = gravitational constant = 6.67×10^{-11} m^3 kg^{-1} s^{-2})

$$M_p = \left(\frac{\hbar c}{G_N}\right)^{\frac{1}{2}} = 2.1 \times 10^{-8} \text{ kg} \qquad E_p = M_p c^2 = 1.2 \times 10^{19} \text{ GeV}$$

and the corresponding Compton wavelength

$$l_p = \frac{\hbar}{M_p c} = 1.6 \times 10^{-35} \text{ m}$$

The corresponding time and temperature which we shall refer to in chapter 15 are then

$$t_p = \left(\frac{\hbar G_N}{c^5}\right)^{\frac{1}{2}} = 5.4 \times 10^{-44} \text{ s} \quad \text{and} \quad T_p = \frac{1}{k}\left(\frac{\hbar c^5}{G_N}\right)^{\frac{1}{2}} = 1.4 \times 10^{32} \text{ K}$$

The string length is thus expected to be $\sim 10^{-35}$ m which is clearly beyond the reach of investigation of any conceivable particle accelerator although possibly relevant in the first instants following the Big Bang. For this reason it will not be possible to achieve *direct* tests of superstring theory. This does not preclude indirect tests at attainable energies although in view of the energy scale of 10^{19} GeV it seems likely that only a very complete formulation of the theory will allow predictions within the range of foreseeable accelerators. At present the theory is far from such a developed state and there are no such predictions.

Superstring theory differs from most physical theories in that it is a 'top down' approach where the guiding principles are the universality and the beauty of the mathematical formulation. In most cases the development of a theory to a realistic and successful conclusion has depended on experiment to provide clues when difficulties are encountered, to select among possible development tracks and to test its predictions. Without benefit of experiment it seems likely that superstring theory will take many years to bear fruit – even if it is true.

15

Particle physics and cosmology

15.1 Introduction

There have always been close links between physics and astronomy and physics and cosmology. This is to be expected since astronomy and cosmology are fields of science which apply physics in an attempt to understand observations of the cosmos and to comprehend the structure, origin and evolution of the universe. Equally our understanding of physical laws and in particular their apparent universality in space and time, has in a number of instances depended crucially on astronomical data.

In the first half of the twentieth century a number of common areas of interest – relativity theory, nuclear reactions and stellar evolution, and cosmic rays, among others – provided fruitful points of contact between particle and nuclear physics on the one hand and cosmology on the other. After a period of less interaction common interests have increased dramatically over the last 10–15 years in a way which has been highly stimulating for both subjects.

Most fundamentally the development and general acceptance of the Big Bang model implies energy densities during the first instants after the Big Bang which can now be even distantly approached only in high energy particle collisions. On the other hand it is clear that no experiments we can ever do will attain the energies at which we might expect unification of all four forces including gravity (energies $\sim$ Planck mass $\sim 10^{19}$ GeV/c^2) or even Grand Unification energies $\sim 10^{15}$ GeV, so that the only laboratory for direct study of these phenomena existed in the first instants after the Big Bang. This symbiotic relationship between particle physics and cosmology is further highlighted when we realise that during the first 10^5 years after the Big Bang the universe was opaque to photons so that even the most powerful telescopes probing the limits of the universe can not penetrate beyond this time for an understanding of which we must rely on particle physics theory tested by its observable consequences in the laboratory and in the cosmos.

The Big Bang model itself relates to fundamental questions in particle physics – in particular the breaking of charge conjugation symmetry and baryon number conservation at GUT energies to account for the dominance of matter over anti-matter in the universe; the existence of monopoles; the existence of new particle species and/or massive neutrinos in relation to the closed or open nature of the universe. This latter question of the nature of the 'dark matter' may be open to direct experimental investigation using particle physics techniques, as well as to theoretical speculation.

Perhaps the most fruitful area of common interest currently concerns neutrinos where terrestrial measurements of neutrinos from the sun and from supernova 1987A are important both for the study of neutrino properties and for the study of the neutrino sources.

This book is not a textbook on cosmology and in this chapter we treat specifically only the topics with an obvious relation to particle physics. However for the benefit of readers not familiar with the cosmological background we start with a short account of the Big Bang model, sufficient to place these topics in perspective.

15.2 The Big Bang

As the name implies, in the Big Bang model the universe started from the release of a very great quantity of energy in a singularity in space-time. In fact present theory does not allow us to extrapolate to the singularity since at an energy $\sim 10^{19}$ GeV occurring 10^{-43} seconds after the Big Bang the quantum aspects of gravity must be important and a theory of quantum gravity does not yet exist so that the nature of space-time itself is not known before this instant. The fact that the model has nothing to say about the nature of the singularity (if it exists), or whether there is meaning to the question of what could have preceded it, is aesthetically and psychologically unsatisfactory but, given the initial assumption, it has proved an immensely powerful theory.

Following the initial explosion the universe expanded and cooled. In the early instants the temperature was sufficiently high that thermal equilibrium existed between all kinds of particles including the heavy bosons exchanged in the GUT regime, the Higgs bosons, all the quarks and antiquarks, the other particles with which we are familiar and any other particles which we have not yet discovered. As the temperature dropped various energy thresholds were crossed resulting in the decoupling of components, such as for instance electromagnetic radiation, which thereafter did not interact significantly with the rest of the material of the universe. Symmetry breaking transitions had the character of phase changes. The history of the universe in the Big Bang model is illustrated on a logarithmic scale in fig. 15.1.

The experimental justification of the Big Bang model rests on three principal pieces of evidence:

(*i*) the expansion of the universe based on the recession of the galaxies and Hubble's law,

(*ii*) the discovery of the 2.7 K microwave background,

(*iii*) the cosmic abundance of the very light elements.

We discuss these features in the following sections.

15.2.1   *The expansion of the universe and Hubble's law*

The light from stars and galaxies exhibits characteristic line spectra for many elements. It was observed early in the study of such spectra that although the ratios of the frequencies of the lines from a given source were always the same so that the characteristic patterns could be identified, the absolute frequencies were shifted and in most cases (the

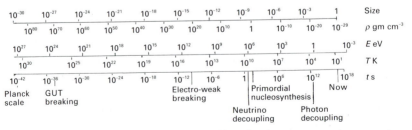

Fig. 15.1. Big Bang chronology. Size, density, energy and temperature following the Big Bang as a function of time. The main transitions in the universe are indicated.

exceptions being mostly certain nearer objects) the shift was toward the red. Although there are processes such as scattering which can redden light it is generally accepted that the shifts in spectral lines from distant sources are due to the Doppler shift associated with rapidly receding sources, which may include an expansion of space during the transmission of the light.

By a series of ingenious steps astronomers have been able to estimate the distances of many of the stellar and galactic sources and in 1929 Hubble published the results of a study of the redshift as a function of the source distance from which he concluded that the recession velocity is proportional to the distance of the source, a relationship now known as Hubble's law:

$$v = Hd$$

where H is Hubble's 'constant', the inverted commas indicating that in fact H is not a constant but changes as the universe evolves. The data available to Hubble were quite limited compared to those available now which extend to much greater distances (up to redshift $z = 4$ where $z = (v - v_s)/v_s$ and v and v_s are the true and shifted frequencies). The early data also suffered from errors in some distance estimations which were later resolved. Although the actual value of the constant has thus changed since Hubble's original work the new data have confirmed the form of the law. The measured value of the constant is known only within a factor of about two. It is

$$H_0 = 0.5 - 1.0/10^{10} \, y$$
$$= 15 - 30 \text{ km s}^{-1}/10^6 \text{ light years}$$
$$= 0.5 - 1.0 \times 100 \text{ km s}^{-1} \text{ Mpc}^{-1}$$

(1 parsec $= 3.09 \times 10^{16}$ m, 1 year $= 3.16 \times 10^7$ s).

The suffix zero on H_0 indicates that this is the value at the present time.

15.2.2   *The Friedmann equation*

A rigorous treatment of the expansion of the universe naturally rests on the appropriate theory of gravitation which is the general relativity theory of Einstein in which gravitational accelerations are related to the curvature of space-time itself. With a few plausible assertions however we can understand the main features of the motion on the basis of elementary non-relativistic dynamics.

Consider a spherical universe in which a shell at radius r is expanding with velocity v. The kinetic energy of the shell is $\frac{1}{2}v^2$ per unit mass. The gravitational potential energy of the shell is $-G_N M/r$ per unit mass where M is the mass enclosed by the shell. If the average density of the matter enclosed is ρ (for the present we neglect the effect of radiation for which the energy density is negligible compared with the matter density in the present era; we return to this point later) then energy conservation yields

$$\tfrac{1}{2}v^2 - \tfrac{4}{3}\pi G_N \rho r^2 = \text{constant}$$

If we use the Hubble law $v = Hr$ we obtain

$$\tfrac{1}{2}H^2 r^2 - \frac{4\pi}{3} G_N \rho r^2 = \text{constant}$$

We can write r in terms of a scale R_0 as $r = R(t) \cdot R_0$ where R_0 is the radius at the present time. We then write the constant on the right-hand side (the 'time independent curvature') as $-\tfrac{1}{2}kc^2 R_0^2$. We shall see that k is the parameter which defines the long term fate of the universe. In these terms

$$H^2 - \frac{8\pi}{3} G_N \rho = -kc^2 \left(\frac{R_0}{r}\right)^2 = -kc^2 R^{-2}$$

This is known after its originator as the Friedmann equation. We can write the equation as

$$H^2 = \left(\frac{\dot{R}}{R}\right)^2 = \frac{8\pi}{3} G_N \rho - kc^2 R^{-2} \qquad \textbf{15.1}$$

If we compare this equation with the corresponding expression from general relativity theory we see that it is the same apart from an additional term in the Einstein expression corresponding to a constant background energy density. In this case

$$\left(\frac{\dot{R}}{R}\right)^2 = \frac{8\pi}{3} G_N \rho - kc^2 R^{-2} + \frac{\Lambda}{3} \qquad \textbf{15.2}$$

where the quantity Λ is known as the 'cosmological constant'. Einstein formulated the theory of general relativity in the era before the discovery of Hubble's law and needed a mechanism to ensure that the universe did not collapse under the mutual gravitational attraction of its parts. Inclusion of the cosmological constant provided such a mechanism although the origin of the constant was not known. The measured data yield a value for the cosmological constant which is very small indeed and consistent with zero. As we shall see below this now raises serious problems and it seems likely that the cosmological constant may not have been small at an earlier epoch in the expansion. We shall return to this question in section 15.2.7.

Neglecting the constant and returning to **15.1** we write

$$\dot{R}^2 = \frac{8\pi}{3}G_N(\rho R^3)\frac{1}{R} - kc^2$$

where, since the total amount of matter in a given expanding volume of space remains unchanged, (ρR^3) is constant so that, differentiating with respect to time we have

$$2\dot{R}\ddot{R} = -\frac{8\pi}{3}G_N(\rho R^3)\frac{\dot{R}}{R^2}$$

or

$$\frac{\ddot{R}}{R} = -\frac{4\pi}{3}G_N\rho$$

We note that of course H is not in general constant on the cosmic time scale.

From the above equations we can readily see the significance of the parameter k which defines the sign of the total energy. If $k = -1$ the gravitational term is less than the expansion term, $\dot{R}$ increases asymptotically towards the speed of light, expansion goes on indefinitely and the universe is said to be 'open'. If $k = +1$ then R reaches a maximum at the first zero of $(\dot{R}/R)^2$, the universe subsequently collapses and is said to be 'closed'. If $k = 0$ then

$$\rho = \rho_c = \frac{3H^2}{8\pi G_N}$$

where ρ_c is the 'critical' density corresponding to a 'flat' universe. Substituting values for the measured value of H_0 ($H_0 = H$ at the present time t_0) we obtain for the present era $\rho_{0c} = 4.5 \times 10^{-30}$ g cm^{-3}. The development of the radius as a function of time is conventionally illustrated in fig. 15.2.

For the flat universe (Einstein–de Sitter universe) with $k = 0$ if we neglect the cosmological constant the Friedmann equation becomes

$$\dot{R}^2 = \left(\frac{8\pi G_N}{3}\right)\rho_{0c}\frac{R_0^3}{R}$$

where we have assumed matter conservation so that

$$\rho_{0c}R_0^3 = \rho_c R^3$$

Integrating we have

$$t = (6\pi G_N\rho_{0c}R_0^3)^{-\frac{1}{2}} \cdot R^{\frac{3}{2}}$$

or

$$R(t) = (6\pi G_{\mathrm{N}} \rho_{0\mathrm{c}} R_0^3)^{\frac{1}{3}} \cdot t^{\frac{2}{3}}$$ **15.3**

Then

$$\rho_{\mathrm{c}}(t) = (6\pi G_{\mathrm{N}} t^2)^{-1}$$

and in terms of H

$$H = \frac{2}{3t}$$

A commonly used parameter is $\Omega (\Omega = \Omega_0$ at the present time) the ratio of the density to the critical density:

$$\Omega = \frac{\rho}{\rho_{\mathrm{c}}} = 1 + \frac{kc^2}{H^2 R^2}$$

The observed luminous matter density (ρ_{L}) is much less than the critical density; $\rho_{\mathrm{L}} \sim 0.01 \rho_{\mathrm{c}}$. We shall see in section 15.3 however that there is strong evidence for the existence of great quantities of 'dark matter' in the universe so that the critical density is not excluded. In addition the nature and age of the universe constrain the value of Ω to be not too different from 1. Had the density been substantially different from ρ_{c} the heavy elements, stars and galaxies could not have formed. For $\rho \ll \rho_{\mathrm{c}}$ the universe would have expanded too rapidly for stars to form while for $\rho \gg \rho_{\mathrm{c}}$ the universe would have recollapsed before star formation could have

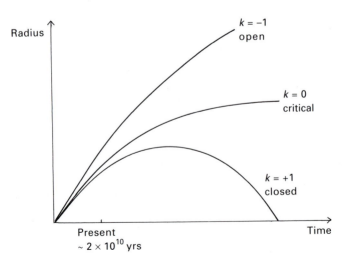

Fig. 15.2. Curves illustrating the development of the size of the universe R, as a function of time for different values of the curvature constant k.

been achieved. As we shall see below the fine tuning required to maintain Ω *near* to unity over the required period turns out to be so extremely delicate that some specific mechanism is needed which probably leads to $\Omega = 1$.

15.2.3 *The early universe – content and coupling*

In the very early universe the density and temperature were sufficiently high that the universe was surely in thermal equilibrium. Matter density ρ_m is proportional to R^{-3} but the expansion of the universe produces a redshift in the wavelength of radiation proportional to R^{-1} so that the radiation density ρ_r is overall proportional to R^{-4} (photon energy $= \hbar c/\lambda$). Thus at early times when R was very small we expect ρ_r to dominate the dynamics. In this era

$$H^2 = \left(\frac{\dot{R}}{R}\right)^2 = \frac{8\pi}{3} G_N \rho_r \propto R^{-4}$$

since the curvature term proportional to R^{-2} can be neglected.

Writing ρ_r at the present time as ρ_{0r} we have

$$(\dot{R})^2 = \frac{8\pi}{3} G_N \rho_{0r} R_0^4 R^{-2}$$

$$\dot{R} = \left(\frac{8\pi}{3} G_N \rho_{0r} R_0^4\right)^{\frac{1}{2}} R^{-1}$$

$$t = \frac{1}{2}\left(\frac{8\pi}{3} G_N \rho_{0r} R_0^4\right)^{-\frac{1}{2}} R^2$$

$$R = \left(\frac{32\pi}{3} G_N \rho_{0r} R_0^4\right)^{\frac{1}{4}} \cdot t^{\frac{1}{2}} \qquad \textbf{15.4}$$

R is thus proportional to $t^{\frac{1}{2}}$ and $H = (2t)^{-1}$ while

$$\rho_r = \frac{3}{32\pi G_N} t^{-2}$$

In terms of temperature T we use the statistical mechanics result that the energy density (for photons) is given by

$$\rho_r = aT^4$$

where $a = 4\sigma/c = 7.5659 \times 10^{-16} \ \text{Jm}^{-3} \ \text{K}^{-4}$ where σ is the Stefan–Boltzmann constant. Thus

$$T = \left(\frac{3c^2}{32\pi G_N a}\right)^{\frac{1}{4}} \cdot t^{-\frac{1}{2}} = \frac{10^{10}}{t^{\frac{1}{2}}} \text{K} \qquad \textbf{15.5}$$

Thus as t and $R \to 0$, $T \to \infty$ and the universe must have started from a hot big bang

The constant of proportionality linking T and $t^{-\frac{1}{2}}$ is modified if particles other than photons are present and depends on the number of particle species and their spins.

At $t = 1$ s, $T \sim 10^{10}$ K and the mean photon energy can be readily calculated to be $\tilde{E}_\gamma = 2.7kT$ (k = Boltzmann's constant) $= 2.6$ MeV so that at this time electron–positron pair production was possible. At about $t = 1.5$ s, $\tilde{E}_\gamma \sim 2m_e$ and e$^+$–e$^-$ pair creation begins to fade out. Prior to this time various other thresholds were crossed. In particular phase changes at the electroweak and possibly at the GUT breaking energies, took place.

15.2.4   *Decoupling and the 2.7 K cosmic background radiation*

We are aware from the experimental and theoretical picture developed in earlier chapters that there are many established particles ranging in mass from W and Z to neutrinos, as well as various hypothetical particles such as the Higgs, top quark and supersymmetric particles. The particles interact through the forces to which they couple and will remain in thermal equilibrium as long as the interaction rate Γ is sufficiently high that Γ^{-1} is much less than the age of the universe

$$\Gamma \gg \frac{\dot{R}}{R} = H = \frac{1}{2t}$$

The average interaction rate is given by

$$\Gamma = \langle n\sigma v \rangle$$

where n is the number density of the particle species in question, σ is the interaction cross section and v is the particle velocity, averages being over the thermal distribution. As long as n and v are high, equilibrium is maintained but as the temperature drops reaction thresholds are crossed. As the distribution crosses the threshold the cross section disappears and the particle in question 'decouples'.

Such decoupling in the case of photons gives rise to the cosmic microwave background the discovery of which by Penzias and Wilson in 1965 is one of the most compelling pieces of evidence supporting the Big Bang picture of the origin of the universe.

As long as the photon energy is sufficiently high and in the presence of a high density of electrons and photons we expect an equilibrium balance in the ionisation and recombination processes

$$\gamma + H \leftrightarrow p + e^-$$

The ionisation potential of hydrogen is 13.6 eV which corresponds to a temperature of 1.6×10^5 K. The equilibrium between ionisation and recombination is described by the 'Saha equation' where, if x is the ionised fraction

$$\frac{x^2}{1-x} = \frac{(2\pi m_e kT)^{\frac{3}{2}}}{n h^3} \exp(-I/kT)$$

where I is the ionisation potential for hydrogen. This relation yields a sharp change in x for $T \sim 4 \times 10^3$ K or $T \sim 3 \times 10^3$ K if ionisation from the first excited level is included. When the energy of most of the photons dropped below that corresponding to 3×10^3 K then the interaction rate dropped sharply, matter became transparent to the radiation and the photons decoupled. Using the (radiation dominated) relation $T \propto t^{-\frac{1}{2}}$, we see that this took place when the age of the universe was $\sim 3.5 \times 10^5$ years. In the 10–20 billion years since that time the universe has expanded by a factor of more than 10^3 and the decoupled photons will have cooled by redshift to ~ 3 K without significant interaction so retaining their black body spectrum.

Only a rather limited spectral range can be measured in experiments on the surface of the earth due to the absorption of the radiation by the atmosphere. Beautiful results have however recently been obtained by Mather *et al.* (1990) by means of the *Cosmic Background Explorer* satellite (COBE) launched into earth orbit at the end of 1989. Spectral measurements made using a Michelson interferometer cover the wavelength range from 0.5 mm to 1 cm. The results of just nine minutes of observation are shown in fig. 15.3. The fit to the expected black body spectrum is good to within 1% and the temperature measured is 2.735 ± 0.06 K.

COBE has also yielded data from differential radiometers which map variations in the sky temperature. These data show a 0.1% 'dipole' asymmetry attributed to a Doppler shift due to the Earth's motion relative to the comoving reference frame in which the cosmic microwave background is apparently isotropic. At a present precision of $\Delta T/T \leqslant 10^{-4}$ no other asymmetry is detected. These observations are at relatively coarse resolution (angular field $\sim 7°$) but cover a large area of the sky. Earth based studies show no anisotropy at a level of a few parts in 10^5 but extend over rather limited areas.

15.2.5   *Primordial abundance of the elements*

The decoupling phenomenon allows predictions of the present abundance of the light elements so that comparison of such predictions with observation provides one of the canonical tests of the Big Bang theory. The abundances are a function of the ratio of baryon to photon density so that such measurements also allow the most stringent determination of this quantity which is of great importance in other aspects of cosmology. The abundance of ^{4}He is also sensitive to the number of light neutrino species and the experimental limits on abundance required $N_\nu = 3 \pm 1$. Now that N_ν has been unambiguously determined to be three from measurements on the Z° (see section 11.6.5) this number can be used in the cosmological argument rather than the other way round. If the abundances could not be reconciled with the Big Bang picture then it might be necessary to invoke exotic effects due to new particles to account for the discrepancy. Since this is not so a discussion of abundances is primarily of relevance for cosmology rather than for particle physics. For this reason the treatment given below is brief and those interested in pursuing this important and fascinating subject are referred to other texts (e.g. Audouze, 1987 or any modern text on cosmology).

The most important element in the study of primordial abundances is

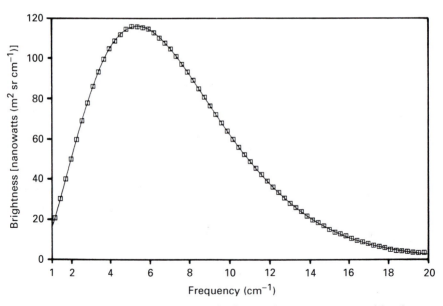

Fig. 15.3. Cosmic microwave background spectrum measured by the far-infrared absolute spectrometer aboard the COBE satellite. The fit is to a black body spectrum for a temperature of 2.735 ± 0.06 K (Mather *et al.*, 1990).

^{4}He. According to the Big Bang model nearly all the ^{4}He in the universe was formed at an early time by the following processes:

$$p+n \leftrightarrow {}^2H + \gamma \quad {}^2H + n \leftrightarrow {}^3H + \gamma \quad {}^3H + p \leftrightarrow {}^4He + \gamma$$
$$^2H + p \leftrightarrow {}^3He + \gamma \quad {}^3He + n \leftrightarrow {}^4He + \gamma$$

The binding energy of deuterium is only 2.2 MeV so the deuterium is photodisintegrated until the temperature falls below about 10^9 K (about 4' after the Big Bang) at which point *all the neutrons are bound* and the ratio of neutrons to protons is frozen.

In terms of temperature the N_n/N_p ratio is given simply by the ratio of the Boltzmann factors:

$$N_n/N_p = \exp(-m_n c^2/kT)/\exp(-m_p c^2/kT)$$
$$= \exp(-\Delta mc^2/kT)$$

where $\Delta m = m_n - m_p = 1.3$ MeV/c^2. The effective temperature T at which deuterium ceases to be disintegrated as fast as it is produced clearly depends on the baryon/photon ratio η. For $\eta = 10^{-9}$ nucleosynthesis wins over disintegration at $T \sim 9 \times 10^9$ K and $N_n/N_p = \frac{1}{6}$ reducing to $\sim \frac{1}{7}$ by the time the ^{4}He production chain is effective due to the reduction of N_n by neutron decay (mean life 15'). We can then easily calculate the ^{4}He abundance when all the neutrons are eaten up:

$$\frac{\text{No. of } {}^4\text{He}}{\text{No. of p}} = \frac{1}{12} \qquad \frac{\text{Mass of } {}^4\text{He}}{\text{Mass of p}} = \frac{1}{3}$$

$$\frac{\text{Mass of } {}^4\text{He}}{\text{Total Mass}} = \frac{1}{4}$$

Neglect d, t, He^3 and any other nuclear species

More elaborate models also result in an abundance very close to 25%.

The ^{4}He abundance has been measured in sites including the Sun, Jupiter, the solar wind, stellar atmospheres, planetary nebulae, globular clusters and blue compact galaxies. The results are of different precisions but are all consistent. It is difficult to know how these various measurements should be combined to yield a representative value but $24 \pm 1\%$ is generally accepted and the error of $\pm 1\%$ may even overestimate the uncertainty.

The gaps in the series of stable nuclei at $N = 5$ and 8 inhibit higher mass nuclear formation but small amounts of primordial ^{7}Li and ^{8}Be were produced via the reactions

$$^3He + {}^4He \rightarrow {}^7Be + \gamma \quad \text{and} \quad {}^4He + {}^3H \rightarrow {}^7Li + \gamma$$

The expected amounts of these nuclei and of the residual ^{2}H, ^{3}H and ^{3}He are more model dependent than that for ^{4}He. The abundances are given as a function of the baryon to photon ratio η in fig. 15.4 which is based on a computation taking into account the effects of galactic evolution

(Delbourgo-Salvador *et al.* 1985). Comparison with observation shows that consistency is possible for $(1.2 \leqslant \eta \leqslant 4.5) \times 10^{-10}$ corresponding to a density ratio for baryons $0.004 \leqslant \Omega_B \leqslant 0.06$.

15.2.6   *The matter–antimatter asymmetry*

We have seen above that the ratio of the number of baryons to photons at decoupling was $\sim 10^{-10}$. The density of photons as measured

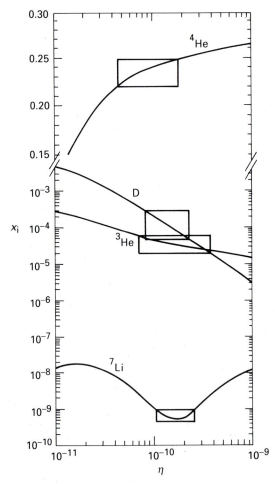

Fig. 15.4. Abundance by weight of ^{4}He, D, ^{3}He and ^{7}Li as a function of the baryon to photon ratio η calculated on the basis of the Big Bang model taking into account the effects of stellar evolution (Delbourgo-Salvador *et al.*, 1985). The curves are for $N_\nu = 3$. The boxes give the measured limits and the corresponding variations for η.

from the $2.7\,\mathrm{K}$ background is $\sim 400\,\mathrm{cm}^{-3}$ yielding a baryon density $\sim 10^{-8}\,\mathrm{cm}^{-3}$. The observed baryon density today is $\rho_{0\mathrm{B}} \sim 10^{-31}\,\mathrm{gm\,cm}^{-3}$ which also yields a number density $\sim 10^{-8}\,\mathrm{cm}^{-3}$.

On the other hand we have practically no evidence for the existence of *antimatter* in the universe. Cataclysmic events resulting from collisions of matter and antimatter galaxies are not observed, the high energy gamma rays which would result if matter–antimatter annihilation occurred to a significant extent are not seen and the proportion of antimatter in the cosmic radiation is very small ($\sim 10^{-4}$).

If baryon number is conserved this implies that

$$\frac{n_{\mathrm{B}} - n_{\bar{\mathrm{B}}}}{n_{\gamma}} \geqslant 10^{-10}$$

On the other hand if we start from a Big Bang which is symmetric with respect to matter and antimatter and require baryon number conservation then we expect (Steigmann, 1976) at present

$$\frac{n_{\mathrm{B}} - n_{\bar{\mathrm{B}}}}{n_{\gamma}} \leqslant 10^{-19}$$

Attempts to explain the asymmetry by mechanisms which *separate* the matter and antimatter in space have been unsuccessful. No mechanism is known which would *generate* one extra baryon for every 10^{10} baryon–antibaryon pairs produced. The way out of the dilemma apparently lies in relaxing the assumption of baryon conservation. Such an explanation, as first pointed out by Sakharov (1967), has three ingredients:

(*i*) baryon number violating forces

(*ii*) asymmetry under the C and CP operations to avoid the asymmetry created by the baryon number violating forces being cancelled by the anti process

(*iii*) freeze-out from thermal equilibrium.

We have seen in section 14.1 that Grand Unification Theories imply baryon number non-conservation. C and CP violation is familiar to us in the weak interaction involved in $K^0 - \bar{K}^0$ mixing and is also a natural feature in GUTs. In the discussion of GUTs we also saw that due to the variation of couplings with energy we should expect interaction rates for baryon violating processes to vary with energy, with GUT symmetry breaking at $\sim 10^{15}\,\mathrm{GeV}$ providing a natural mechanism for decoupling from thermal equilibrium. Thus, at least qualitatively, Grand Unification Theories offer a possible solution to the matter–antimatter asymmetry problem.

The quantitative development of these ideas is not simple. For instance the low energy CP violation $\sim 10^{-3}$ observed in K^0–$\bar{K}^0$ will not in general be the same in the GUT regime. GUTs currently do not predict the asymmetry at all precisely but they can certainly accommodate the observed baryon dominance.

It is a startling fact that apparently in the very early universe nearly all matter annihilated with antimatter into photons while only the existence of a *tiny asymmetry* in CP allowed the survival of the matter from which all the structure in our universe is made.

15.2.7   *Inflation*

The Big Bang model is highly successful in accounting for some of the most striking characteristics of the universe – the expansion, the microwave background and the light element abundance. It does however suffer from some serious general difficulties the solution of which demands an important addition or modification to the basic model.

The various problems are conventionally known as:

(a) *The horizon problem.* We have seen that the universe is remarkably isotropic and uniform on the large scale. In particular the microwave background is uniform within $\Delta T/T \leqslant 10^{-4}$ measured over large and small areas of the sky (section 15.2.4). Yet the radiation reaching us from opposite poles can have had no mutual interaction since the decoupling time at $t \sim 0.5 \times 10^{-13}$ s and in general there can have been no opportunity to establish or maintain equilibrium between causally disconnected regions. We may note that our horizon, i.e. the distance out to which we can see, increases with time since the universe is expanding at a speed less than the speed of light.

(b) *The flatness problem.* We know from observation that the universe is at least fairly 'flat'

$$\left| \frac{\rho - \rho_{\mathrm{c}}}{\rho} \right| \lesssim \text{almost } 1$$

Neglecting Λ we have (from (15.2))

$$\frac{\rho - \rho_{\mathrm{c}}}{\rho} = \frac{3kc^2}{8\pi G_{\mathrm{N}} \rho R^2}$$

We can then extrapolate to find $(\rho - \rho_{\mathrm{c}})/\rho$ at earlier times since we know that $R \propto t^{\frac{2}{3}}$ or $t^{\frac{1}{2}}$ and $\rho \propto R^{-3}$ or R^{-4} in the matter and radiation phases respectively (equations 15.3 and 15.4). Thus if $(\rho - \rho_c)/\rho$ is presently of the order of unity it must have been $\sim 10^{-60}$ at the Planck time $t = t_p = (\hbar G_N/c^5)^{\frac{1}{2}} = 5.4 \times 10^{-44}$ s. Without inflation a deviation from ρ_c of this order would have been magnified to a level incompatible with present observation. Thus very fine tuning to one part in 10^{60} is needed to keep $\rho \sim \rho_c$ or $\Omega \sim 1$ over such a long time. This problem is sometimes expressed as the problem of how the universe has come to last so long. This feature is highlighted if we bear in mind that the 'natural' time scale for the expansion is the Planck time $\sim 10^{-43}$ s. Rather than demanding such fine tuning it seems more likely that there should be an underlying reason for k to be identically zero and $\Omega = 1$.

(c) *The cosmological constant problem.* We have seen (section 11.4.4) that renormalisable gauge theories need spontaneous symmetry breakdown which in turn requires a field such as the Higgs field implying a vacuum energy density. When gravity is included in the overall theory such a vacuum energy density generates an effective cosmological constant. In fact if ρ_E is the energy density then $\Lambda = 8\pi G_N \rho_E$. A proper analysis shows that if spontaneous symmetry breakdown occurs at a mass scale M then the corresponding vacuum energy density $\rho_E \propto M^4$. For the GUT phase transition $M_G \sim 10^{15}$ GeV/c^2. However we know that $\Lambda_0 \lesssim 10^{-50}$ GeV^4 so that

$$\frac{\Lambda_0}{M_G^4} \leqslant 10^{-110}$$

One could imagine that in addition to an energy density generated by a Higgs type field there was a pre-existing density which precisely cancelled the Higgs density but the fine tuning required is again implausible unless there is some underlying mechanism which ensures absolute cancellation.

(d) *The problem of monopoles* (or other topological singularities). Grand Unified theories require the existence of magnetic monopoles. Monopoles are produced at the symmetry breaking transition and in the standard cosmology the monopole density in most GUTs is completely incompatible with the known subsequent evolution of the universe and the observed lack of monopoles in experiments (see section 14.3).

Magnetic monopoles arise at symmetry breaking in the form of topological discontinuities. For instance the direction of the Higgs field is arbitrary and will in general be different in causally unconnected domains. Between such domains there will be regions where the field $\langle \phi \rangle = 0$

instead of the value $\langle \phi \rangle = \sigma$ which minimises the potential energy. The situation is analogous to magnetisation of a material where the local domains are not aligned. These regions of high potential energy are monopoles with mass $M_M \sim M_x/\alpha$ where M_x is the symmetry breaking mass (mass of the X boson) and α is the unified coupling constant so that $M_M \sim 10^{16}\ \mathrm{GeV}/c^2 = 0.02\ \mu\mathrm{g}$.

There should be about one monopole per horizon volume at the time when symmetry breaking occurred i.e. when $kT = M_x c^2$ (k = Boltzmann's constant). Substituting this value of T in **15.5** we obtain for t

$$t \sim (10^{20}k^2/M_x^2 c^4)$$

The number density of monopoles is then

$$n_M = \frac{1}{(ct)^3} \sim \left(\frac{M_x^2 c^4}{10^{20}k^2 c}\right)^3 = \frac{c^9 M_x^6}{10^{60}k^6}$$

to be compared with a photon density $n_\gamma \sim (M_x c/\hbar)^3$ so that

$$\frac{n_M}{n_\gamma} \gtrsim 10^{-11} \sim 10^{-2}\frac{n_B}{n_\gamma}$$

since annihilation will have produced negligible reduction. The monopole mass density is then

$$\rho_M = M_M n_M = \frac{M_x}{\alpha} 10^{-2} n_B \simeq 10^{13} \rho_c$$

a number totally inconsistent with the age of the universe.

Other topological discontinuities such as strings and domain walls may also give rise to similar problems.

(e) *The small scale inhomogeneity problem.* Our very existence and most of our knowledge of the universe depends on the fact that on the 'small' scale the universe is not uniform but consists of stars, galaxies, gas clouds, galaxy clusters and other interesting objects. Such structures require density perturbations of amplitude $\sim 10^{-3}$ on the scale of galaxies at decoupling which would thereafter grow under gravitational forces. The standard Big Bang cosmology does not account for such primordial density perturbations. Despite intense theoretical effort this is still an outstanding problem for cosmology. This question is related to the problem of the nature of the dark matter and we return to this aspect in section 15.3.

The nature of the problems outlined above suggests that there must have been a period in the history of the universe when the rather smooth development we have discussed so far was interrupted by a period of much more rapid and dramatic change. This idea is the basis of the '*inflation*' hypothesis first proposed by Guth (1981) and Kazanas (1980) and subsequently developed by a number of authors. 'Inflation' requires a period in the early history of the universe when the expansion was *exponential*. If at some time the cosmological constant was large and dominant then the Friedmann equation **15.2** becomes

$$\frac{\ddot{R}}{R} = \left(\frac{\dot{R}}{R}\right)^2 = \frac{\Lambda}{3} = H_\Lambda^2$$

and

$$R(t) = R_1 \exp\left(H_\Lambda(t - t_1)\right)$$

where $R(t) = R_1$ at time $t = t_1$. If the duration of the exponential expansion was Δt then during the inflation the universe expanded by a factor

$$z = \exp\left(H_\Lambda \Delta t\right)$$

and the curvature reduced by a factor z^{-2}.

The flatness problem arises from the smallness of the ratio of the curvature term to the energy density term. During the inflation the energy density remains approximately constant but the curvature term which is proportional to R has decreased by a factor z^{-2} so that the problem is solved if $z \gtrsim 10^{30}$. After the inflation Λ must decrease to its present value of zero or near zero and so Ω becomes equal to unity within some small limits.

The horizon problem is also solved. Our observable universe at the present time has a size $\sim 10^{26}$ cm. At the start of the inflation at a temperature corresponding to about 10^{14} GeV ($t = 10^{-34}$ s) the size was only about 10 cm within which volume thermal equilibrium reigned. Inflation models are currently grappling with the small scale inhomogeneity problem. Quantum fluctuations may possibly seed the inhomogeneities though the mechanisms required to account for the observed situation are not established.

The monopole problem is also solved since within the small volume region at the start of inflation the Higgs field would have been aligned in the lowest energy configuration so that there would have been $\leqslant$ one monopole within the universe.

As we shall see below inflation does not offer a solution to the cosmological constant problem.

A variety of models exists which generate inflation by means of evolution of a weakly coupled scalar field which was initially displaced from the minimum of its potential. A mechanism of this kind was proposed by the originators of the inflation idea who suggested that at the GUT phase transition a suitable choice of potential leads to the universe being trapped in the symmetric phase for a period before passing to the broken symmetry $SU(3) \times SU(2) \times U(1)$ phase. The universe is thus 'supercooled'. The situation is illustrated in fig. 15.5 where the free energy density is plotted as a function of the field ϕ for different temperatures T. Before inflation the vacuum energy density has a minimum at $\phi = 0$ with a value corresponding to $V(\phi = 0) - V(\phi = \phi_{\min})$ which is proportional to M_G^4, with a correspondingly large cosmological term leading to inflation. If the phase change occurs at $M_G \sim 10^{14}$ GeV then $V \sim 10^{56}$ GeV^4 and a sufficient expansion will take place if the inflation lasted for a time of $\gtrsim 10^{-31}$ s. Inflation gives the universe a 'fresh start' at $t = 10^{-31}$ s and the nature of the original configuration becomes largely irrelevant to its subsequent development.

At $T = T_c$ the potential has developed a second minimum degenerate with the minimum at $\phi = 0$. As T falls to $\ll T_c$ the asymmetric minimum at $\phi = \phi_{\min}$ becomes the true minimum and the symmetric minimum at $\phi = 0$ is now metastable. Inflation occurs when the 'false vacuum' or supercooled state decays by penetrating the potential barrier by means of quantum tunnelling or as a result of thermal fluctuations.

After the phase transition Λ apparently reverts to zero, the latent heat released reheats the universe to $T \leqslant T_c$ and the hot Big Bang model development continues as before (fig. 15.6). The effective source of the matter and radiation in the universe is thus the expansion of the false vacuum. Gravitational energy is transferred into the energy of the false vacuum which is in turn converted into matter and radiation following the phase transition.

There are several classical analogies for this picture. For instance we can envisage bubbles of the broken symmetry $SU(3) \times SU(2) \times U(1)$ starting within the symmetric phase. The bubbles grow and fill up the symmetric phase, coalescing and yielding up their surface energy to thermal energy which then reheats the universe back to nearly T_c at which the phase change began.

Potentials of the form of fig. 15.5 however prove not to yield the required behaviour in that the bubble expansion rate turns out to be too small compared with the exponential rate of the environment, the bubbles of the broken phase do not coalesce and the transition is never completed. This problem is avoided in more recent models by Linde (1982), Albrecht and Steinhardt (1982) and others. In these models the scalar potential has a small barrier near the origin followed by a long rather flat region (fig. 15.7). This results in the field taking a long time to reach the vacuum expectation value. Inside a single bubble the vacuum energy density remains approximately constant, the bubble expands exponentially along with the rest of the universe and a single bubble fills up the whole of our visible universe thus avoiding the problem of topological singularities including monopoles, since there would be just one monopole in the visible universe.

At present the particle physics origin of a potential of the required form

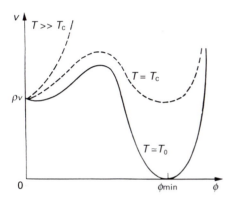

Fig. 15.5. Illustrative potentials for an inflationary model as a function of the field, for different temperatures.

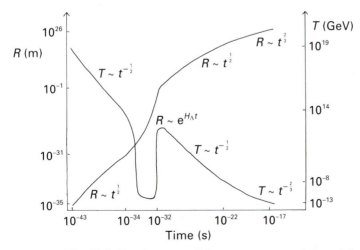

Fig. 15.6. Development of the temperature and size of the universe as a function of time in the inflationary scenario.

is not clear. The Higgs potential is not suitable nor is that of the SU(5) GUT. Other theories beyond the standard model, and in particular supersymmetric theories, may prove more successful but none is presently wholly satisfactory.

An outstanding question is why the vacuum energy minimum and thus the cosmological constant should be zero or at least extraordinarily small. It is an intriguing fact that since Λ and the curvature terms are both zero or very small the net energy of the universe is apparently close to zero. We also believe that the net amount of other strictly conserved quantities such as charge is zero. This suggests that the universe might have been created from nothing by a vacuum quantum fluctuation with the matter of the universe as we know it being the result of tiny violations of perfect symmetries in C, P, T, B and L.

Inflation must surely be considered at present as no more than a very attractive hypothesis which offers solutions to most of the outstanding problems of Big Bang cosmology at the cost of a potential for which there is as yet no very strong theoretical foundation in particle physics coupled with an unexplained value for the cosmological constant. However inflation models do have some predictions concerning which experimental information will surely improve over the next few years. Most important is the requirement that $\Omega = 1$. We shall return to this question in the following section of this chapter. Detailed studies also make predictions concerning the density perturbations and suggest anisotropies $\sim$ a few parts in 10^6, a level which should be within the experimental range of the COBE satellite instruments (section 15.2.4). In addition the predictions of detailed models concerning structure in the universe are sensitive to the nature of the dark matter, searches for which are discussed in the

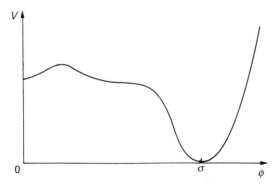

Fig. 15.7. Effective potential for a field ϕ of the type corresponding to inflation models with a 'slow roll-over'.

following section. We shall have to wait and see if the great theoretical structure of inflation survives the ultimate test of experiment.

15.3   Dark Matter

One of the most intriguing and important questions for cosmology at the present time concerns the existence of 'dark matter'.

Direct observations of extraterrestrial phenomena depend very largely on electromagnetic radiation emitted by the material concerned. The radiation extends from γ-rays through the complete spectrum to radio wavelengths. We refer to such material as luminous matter. If matter is not itself luminous we may sometimes infer its existence from its absorption of light from other luminous bodies or from its gravitational effect on luminous matter. This latter effect provides evidence that there exists in the universe an amount of dark or non-luminous matter, which probably exceeds the amount of luminous matter by a factor of at least 10 and maybe by as much as 100.

While the luminous matter gives a density ~ 0.01 of the critical density for closure ρ_c, the additional dark matter clearly associated with galaxies raises this fraction to $(0.2 \pm 0.1) \times \rho_c$. This is consistent with the baryon density inferred from the data on light element abundance and the Big Bang model for primordial nucleosynthesis (section 15.2.5) which resulted in a baryon/photon number ratio $\eta \lesssim 10^{-9}$ equivalent to a baryon number density $\lesssim 10^{-7}$ cm^{-3}. Some estimates of total dark matter in galaxies and clusters would even allow upper limits consistent with a density ρ_c as required by the inflation hypothesis. However it seems more likely that if $\Omega = 1.0$ this will require an extended or smooth distribution of non-baryonic dark matter which will not give rise to local dynamical effects, in addition to the dark matter clumped around galaxies. The nature of this matter is of great interest for both particle physics and cosmology. In this section we review the direct evidence for dark matter, speculate as to its nature and describe some proposals as to how it might be detected in terrestrial experiments.

15.3.1   *Evidence for dark matter*

The most convincing evidence of the existence of dark matter comes from the observation of the velocity of the luminous matter as a function of distance from the centre (the 'rotation curve') for spiral galaxies. Spiral galaxies are chosen which happen to have their planes tilted at an angle which allows measurement of the circumferential velocity along the line of sight from Doppler shifts.

The luminosity falls off exponentially with radial distance from the centre of the galaxy. Beyond the central region the measured velocity is found to be nearly constant out to radii of 30 kpc or more (fig. 15.8). Newtonian gravity gives

$$\frac{mv^2}{r} = \frac{GmM(r)}{r^2}$$

where $M(r)$ is the mass contained within the radius r. If the mass density $\rho(r)$ fell off exponentially like the luminosity so that practically all the mass was contained within the central region then we would expect

$$v \propto r^{-\frac{1}{2}}$$

For constant v on the other hand M must be proportional to r since

$$M(r) = \left(\frac{v^2}{G}\right) r$$

and the mass density $\rho(r) \propto r^{-2}$.

The natural interpretation of the observed variation of v as a function of r is that there is a halo of dark matter around the galaxy and that this

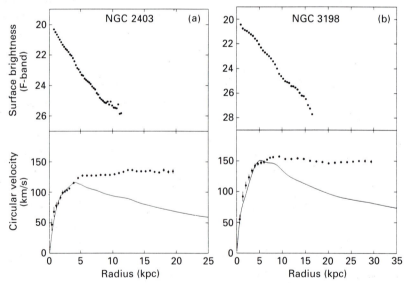

Fig. 15.8. Galactic luminosities and velocities as a function of distance from the centre (light profiles and rotation curves for two galaxies (a) NGC2403 and (b) NGC3198) (*see* Sancisi and Van Albada, 1987). The solid curve on the lower diagrams is the rotation curve calculated from the light profile.

halo extends to a large radius with the ratio of luminous to dark matter decreasing with increasing r.

Evidence for the existence of dark matter also comes from the relative motion of the components in apparently bound galaxy clusters. These studies require masses greater by a factor of ~ 10 than the masses inferred from the internal dynamics of the luminous parts of the galaxies.

A further indicator of the existence of dark matter on the large scale is the ratio of mass to luminosity. The luminosity is generally measured in a standard spectral band (blue band-pass luminosity, L) and the mass of the system (galaxy, binary galaxy, galaxy cluster or supercluster) may be estimated by application of the virial theorem relating average kinetic and potential energies ($\bar{T} = -\frac{1}{2}\bar{V}$) using velocities from measured Doppler shifts. If we normalise to the Sun, $M_\odot / L_\odot = 1$, then for our galaxy we have $L_G \simeq 10^{11} L_\odot$ and $M_G \simeq 10^{12} M_\odot$ so that $M_G / L_G \simeq 10$. Some typical values for larger scales are given in table 15.1. Also given in the table are corresponding values of the ratio of measured density to critical density which appear to show that estimates of Ω increase with increasing scale.

15.3.2 *Nature of the dark matter*

There seems little doubt that the dark matter exists and indeed dominates the mass of the universe. In this section we discuss candidates for its constituents.

One naturally first asks whether the matter could be accounted for by dust clouds or heavy star remnants such as white or brown dwarfs, neutron stars or black holes. Astronomers reject this explanation on the ground that the existence of such matter implies an abundance of high Z matter inconsistent with the observed composition of the visible stars.

Another hypothesis in terms of known phenomena explains the dark matter as consisting of very large numbers of low mass, low luminosity stars or objects with mass substantially less than $M_\odot$ such that they do not form stars ('Jupiters'). Such an explanation however implies on any reasonable model, the existence of a distribution including large numbers of similar objects at somewhat larger, observable, masses and luminosities. Such objects are not found.

Primordial black holes have also been proposed as a solution to the problem. Small black holes would have evaporated by the present era and there is no evidence for the necessary number of large black holes to explain the observations.

Although none of the above explanations is a plausible solution to the dark matter problem on its own each of them may make some contribution. Nevertheless the phenomenon seems to demand a more radical and unconventional explanation. We shall consider three types of candidate for non-baryonic dark matter: massive neutrinos, axions, and SUSY particles or other Weakly Interacting Massive Particles (WIMPs). This list is not exhaustive but includes the most promising candidates. These particles also illustrate a variety of problems in the actual experimental detection of dark matter.

Massive neutrinos

Neutrinos with mass seemed at one time an attractive candidate for the non-baryonic dark matter on the grounds that (unlike the other candidates) we at least know that neutrinos exist and the neutrino mass is not required to be zero by any over-riding theoretical principle. Indeed quite recently measurements indicated that the electron neutrino mass was $\sim 30\ eV/c^2$ although more recent measurements have not confirmed this result (section 15.4.6).

Although we shall see below that there are reasons to believe that neutrinos do not constitute the dark matter nevertheless it is instructive to consider briefly the arguments concerning neutrinos especially since some of the same arguments apply to other candidate particles.

If we apply the same treatment concerning decoupling which we used for photons in section 15.2.4 then the interaction rate Γ is given by

$$\Gamma = \langle n\sigma v \rangle \gg H = (2t)^{-1}$$

Table 15.1. *Order of magnitude values for mass/luminosity ratios for systems of various size*

System	Mass (Units of $M_\odot$)	Scale (kpc)	M/L (units of M/L for Sun)	Corresponding Ω
Galaxy – visible	10^9–10^{11}	10	1–2	0.002–0.006
Dynamical:				
Galaxy	10^{10}–10^{12}	25	10	0.01–0.03
Binary galaxies	10^{10}–10^{13}	10–100	20–80	0.05
Galaxy cluster	10^{14}	~ 500	50–200	0.2
Galaxy supercluster	10^{15}	10^4–10^5	50–300	0.3

with $v \sim c$. σ is the weak interaction cross section which we know (section 11.2) and which is proportional to $E_v^2 = (kT)^2$. n is the neutrino density which we can express in terms of the temperature at decoupling from statistical mechanics (radiation era) and which is proportional to $(kT)^3$ so that $\Gamma \propto T^5$. The result for three neutrino families is a decoupling temperature $T_v \sim 3$ MeV at time $t \sim 1$ s. This leads to a present neutrino temperature ~ 1.9 K. A spin factor and a factor arising from additional production of photons by $e^+ e^-$ annihilation lead to a difference between the number densities of neutrinos and photons so that

$$n_v = (3/11) n_\gamma \sim 100 \text{ cm}^{-3}.$$

If the three species of neutrino have masses m_i ($i = 1, 2, 3$) then for massive neutrinos we have a mass density

$$\rho_v = n_v \sum_i m_{v_i} = (3/11) n_\gamma \sum_i m_{v_i} = 2.10^{-31} \sum_i m_i (\text{eV}) \text{ gm cm}^{-3}$$

and neutrinos can close the universe if $\rho_v = \rho_c$ and $\Sigma m_{v_i} = 25\text{--}100 \text{ eV}/c^2$ ($H_0 = 50\text{--}100 \text{ Km s}^{-1} \text{ Mpc}^{-1}$).

Limits on m_{v_e} (section 15.4.6) now appear to exclude the electron neutrino from filling this role but the mass limits for muon and tau neutrinos are such that these particles could account for the dark matter.

There is however another argument which appears to disfavour this solution to the dark matter problem. This argument is related to the complicated problem of galaxy formation – a process at present not fully understood. The neutrinos were certainly relativistic at decoupling and are 'hot dark matter' (HDM) candidates as opposed to the 'cold dark matter' (CDM) slow particles discussed below. Galaxy formation in an otherwise uniform universe must depend on 'seeding' by density fluctuations. A sufficiently large mass ($\Delta\rho/\rho \sim 10^{-3}$) will then grow by gravitational attraction. HDM such as neutrinos, 'streaming' through the universe with only gravitational interactions after decoupling will however wash out any incipient fluctuations.

We can estimate the size of structure which can result from seeding by HDM particles by estimating the distance travelled by such particles before they slow down to non-relativistic velocities when $mc^2 \sim kT$. Using 15.5 to relate temperature T to time t we can then determine the distance travelled, $d \sim ct$. The total mass in this volume is then $M = d^3 mn$ where n is the number density at temperature T. For $m_v \sim 25 \text{ eV}/c^2$, $t \sim 10^4$ y and $M \sim 2.5 \times 10^{15} M_\odot$ which is the mass of typical galaxy superclusters. In order to account for the observed situation we then need a mechanism to break down such large structures into smaller components. This 'top-down' scenario has proved difficult to reproduce in current models.

The same arguments can be applied to other particles which decoupled earlier than the neutrinos. Such particles would have lower number density today and could be more massive ($\sim 1 \text{ keV}/c^2$) thus becoming non-relativistic more quickly than massive neutrinos. They would then form smaller structures ($\sim 10^{11} M_\odot$) which are still however of galactic size.

Axions or other light bosons

Axions are hypothetical particles which result from a fairly arcane argument concerning the suppression of *CP* violation in strong interactions. *CP* violation appears naturally in weak interactions when there are at least three generations (section 12.8). Non-perturbative effects can then couple *CP* violation into QCD but such *CP* violation is not observed. In particular we would expect a neutron electric dipole moment much larger than the experimental upper limit. A complex pseudopotential arising from breaking U(1) global symmetry (Peccei and Quinn 1977) can be introduced to cancel the QCD *CP* violating effects but at the cost of requiring an almost massless, neutral pseudoscalar boson, the axion. Other hypothetical particles with similar properties are 'familons' and 'Majorons' associated respectively with spontaneously broken family and lepton number symmetries.

The best limits on the axion mass are set by cosmological constraints. If the axion mass density is not to exceed the critical density then $m_A > 10^{-5} \text{ eV}/c^2$ otherwise axions would be too copiously produced in the Big Bang. At the other limit, since the axions could readily carry energy away from the centres of stars, the observed longevity of helium burning stars sets a limit $m_A < 0.1 \text{ eV}/c^2$.

Axions couple only to net spin with dimensionless coupling constant to fermions, g_{Aff}, corresponding to the above masses in the range 10^{-16}–10^{-12} which is at best some 11 orders of magnitude weaker than the weak interaction coupling. The axions are expected to be very 'cold' with kinetic energy $\sim 10^{-6}$ of rest energy and $\beta \sim 10^{-3}$.

SUSY particles or other weakly interacting massive particles – WIMPs.

We have already discussed the supersymmetric hypothesis in section 14.4. We may expect that at least the lightest SUSY particle should be stable. This could be the gravitino, the *s*-neutrino or the photino. The absence of exotic isotopes demonstrates that the stable lightest SUSY particle cannot interact with ordinary matter through electromagnetic or strong interactions. The most probable candidate for the lightest SUSY particle is the photino, $\tilde{\gamma}$.

The SUSY particle masses are not predicted by theory and only informed guestimates are available. For the photino the mass range is ~ 5–100 GeV/c^2. The average velocity will be close to the typical velocity of visible matter in the solar neighbourhood, $\beta \sim 10^{-3}$. The galactic escape velocity, $\sim 2 \times 10^{-3}c$, sets an upper bound while the extent of dark matter halos constrains the lower bound. Such WIMPS are certainly 'cold' dark matter.

We would expect a residue of $\tilde{\gamma}$ from the Big Bang with number density suppressed relative to the photon density by a Boltzmann factor dependent on the $\tilde{\gamma}$ mass. The residual number then depends also on the annihilation cross section.

15.3.3 *Detection of axions and WIMPS*

Although a number of ingenious ideas have been studied there seems at present to be no practicable way to detect the very low energy (kinetic energy $\sim 10^{-5}$ eV) massive neutrinos which might possibly be candidate particles for dark matter. We have however seen that massive neutrino HDM candidates are disfavoured for other cosmological reasons.

Axions

A number of direct and indirect methods have been suggested for the detection of dark matter. Indirect methods involve a primary interaction, for instance in the Sun, to produce secondary particles which can then be detected on the Earth. We shall not discuss these here but rather the possibility of direct detection in terrestrial experiments.

It has been proposed by Sikivie (1983) that axions could be best detected by means of their interaction with a strong magnetic field. The axion interacts with a photon in the field via a fermion vacuum loop (fig. 15.9). The rate for such a process is given by

$$R = 10k_1^2(B/8T)^2(V/10^{-3}\text{ m}^3) \text{ events per day}$$

where B is the magnetic field in Tesla, V the detector volume in m^3 and k_1 is a constant equal to 0.25 or 1.0 respectively according to whether or not the axion also couples to the two photons via the π^0. This result gives usable event rates for possible values of B and V. The experiment requires to be carried out at low temperatures (10 mK) in order to reduce the normal black body photon background in the same energy range ($100\,\text{mK} = 10^{-5}\,\text{eV}$) to a negligible level.

Unfortunately no techniques currently exist to detect single photons of energy corresponding to the expected axion mass. There is still however the possibility that instead of a broad band experiment the measurement could be made in a high Q resonant cavity placed in a high magnetic field. For photons with energy corresponding to the resonant frequency of the cavity the photons would accumulate to give a detectable signal. Axion mass in the range 10^{-5}–$10^{-1}\,\text{eV}/c^2$ corresponds to a frequency range 2–20000 GHz with a very narrow line width corresponding to the small kinetic energy. Since the axion mass is unknown a very fine systematic frequency scan is required to search the likely range.

A pilot experiment of this kind has already been performed. It covered the mass range $(0.45\text{–}1) \times 10^{-5}\,\text{eV}/c^2$ at a level ~ 10 times less sensitive than the expected coupling and no signal was detected. (De Panfilis *et al.*, 1987). Further experiments of this kind with higher sensitivity are now in progress.

WIMPs

The challenge in this case is to detect particles which interact only through the weak interaction and which have kinetic energies of 1 keV or even less. We expect the particles to have mass in the range 5–100 GeV/c^2 i.e. comparable with light or middle weight nuclei.

Scattering on electrons will produce direct ionisation but for photinos the cross section for this process is too small to be useful. More promising is scattering on nuclei the recoil nucleus being then detected either by

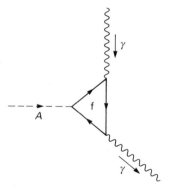

Fig. 15.9. Feynman diagram for the interaction of an axion A, with a photon from a magnetic field to produce a real photon. The interaction is via a vacuum fermion loop.

ionisation or through the larger fraction (90 %) of its energy converted to phonons. The phonons are detected either directly or as small temperature pulses.

For photinos the interaction is mainly spin dependent so does not benefit from the coherent enhancement arising from the addition of amplitudes for scattering off many nucleons spanned by the incident wave corresponding for instance to an incident heavy neutrino.

The nuclear recoil energy is given by simple kinematics as

$$E_R = \frac{E_I 2 M_R M_I}{(M_R + M_I)^2}(1 - \cos\theta)$$

with maximum value $4 M_R M_I (M_R + M_I)^{-2} E_I$, where M_R, E_R and M_I, E_I are the mass and kinetic energy of the recoil and incident particles respectively and θ is the c.m. scattering angle. Thus in the optimum case when $M_I = M_R$ the maximum recoil energy is equal to the incident energy. We assume this low energy scattering process to be isotropic in the c.m. so if the detector threshold is known we can calculate the fraction of events detected for any given incident energy.

For dark matter associated with galaxies we can estimate the $\tilde{\gamma}$ density, ρ_{DM}, necessary to account for all the dark matter on the basis of the rotation curve. This yields a value of about 10^{-24} gm cm^{-3}. The number density is then $n_{DM} = \rho_{DM}/M_I \sim (0.5 \text{ GeV}/c^2 \text{ cm}^{-3})/M_I$. Using this value for n_{DM} and assuming a weak interaction axial coupling to obtain the cross section we arrive at an event rate

$$R \simeq 1.2 r C \lambda^2 J(J+1) \text{ events kg}^{-1} \text{ day}^{-1}$$

r is the kinematic factor $4 M_R M_I/(M_R + M_I)^{-2}$, C is the squared hadronic matrix element for the $\tilde{\gamma}$-nucleon interaction, λ^2 is an easily evaluated shell model factor depending on the spin configuration and J is the total nuclear spin. $C\lambda^2 J(J+1)$ is always less than 1 but in favourable cases may be greater than 0.5. This formula leads to rates ~ 0.1–1 event kg^{-1} day^{-1} although some models lead to rates 10 or even 100 times greater arising from a coherent term or from a zino component mixing with the photino.

For detection of the phonons produced by the recoiling nucleus it is essential to use a detector such as a superconductor or a very pure crystalline material at a sufficiently low temperature (< 0.1 K) that the electronic contribution to the specific heat is negligible and the lattice specific heat is proportional to T^3. For instance a photino of mass 10 GeV/c^2 and velocity $\beta = 10^{-3}$ will yield an average nuclear recoil energy of 2 keV in silicon. For a 0.5 mm side silicon cube this gives a 20 mK temperature rise.

Prototype detectors of this kind are under development using a stack of thin wafers each with a deposited thin film thermometer circuit on its surface. Wafers of several materials can extend the mass range of particles studied and allow the possibility of mass estimation. Materials such as GaAs, LiF and TlBr are interleaved with silicon, which is itself likely to be a poor detector for axial coupling particles such as photinos since the only silicon isotope with nuclear spin is present only at an abundance of 5%.

As always with experiments studying very low rate processes, background minimisation is a serious problem. Use of external shields and pure materials free of radioactive contamination is essential. Genuine signals will be confined to a single wafer in the stack. A further characteristic which will distinguish genuine signals is modulation of the rate as the Earth moves round the Sun. This motion must give rise to a variation in the incident particle velocity and hence to a variation in both E (proportional to v^2) and in the incident flux (proportional to v). The Earth's orbital velocity $\sim 10^{-4} c$ oriented at $\sim 60°$ to the galactic plane produces a 5% variation in galactic velocity and thus a $\pm 15\%$ change between the maximum in June and the minimum in December.

15.4   Neutrinos

15.4.1   *Neutrinos from the Sun*

As we shall see below the Sun's energy generation must be accompanied by copious production of electron neutrinos. This neutrino source is of great interest for two reasons: on the one hand it provides information about the innermost parts of the Sun and on the other it allows measurements concerning the neutrino masses and magnetic moments and interactions, which are not possible on Earth. These two areas of study are not of course independent and only if the observations fitted the models would the interpretation of the data be straightforward. In fact the solar neutrino flux turns out to be only about one third to one half of the expected value. The following sections are devoted to the presentation of the expectations from the *Standard Solar Model*, the measured data and the possible interpretations of the measurements.

The Standard Solar Model (SSM)

The Sun is a middle-aged main sequence star the characteristics of which are believed to be very well understood. The principal assumptions (or approximations) of the SSM are:

(*i*) the Sun is a plasma in hydrostatic equilibrium where the radiative pressure (and, to a minor extent, particle pressure) balance gravity.

(*ii*) energy transport is mainly radiative but also by convective motion.

(*iii*) energy generation is by nuclear reactions and gravitational effects are small.

(*iv*) the relative nuclear abundances change only as a result of nuclear reactions.

The central temperature of the Sun is $\simeq 1.5 \times 10^7$ K equivalent to ~ 1 keV and the central density is ~ 150 g cm^{-3}. The age of the Sun is $\simeq 4.6 \times 10^9$ y.

The SSM gives a good account of the rate of energy production and the solar elemental composition. The neutrino data are the main data concerning the Sun which are clearly inconsistent with the SSM.

The main features of the nuclear fusion processes which fuel the Sun were elucidated by Bethe in 1939 and although the model has become much more refined as a result of both theoretical and experimental work since that time the primary features remain unchanged. The *overall* energy producing process is

$$4p \rightarrow \alpha + 2e^+ + 2\nu_e + 25 \text{ MeV}$$

Table 15.2. *Solar energy producing processes*

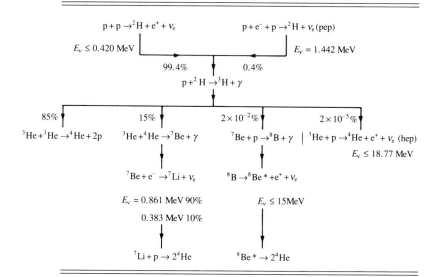

but this process takes place through a series of intermediate steps which define the energy spectrum. These reactions are listed in table 15.2 which also gives the neutrino energy where appropriate and the fractions of processes in the various branches. The corresponding energy spectra are shown in fig. 15.10. An alternative track by which four protons fuse to produce helium is by means of reactions involving carbon, nitrogen and oxygen – the CNO cycle. In the Sun however conditions are such that in the SSM this cycle contributes only 1.5% of the total luminosity.

Figure 15.10 shows clearly that the neutrino flux measured by any detector (see section 15.4.5) will depend on its energy threshold and on the cross section for the detection process. The total neutrino flux at the Earth is $6 \times 10^{10}/cm^2/s$ and is dominated by the low energy neutrinos from the pp interaction (table 15.2). Measurements to date have detected only the higher energy neutrinos from ^8B, ^7Be, pep and hep processes (see table 15.2) which form a small part of the total flux.

It has become customary to measure rates in units known as *solar neutrino units* or SNUs. The SNU is the product of the calculated neutrino flux and a theoretical cross section for the detection process in question. One SNU is defined as 10^{-36} events per target atom per second. For a given detector the expected signal is calculated by convoluting the

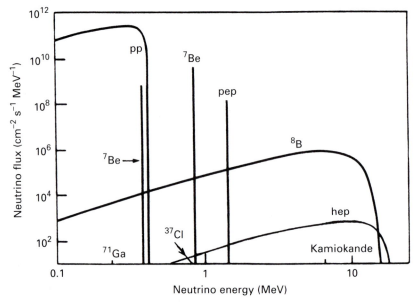

Fig. 15.10. Energy spectra of neutrinos from nuclear reactions in the Sun. The threshold energies are indicated for various detectors.

expected flux with the appropriate interaction cross section all integrated over the accessible energy range.

The Solar neutrino problem

Two measurements of the solar neutrino flux are presently available. First we have the data from the experiment of Davis and collaborators taken over a period of 20 years using absorption of the neutrinos by ^{37}Cl to produce ^{37}Ar (section 15.4.5). The threshold is 0.814 MeV so the detector is sensitive to neutrinos from ^{7}Be, ^{8}B, pep and hep but not pp. The predicted rate is 7.9 ± 2.6 SNU (Bahcall & Ulrich, 1988; Bahcall, 1989) where the authors choose to quote an error of three standard deviations. The measured value averaged over the period 1970–84 is 2.1 ± 0.3 SNU. The 10 most recent runs average 3.6 ± 0.7 SNU and the overall average is 2.2 ± 0.3 SNU. In 1989 the Kamiokande II experiment reported first results based on neutrino–electron scattering events with a threshold cut of 9.3 MeV so that only the ^{8}B and hep neutrinos are detected. The ratio of the observed flux to that expected on the basis of the SSM is $0.46 \pm 0.13(\text{stat}) \pm 0.08(\text{syst})$. This result is fully consistent with the ^{37}Cl data taken over the same period.

The difference between the expected and measured values given above constitutes the 'Solar Neutrino Problem'. In principle the problem may arise either from:

(*a*) something wrong with the Standard Solar Model or
(*b*) something happening to the neutrinos on their way from the Sun to the Earth.

In either case the resolution of the problem is of great interest. The neutrinos are our only *direct* source of information concerning the core of the Sun but since all other data are well accounted for by the SSM it seems unlikely that the SSM can be seriously in error. As discussed in section 15.4.2, mechanisms of neutrino oscillation coupled with interaction on the way through the Sun may account for the observations. In this case there is important particle physics information to be gained concerning neutrino masses and mixing angles.

A further aspect of the ^{37}Cl data deserves mention here. The data are shown as a function of date in fig. 15.11 which shows some evidence for maxima in 1977 and 1988 with a minimum in 1980. These maxima correspond to minima in sunspot activity with the intermediate neutrino minimum corresponding to a sunspot maximum. If neutrinos have a magnetic moment of sufficient magnitude then spin flip in the strong magnetic fields associated with sunspots might account for the effect (see section 15.4.4). The significance or otherwise of the effect should become clearer with observations over the next sunspot maximum around 1991.

In order to test fully the possible solutions to the solar neutrino problem or perhaps to uncover unsuspected phenomena we would ideally like to have complete data on the neutrino flux and flavour composition at all energies. As we shall see in section 15.4.5 different detectors have different capabilities in these respects. For instance the electron scattering process is 'flavour blind' while the ^{37}Cl absorption is sensitive only to electron neutrinos. Before discussing the detectors however we shall examine the fascinating phenomena of oscillation which may have a profound effect on the neutrinos arriving at the Earth.

15.4.2   *Neutrino Oscillations*

We have already met the idea of particle mixing in connection with K_1^0, K_2^0 (section 8.2–8.4), singlet-octet mixing in the meson nonets (section 10.6) and quark mixing (sections 12.1 and 12.8). In the latter two cases we have seen that a particularly convenient way to parameterise the mixing is in terms of a *mixing angle*. In the case of K_1^0, K_2^0 we saw that the mixing and oscillation phenomena provided an amazingly sensitive

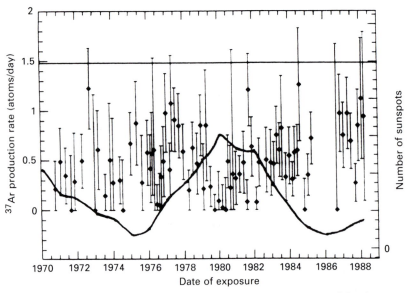

Fig. 15.11. Solar neutrino flux from the ^{37}Cl experiment of Davis *et al.* as a function of date. The solid line shows the sunspot rate.

measure of the mass difference. We now consider the possibility of mixing between different neutrino flavours v_e, v_μ, and v_τ, by non-flavour-conserving forces. As before the mixing is a function of the mass difference. If, as in the Standard Model, all neutrino masses are identically zero, then we will observe no mixing. Alternatively if mixing is observed we may hope to get some measure of the mass difference.

We may note that in general GUTs *do* give rise to non-zero neutrino masses. For instance in a GUT based on the group SO(10) the neutrino mass is given by the so called 'see-saw' formula

$$m_v = (\text{'normal mass'})^2/M$$

where M is the Grand Unification mass $\sim 10^{15}$ GeV/c^2 and where 'normal mass' is the mass of the electron, muon or tau, depending on the neutrino in question, or possibly the mass of the corresponding quark. This formula gives

$$m_{v_e} \sim 10^{-14}\text{--}10^{-7} \text{ eV}/c^2 \quad \text{and} \quad m_{v_\tau} \sim 10^{-10}\text{--}10^{-3} \text{ eV}/c^2$$

with

$$m(v_e):m(v_\mu):m(v_\tau) \simeq m^2(e):m^2(\mu):m^2(\tau).$$

The treatment is very similar to that employed in our earlier examples of mixing. We can write the flavour eigenstates $|v_f\rangle$ as linear super-positions of the mass eigenstates $|v_1\rangle, \ldots$

$$|v_f\rangle_t = \sum_j U_{fj} \exp(-iE_j t)|j\rangle \quad \hbar = c = 1$$

where we take the momenta of the mass eigenstates to be the same so that the energies E_j will differ by small amounts depending on the mass differences. U is a unitary matrix (which can be arranged to be real if *CP* conservation holds) to be compared with the Kobayashi–Masakawa matrix of section 12.8.

Since there are three flavours of leptons we can write the flavour eigenstate for the electron neutrino as a mixture of the mass eigenstates v_1, v_2, v_3:

$$|v_e\rangle = U_{e1}|v_1\rangle + U_{e2}|v_2\rangle + U_{e3}|v_3\rangle$$

with corresponding expressions for $|v_\mu\rangle$ and $|v_\tau\rangle$. In the vacuum the mass eigenstates $|v_1\rangle \ldots$ propagate independently and do not transform into each other.

The mixing is conveniently parameterised in this case as elsewhere in terms of mixing angles. The characteristic features are most readily seen if we consider the simplified case of two rather than three flavours. In this case there is one mixing angle, θ_V, the V indicating that the propagation is in vacuum, i.e. there are no interactions with matter to be considered. U is then a 2×2 matrix

$$U = \begin{pmatrix} \cos \theta_v & \sin \theta_v \\ -\sin \theta_v & \cos \theta_v \end{pmatrix}$$

The time evolution for v_e and v_μ is given by

$$|v_e\rangle_t = \cos \theta_v \exp(-iE_1 t)|v_1\rangle + \sin \theta_v \exp(-iE_2 t)|v_2\rangle$$
$$|v_\mu\rangle_t = -\sin \theta_v \exp(-iE_1 t)|v_1\rangle + \cos \theta_v \exp(-iE_2 t)|v_2\rangle$$

Due to the difference in E_1 and E_2 the relative phase of the components changes as a function of time or distance and we thus get changes in the composition of the beam. For an initially pure v_e beam the electron neutrino amplitude after a time t is given by

$$\langle v_e|v_e\rangle_t = \cos^2 \theta_v \exp(-iE_1 t) + \sin^2 \theta_v \exp(-iE_2 t)$$

The muon neutrino amplitude is

$$\langle v_\mu|v_e\rangle_t = \sin \theta_v \cos \theta_v (\exp(-iE_2 t) - \exp(-iE_1 t))$$

The corresponding intensities are then

$$|\langle v_e|v_e\rangle_t|^2 = 1 - \sin^2 2\theta_v \sin^2(\tfrac{1}{2}(E_2 - E_1)t)$$
$$|\langle v_\mu|v_e\rangle_t|^2 = \sin^2 2\theta_v \sin^2(\tfrac{1}{2}(E_2 - E_1)t)$$

In terms of masses $E^2 = m^2 + p^2$ and we assume that $p \gg m$ so that

$$E \simeq p(1 + m^2/2p^2) \quad \text{and} \quad E_2 - E_1 = \Delta m^2/2E$$

where $\Delta m^2 = |m_1^2 - m_2^2|$ and $E \simeq p$. Thus

$$|\langle v_\mu|v_e\rangle_t|^2 = \sin^2 2\theta_v \sin^2 \left(\frac{1}{2} \frac{\Delta m^2}{2E} t \right)$$

The oscillatory term is frequently written in terms of the distance travelled L, and a 'vacuum oscillation length' $L_v = 4\pi E/\Delta m^2$. L_v is the distance over which the neutrino returns to its initial state. The probability to see a flavour-changed neutrino after a distance L, is then

$$\sin^2 2\theta_v \sin^2 (\pi L/L_v)$$

since the neutrino is taken to be travelling with velocity $\simeq c = 1$. The vacuum oscillation length can be written in convenient units:

$$L_v = \frac{4\pi E h}{\Delta m^2 c^3} = 2.48 E/\Delta m^2 \text{ metres}$$

where E is in MeV and Δm^2 in $(\text{eV}/c^2)^2$. We have no *a priori* knowledge of the value of θ_v but Grand Unification ideas might suggest that it should be of the same order as the quark mixing angle θ_c for which $\sin^2 \theta_c \sim 0.05$. The probability of transformation of the original neutrino is then proportional to $\sin^2 2\theta_c = 0.2$ giving a minimum survival probability of 80% and an average survival probability $\sim 90\%$. The smaller the mass difference the larger is L_v and the larger must be L in order to have a finite probability to obtain flavour-transformed neutrinos. Extraterrestrial sources may thus provide the only laboratory for observations of neutrino oscillations if there is a very small mass difference between the eigenstates.

We may note that we can only hope to observe variations with distance (or time) if the neutrino source is localised within a distance $\lesssim L_v$ otherwise the neutrinos from different parts of the source will smear the effect. As Δm^2 decreases L_v increases, the tolerable source size increases and the distance necessary to observe significant oscillation effects also increases. For $L_v \sim 0.5$ solar radii (includes most of the mass of the Sun) $\Delta m^2 \simeq 6 \times 10^{-8}$ $(eV/c^2)^2$ for 8 MeV neutrinos. Most of the ^8B neutrinos are produced within a region near the core of the Sun at smaller radii than $0.5R_\odot$.

Transmission through matter – the MSW effect
An important modification to the theory of neutrino propagation *in matter* was pointed out by Mikeyev and Smirnov (1986a, b, c) based on earlier work by Wolfenstein (1978). This is now generally referred to as the MSW effect.

In order to include the effects of interactions with matter it is usual to use the formalism of the *mass matrix*. Including the effects of matter we have phase differences arising from

    (*i*) the mass difference between v_1 and v_2, as in the vacuum case
    (*ii*) the difference in the interactions of v_1 and v_2.

We have already met the mass matrix formalism in our discussion of singlet-octet mixing in the meson nonets (section 10.6). The treatment for neutrinos in vacuum is formally exactly the same except that in this case the neutrinos are highly relativistic so that $E \simeq p + m^2/2p$.

We start with transmission in vacuum and write the mass eigenstates as before as $|v_1\rangle$ and $|v_2\rangle$. The Schrödinger equation is

$$i\frac{d}{dt}\begin{bmatrix} v_1 \\ v_2 \end{bmatrix} = \begin{bmatrix} E_1^2 & 0 \\ 0 & E_2^2 \end{bmatrix}\begin{bmatrix} v_1 \\ v_2 \end{bmatrix} = \begin{bmatrix} m_1^2/2p & 0 \\ 0 & m_2^2/2p \end{bmatrix}\begin{bmatrix} v_1 \\ v_2 \end{bmatrix}$$

where we have omitted the momentum term which influences the phase of v_1 and v_2 *equally* and does not affect the probability amplitudes.

Expressing the v_1 and v_2 in terms of the flavour base eigenstates we can write

$$i\frac{d}{dt}\begin{bmatrix} v_e \\ v_\mu \end{bmatrix} = H_f\begin{bmatrix} v_e \\ v_\mu \end{bmatrix}$$

where the reader can check that if the algebra is carried through we find

$$H_f = \frac{\Delta_V}{2}\begin{pmatrix} -\cos2\,\theta_v & \sin2\,\theta_v \\ \sin2\,\theta_v & \cos2\,\theta_v \end{pmatrix}$$

where $\Delta_V = E_2 - E_1 = \Delta m^2/2p$ is the difference in the energy eigenvalues of the two mass eigenstates. Here and subsequently we have taken m_2 to be greater than m_1.

In matter the relative phases of the v_f change due to the difference in the interactions of v_e on the one hand and v_μ and v_τ on the other. All three neutrinos share the same neutral current (Z^0 exchange) interaction with electrons but only v_e can interact by $W^\pm$ exchange (fig. 15.12).

The effect of matter can be included by modifying the mass matrix by adding additional terms

$$H = H_f + H_M$$

where H_f is our earlier vacuum expression and H_M is due to scattering on matter. The effective masses of v_1 and v_2 and the mixing angle *become functions of the scattering electron density* since

$$H_M = \sqrt{2}G_F n_e$$

where G_F is the Fermi coupling constant and n_e is the electron density.

For passage through matter we write a modified mass matrix

$$\frac{\Delta_M}{2}\begin{pmatrix} -\cos2\,\theta_M & \sin2\,\theta_M \\ \sin2\,\theta_M & \cos2\,\theta_M \end{pmatrix}$$

As before Δ_M corresponds to the difference between the energies of the two mass eigenstates but now taking account of the effect of the difference in scattering.

The relations between Δ_M and θ_M and the corresponding vacuum quantities are obtained by equating the 'matter' quantities to the values obtained by substituting the explicit formulae for H_f and H_M in the vacuum expression. Δ_M is then given by the expression

$$\Delta_M = \left[\left(\frac{\Delta m^2}{p}\cos 2\,\theta_v - \sqrt{2}G_F n_e\right)^2 + \left(\frac{\Delta m^2}{p}\sin 2\,\theta_v\right)^2\right]^{\frac{1}{2}}$$

Fig. 15.12. Feynman diagrams for v-electron scattering.

The mixing angle θ_M is given by

$$\tan 2\theta_M = \frac{\tan 2\theta_v}{[1 - (L_v/L_e)\sec 2\theta_v]} \qquad \textbf{15.6}$$

where we have introduced a new characteristic length, the neutrino–electron interaction length L_e, which is independent of energy and which is given by

$$L_e = \frac{\sqrt{2}\pi\hbar c}{G_F n_e}$$

Corresponding to the vacuum oscillation length L_v we have a 'matter oscillation length' L_M where $L_M = 2\pi/\Delta_M$ (cf. $L_v = 2\pi/\Delta_v$) and L_M is related to L_v by

$$L_M = L_v(1 + (L_v/L_e)^2 - 2(L_v/L_e)\cos 2\theta_v)^{-\frac{1}{2}}$$

Equation 15.6 allows the possibility of 'resonance' in which the mixing angle θ_M goes through a maximum. If, as we have assumed, $m_2 > m_1$ then the mixing is a maximum ($\theta_M = \pi/4$) if

$$L_v/L_e = \cos 2\theta_v$$

We can write this condition in terms of a 'resonant electron density' $n_{e\,res}$ where

$$n_{e\,res} = \frac{|\Delta m^2|\cos 2\theta_v}{2\sqrt{2}\,G_F E}$$

If its energy exceeds the resonance value for the electron density at the solar centre than an electron neutrino will *always* pass through resonance as it passes through progressively smaller electron densities on its way out of the Sun. Mixing can become very large at resonance since even for very small θ_v, θ_M can reach the resonant value.

It is instructive to examine the behaviour of the flavour and mass eigenstates as a function of electron density. At the large densities near the centre of the Sun θ_M is close to $\pi/2$ and the electron neutrinos are in the upper state $|v_2\rangle$. As the neutrino moves to lower densities there is a probability of a transition to $|v_1\rangle$ where the energy levels approach closely but this probability is very small (see below) so that the neutrinos are still in the state $|v_2\rangle$ when they arrive at the surface. Here however $\theta_M \simeq \theta_v$ which is assumed to be small so that the neutrino is now mainly composed of the flavour state $|v_\mu\rangle$.

The situation is illustrated in fig. 15.13 showing the relative level energies for $|v_1\rangle$ and $|v_2\rangle$ as a function of electron density. In the absence of mixing the v_e effective mass in the presence of a charged current potential in matter grows linearly with electron density while the effective v_μ is independent of density. The eigenvalue curves would cross at the resonance value of n_e where the diagonal values of the neutrino mass matrix are equal were it not for mixing which is responsible for the off-diagonal terms proportional to $\sin 2\theta_v$.

We have mentioned that transformations between $|v_1\rangle$ and $|v_2\rangle$ are expected to be of low probability even at resonance where the levels are closest. A small rate for this transition depends on fulfilment of the 'adiabatic condition'. This requires that the electron density changes sufficiently slowly that the mixed neutrino state has time to adjust to density changes without jumps between $|v_1\rangle$ and $|v_2\rangle$. Near resonance it can be shown that the adiabatic condition is that the resonance width in space Δr should be greater than the matter oscillation length:

$$\Delta r > L_M$$

Using values for the electron density and its gradient corresponding to $0.1 \times R_\odot$ (the region where most of the ^{8}B neutrinos are produced) we obtain for the adiabatic condition

$$\frac{\sin^2 2\theta_v}{\cos 2\theta_v} > 4 \times 10^{-9} E/\Delta m^2$$

where E is in MeV and Δm^2 in $(eV/c^2)^2$. For $\theta_v \simeq \theta_c$ this requires $\Delta m^2 \gtrsim 10^{-3}$ $(eV/c^2)^2$.

An analogy often used to illustrate the mixing process is to compare the

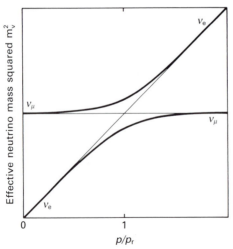

Fig. 15.13. Level energies as a function of the ratio of electron density ρ, to the resonant electron density ρ_r.

v_e, v_μ system to a pair of coupled oscillators (pendulums). The effect of matter is to change the eigenfrequencies of the oscillators but the change is different for the oscillators corresponding to v_e and to v_μ. At resonance the eigenfrequencies of the two oscillators coincide and the energy can pass completely from one to another corresponding to total transformation from v_e to v_μ.

It is clear from the above analysis that there can be regions of $\Delta m^2/\theta_v$ space for which oscillations modified by the MSW effect produce substantial reduction of the solar neutrino flux. Bethe (1986) has made a particularly striking if somewhat simplified calculation of the effect. He argues that all the lost neutrinos are from ^{8}B decay and that all the others with energy less than 2.8 MeV have energies too small to fulfil the resonance condition anywhere in the Sun. As we have seen the ^{8}B neutrinos with energies up to 14 MeV all come from a small central region where the mean density is ~ 130 gm cm^{-3}. Bethe proposes that there exists a critical energy above which all neutrinos from the core pass through a region of resonant density and emerge as muon neutrinos. Bethe uses analytic expressions to express the spectrum and the detector response and calculates the critical energy as ~ 8 MeV. This gives

$$(m_2^2 - m_1^2)\cos 2\theta_v = 1 \times 10^{-4}(\mathrm{eV}/c^2)^2.$$

Assuming that θ_v is small and $m_{v_e} \ll m_{v_\mu}$ then $m_{v_\mu} \sim 10^{-2}\,\mathrm{ev}/c^2$. Detailed calculations arrive at similar conclusions. For example if θ_v is set equal to θ_c then the flux of solar electron neutrinos measured by the ^{37}Cl experiment is reduced from the expected value of 7.9 SNU to $\leqslant 3$ SNU for $10^{-7} \leqslant \Delta m^2 < 10^{-4}(\mathrm{eV}/c^2)^2$ (Parke and Walker (1986)).

In summary neutrino oscillations in matter offer a possible explanation of the solar neutrino puzzle which preserves the standard solar model. This solution requires an extension of the standard electroweak theory to allow neutrinos to have very small masses. There is presently no evidence available to test these ideas but there are clear predictions of the relative fluxes of the three flavours of neutrinos and of their energy spectra. In particular if the Bethe hypothesis is true then the forthcoming ^{71}Ga experiments (section 15.4.5) which have a low energy threshold will see a much smaller flux depletion ($\sim 10\%$) compared with the expected value. The development of a number of new detectors with a variety of different sensitivities should establish or otherwise the truth of this explanation of the puzzle, over the next decade.

15.4.3 *Neutrino oscillation experiments*

As shown in the preceding section the amount by which the original neutrino flux is decreased or by which a different neutrino flavour is produced is proportional to a factor $\sin^2(L\Delta m^2/4E)$. The experimental parameter is thus L/E where L is the distance between source and detector and E is the neutrino energy. If L/E can be varied then oscillations may be detected independent of knowledge of the initial flux. If $L/E \ll 4/\Delta m^2$ then $\sin^2(L\Delta m^2/4E) \sim 0$ and no oscillations will be observed. Thus the area of sensitivity of an experiment in terms of Δm^2 can be defined by the condition.

$$L/E \gtrsim 4/\Delta m^2 \quad \text{or} \quad L/E \gtrsim (\Delta m^2)^{-1}$$

where L is in metres, E in MeV, Δm^2 in $(eV/c^2)^2$

Thus the bigger L and the smaller E the more sensitive is the experiment. On the other hand if $L/E \gg 4/\Delta m^2$ then many cycles of oscillation take place in the L/E range and uncertainties in L (due for instance to an extended source) or in E can smear the result. Some typical values for the relevant quantities are given in table 15.3

Searches for oscillation phenomena have been made using neutrinos from reactors, high energy accelerators, low energy accelerators and interactions of cosmic rays in the atmosphere. None of these has so far given convincing evidence of neutrino oscillations. We discuss below some reactor and high energy accelerator experiments, these being the areas from which have come most of the data. The solar neutrino experiments have already been treated in section 15.4.1.

Reactor experiments

Although reactor experiments have relatively short flight paths compared to the Earth–Sun distance they have the advantage that the flight path can be varied and also that the neutrino fluxes available are very big allowing

Table 15.3. *Characteristics of neutrino oscillation experiments*

Neutrino source	Energy (MeV)	Length (m)	L/E (m/MeV)	Δm^2 (min) $(eV/c^2)^2$
High energy accelerators	10^3–10^5	10^2–10^4	10^{-2}–1	1
Reactors	1–10	10–10^2	1–10^2	10^{-2}
Sun	10^{-1}–10	10^{11}	10^{10}–10^{12}	10^{-12}

the use of mobile detectors small compared with those employed for solar neutrino experiments. Such experiments have played a particularly important role in limiting the possible values of Δm and mixing angle.

Power reactors produce $\sim 1.6 \times 10^{20}$ $\bar{\nu}_e/s/GW$ and often have outputs of 2–4 GW. Fission reactors lead to a complex set of products the precise mix depending on the reactor fuel used. Many of these products are β emitters which generate the flux of *anti*neutrinos.

A number of experiments have been performed. These use one or more of the following processes for detection:

$\bar{\nu}_e p \to e^+ n$	charged current (only ν_e)	threshold 1.8 MeV
$\bar{\nu}_e d \to e^+ n n$	charged current (only ν_e)	threshold 4.0 MeV
$\bar{\nu} d \to \bar{\nu} n p$	neutral current (all flavours)	threshold 2.2 MeV

As an example we take the experiment at the Gösgen reactor in Switzerland which has yielded some of the best data in this field. This reactor has a power output of 2.8 GW. The detector used the $\bar{\nu}_e p \to e^+ n$ process in which the e^+ was detected and its energy measured in an oil based liquid scintillator which was also the antineutrino target. The scintillator was interleaved with layers of ^{3}He filled MWPCs which recorded the neutrons via the process $n + {}^3He \to p + {}^3H + 765$ keV. The detector thus measures both the energy and the position of the absorption. The use of both neutron and positron detection coupled with pulse shape discrimination, achieves a low background of γs, fast neutrons and protons. Measurements of backgrounds not associated with the reactor were made during periods when the reactor was off. The detector was run in three different positions at 37.9, 45.9 and 64.7 m from the reactor. 10 000 neutrino counts were recorded in each position.

Knowing the composition of the reactor it is possible to calculate the energy spectrum of the neutrinos which runs up to a maximum of 8 MeV declining exponentially above 2 MeV. It is difficult to achieve totally reliable values at the top end of the spectrum but measurement of the associated β spectrum allows verification or correction of the calculations. The Gösgen data have been analysed both using the spectra and absolute efficiencies and also using only the measurements from the three different distances. The results agree well and show no evidence for oscillations with the data including the absolute spectra allowing rather more restrictive limits on the Δm^2/mixing angle – plot (fig. 15.14(a)). Some other reactor experiments have from time to time shown some evidence for oscillations but they have generally been shown to have potential weaknesses or have been overtaken by other work showing no such effect.

Fig. 15.14. Exclusion limits for $\sin^2 2\theta$ and Δm^2. The regions to the right of the curves are excluded at 90 % confidence level (or as indicated).

(a) Data from the Gösgen reactor experiment for $\bar{\nu}_e$ (Zacek *et al.*, 1986).

(b), (c), (d), Results of an analysis based on combined data from accelerator experiments (Blümer and Kleinknecht, 1985). (b) is for $\nu_e \leftrightarrow \nu_\mu$ (c) for $\nu_e \leftrightarrow \nu_\tau$ (d) for $\nu_\mu \leftrightarrow \nu_\tau$. Numbered lines correspond to data from different experiments assuming the other mixing angles are zero. The shaded areas are the allowed ranges from the Bugey reactor experiment. The solid lines are the limits from a three flavour analysis with no restriction on the mixing angles.

Accelerator experiments

Reactor experiments are generally 'disappearance experiments' where loss of neutrinos in absolute terms or in the energy spectrum is looked for as a function of distance. At accelerators where the neutrino energies are sufficient to generate their charged lepton counterparts the μ and τ and where we also start with a beam composed primarily of muon neutrinos, we can also have 'appearance' experiments. Only in such experiments can we distinguish mixing between all three different flavours.

The technique for generating a neutrino beam at an accelerator has already been described in our discussion of the 'two neutrino experiment' in section 3.11. The neutrinos in such a beam arise primarily from decay of pions and kaons and are thus muon neutrinos and antineutrinos with a small ($\sim 1\%$) component of electron neutrinos from K_{e3} decay. The neutrinos are detected and their flavours identified by the charged current processes

$$\nu_\mu N \rightarrow e + X$$
$$\nu_\mu N \rightarrow \tau + X$$

where N is a nucleon and X is the residual nucleus.

We may note that L/E is not well defined in such experiments where the origin of the neutrino is not well known.

The most straightforward experiments in this case are *appearance* experiments where transitions of ν_μ to ν_e or ν_τ are indicated by the identification of electrons (beyond the small number expected) or taus in the detector. Disappearance is more difficult to measure since it requires a rather precise knowledge of the incident muon neutrino flux and energy distribution.

A number of experiments of this kind have been carried out using electronic or bubble chamber detectors. There is no evidence for oscillations from these experiments. The results of an analysis of data available up to August 1985 (Blümer and Kleinknecht 1985) are shown in fig. 15.14(b), (c), (d) for the processes $\nu_e \leftrightarrow \nu_\mu$, $\nu_e \leftrightarrow \nu_\tau$ and $\nu_\mu \leftrightarrow \nu_\tau$ respectively. The regions to the right of the line are excluded. More recent data have only slightly extended the forbidden areas.

Accelerator experiments have also been carried out in which the high energy beam is 'dumped' in a heavy target. In this case most of the pions and kaons interact before they decay and the emerging neutrino flux contains a large proportion of neutrinos arising from the decay of the very short lived charmed particles. For these 'prompt' neutrinos the numbers of muon and electron neutrinos are equal. The analysis of prompt neutrino data to reveal oscillations is less direct and subject to greater uncertainties than that for the neutrino beam experiments described above. They do not currently reveal any convincing evidence for oscillations.

15.4.4   *Neutrino magnetic moment*

The Standard Model in particle physics does not endow neutrinos with a magnetic moment but if we allow the neutrino to have mass we should consider the possible consequences if it also has a magnetic moment.

If the neutrino has a magnetic dipole moment (an electric dipole moment would produce similar effects) then traversal of the solar magnetic field may flip the neutrino spin. The spin-flip would change the left-handed neutrino which can interact with ^{37}Cl or ^{71}Ga to a right-handed neutrino which would not be detectable in these experiments. If we allow the neutrinos a mass then conventional electroweak theory gives $\mu_\nu \simeq m_{\nu_e} 10^{-19} \mu_B$ where m_{ν_e} is in eV$/c^2$.

This possibility adds a further dimension to an already complicated picture. Studies of the combined effect of flavour mixing, spin flip and matter interactions lead to the conclusion that spin-flip may also have a resonant character.

The spin-flip probability depends on the product of the magnetic dipole moment μ_ν, and the line integral of the perpendicular component of the magnetic field along the flight path which we can write in terms of an average field B as BL. If the spin is flipped the order of magnitude of the magnetic moment can be estimated by setting

$$\mu_\nu BL \sim \hbar c$$

Estimates of the transverse magnetic field in the Sun can be complicated by the presence of sunspots but values 0.1–1T existing over a distance of 2×10^{10} cm seem plausible. Using these values we find

$$\mu_\nu \sim 10^{-11}\text{–}10^{-10} \mu_B \quad \text{where} \quad \mu_B = e\hbar/2m_e c$$

The small value of such a moment makes it apparently impossible to measure in terrestrial spin precession experiments but some limits may be obtained from scattering experiments at reactors. On the basis of such data Lim and Marciano (1988) have calculated that $\mu_\nu < 2 \times 10^{-10} \mu_B$ but other treatments yield a larger limit.

Resonant effects in a full treatment may well enhance the spin-flip to a level where it is important. The number of parameters in the problem now seems to be so great however that more experimental data are essential in order to make real progress.

15.4.5 *Neutrino detection*

As we have seen (chapters 7 and 11) neutrinos interact only via the weak interaction so that the interaction cross section, especially at low energies, is extremely small ($\lesssim 10^{-44}$ cm^2). Indeed although the cross section increases with energy we have seen (sections 3.11 and 11.3) that even for neutrinos of GeV energies detectors weighing hundreds of tonnes are necessary to achieve useful event rates.

We discuss briefly below the principal detection methods which have been used or are planned, to tackle this very difficult and challenging area of experiment. Most of these techniques are aimed particularly at the study of solar neutrinos with energy of a few MeV. The reader who wishes to pursue this subject in more detail is referred to *Neutrino Astrophysics* by J. N. Bahcall (Cambridge University Press, 1989).

In addition to the common characteristic of large mass, neutrino detectors have to be protected against cosmic ray background and wherever possible are situated in deep mines or tunnels.

The ^{37}Cl detector at the Homestake Mine

The first and longest running solar neutrino experiment is that of Ray Davis and collaborators at the Homestake mine in South Dakota where measurements started in 1968.

The neutrinos are detected by the inverse reaction to the electron capture decay of ^{37}Ar in the process

$$\nu_e + {}^{37}\text{Cl} \rightarrow e^- + {}^{37}\text{Ar}$$

for which the threshold energy is 0.81 MeV. The chlorine is in the form of perchlorethylene, an inexpensive liquid widely used as a cleaning fluid for clothes. In the Homestake experiment 615 tons of the fluid is contained in a tank at a depth of 4850′ (4100 m water equivalent depth). The number of target Cl atoms is 2.2×10^{30}. This detector will yield one atom of argon per day for an electron neutrino flux of 5.35 SNU.

The ^{37}Ar has a half life of 35 days and argon atoms are allowed to accumulate for one to three months before being extracted for counting. The argon exists as a tiny amount of gas dissolved in the liquid and is removed by purging with helium gas which is subsequently passed through a condenser at $-32\,°$C and a molecular sieve at room temperature. The argon is finally absorbed in a charcoal trap at liquid nitrogen temperature. The efficiency of extraction is measured by introducing a precisely known small quantity of isotopically pure ^{36}Ar into the liquid. The amount of this gas which is recovered gives a value for the efficiency of the extraction process, a value which should be identical to that for the ^{37}Ar. The ^{37}Ar is counted by means of its electron capture decay in a low background, gas proportional chamber using special techniques exploiting measurements of both pulse energy and rise time to eliminate background.

In 61 runs between 1970 and 1984 the production rate averaged 0.462 ± 0.040 ^{37}Ar atoms per day including a background of 0.08 ± 0.03 per day. This rate corresponds to a solar neutrino flux of 2.05 ± 0.3 SNU.

There was a break in observation in 1985–6 and results from a further 10 runs in the period 1986–8 have been reported with an average capture rate of 3.6 ± 0.7 SNU.

Other activation experiments

^{71}**Ga**. Two major experiments are starting which use the reaction

$$\nu_e + {}^{71}\text{Ga} \rightarrow e^- + {}^{71}\text{Ge}$$

as the detector. The threshold in this case is 0.2332 MeV so that the measurement is sensitive to neutrinos from the pp process which have an upper energy limit of 0.42 MeV. The ^{71}Ge decays back to ^{71}Ga by electron capture with a half life of 11.43 days. The low threshold energy coupled with a favourable reaction cross section results in a relatively high expected rate of 132^{+20}_{-17} SNU of which 70.8 arises from the pp process.

One of these experiments, known as GALLEX, uses 30 tons of gallium located in a specially constructed laboratory located off the Gran Sasso road tunnel in Italy. The gallium is in the form of a solution of gallium chloride and hydrochloric acid. The germanium is extracted by purging with air or nitrogen and following a series of intermediate processes is collected as the gas GeH_4, known as germane. Pilot tests have given an extraction efficiency of better than 99%. The standard solar model predicts 1.2 events per day for this experiment.

The other gallium experiment (sometimes known as SAGE) uses 60 tons of gallium in an underground laboratory in the Baksan valley in the USSR. In this case the gallium is in the form of metal and the germanium is extracted by first melting the metal (melting point 30 °C) which is then mixed with dilute hydrochloric acid with subsequent extraction similar to that of GALLEX. The experiment will be calibrated using an intense neutrino source from ^{51}Cr produced in a reactor.

Other nuclei. A number of other nuclei have been proposed as target materials for neutrino detection in experiments involving the extraction and counting of the transformed nuclei. Their properties are summarised in table 15.4. The long-lived or stable products afford the possibility of measuring the integrated neutrino flux over periods of millions of years by counting the transformed atoms from deep ore deposits. The shorter lived products (lifetimes of a few days) are generally to be detected by electron capture counting with the exception of ^{7}Be which yields only a 50 eV Auger electron. The longer lived products require spectroscopic detection (mass-spectrometry, resonant ionisation mass spectrometry). Only the main neutrino sources are given.

Neutrino-electron scattering detectors – Kamiokande II

We have earlier discussed the massive detectors built to search for nucleon decay (section 14.2.2). Although no examples of nucleon decay have been so far detected these detectors have an important role as *neutrino telescopes*. The Kamiokande II detector in the Kamioka mine in Japan was originally built as a nucleon decay detector but has been adapted so as to be particularly well suited to the study of solar and other extraterrestrial neutrinos. In comparison with the activation detectors the real-time neutrino–electron scattering detectors give the additional information of neutrino arrival time, some measure of neutrino energy and also the neutrino direction. This latter feature is critical in separation of real events from background. On the other hand the scattering cross section on electrons is some two orders of magnitude smaller than the absorption cross section and the backgrounds are much more severe than in the activation experiments.

Kamiokande II is an imaging water Cerenkov detector which detects ^{8}B neutrinos by neutrino–electron scattering

$$v + e^- \rightarrow v + e^-$$

The detector volume containing 2140 tonnes of pure water is viewed by 948 20″ diameter photo-multipliers (PMs) on a 1×1 m grid around the walls. It is surrounded by an anti-coincidence counter also consisting of a water Cerenkov 1.4 m in thickness which monitors cosmic ray muons, as well as shielding the inner detector from γ-rays from the surrounding rock and ensuring containment of low energy events. The laboratory is at a depth of 1 km (2700 m water equivalent).

The detector is triggered by signals in at least 20 PM tubes within a time window of 100 nsec. A number of event selection cuts are necessary to reduce background of which one is to use a limited fiducial volume of 680 tonnes. Within this volume the trigger currently accepts 6.7 MeV electrons with 50 % efficiency and 8.8 MeV electrons with 90 % efficiency. Fig. 15.15 shows the angular distribution of the electron trajectories with respect to the direction of the Sun for data from a 450 day running period in 1987/88 where the signal in the Sun's direction is clearly visible, particularly for the higher energy data.

The ratio of the measured flux from the Kamiokande detector to that calculated for the Standard Solar Model from these data is

$$0.46 \pm 0.13(\text{stat.}) \pm 0.08(\text{syst.})$$

Other real-time detectors

Real-time counting experiments have been proposed exploiting neutrino interaction or absorption in deuterium in heavy water (Sudbury Neutrino Observatory, SNO), liquid argon in the form of a time projection chamber (ICARUS), ^{115}In, ^{11}B, and ^{19}F and also using coherent scattering on nuclei (see also discussion of WIMP detection, section 15.3.3). These proposals have various thresholds and flavour sensitivities.

A particularly interesting example, now funded for construction, is the Sudbury project to be built in a deep mine (6800′, 5900 m water equivalent) in Ontario. The detector will consist of a transparent tank containing 1000 tonnes of heavy water and viewed by some 2000 PM tubes to detect Cerenkov light. Three different neutrino interaction processes will be studied:

(*i*) $v_e + d \rightarrow p + p + e^-$ threshold 5 MeV   6 SNU
(*ii*) $v + d \rightarrow p + n + v$ threshold 2.2 MeV   2.4 SNU
(*iii*) $v + e^- \rightarrow v + e^-$ threshold 7 MeV   0.04 SNU

Reaction (*i*) is a charged current process and is thus sensitive only to v_e.

Table 15.4. *Characteristics of activation processes for neutrino detection*

Target nucleus	Product nucleus	Threshold MeV	Neutrinos detected	Expected flux SNU	Lifetime of product
^{37}Cl	^{37}Ar	0.814	^{7}Be, ^{8}B	7.9	35 days
^{71}Ga	^{71}Ge	0.233	pp, ^{7}Be, ^{8}B	130	11 days
^{7}Li	^{7}Be	0.862	^{7}Be, ^{8}B	52	53 days
^{81}Br	^{81}Kr	0.470	^{7}Be, ^{8}B	30	2×10^5 years
^{127}I	^{127}Xe	0.789	^{7}Be, ^{8}B	large	36 days
^{98}Mo	^{98}Tc	1.8	^{8}B	18	6×10^6 years
^{205}Tl	^{205}Pb	0.054	pp, ^{7}Be, ^{8}B	263	1.6×10^7 years

Reaction (*ii*) is a neutral current process and is thus 'flavour blind'. Reaction (*iii*) is sensitive to a mixture of charged and neutral currents. The capture rate for (*i*) should be $\sim 10^4/\text{y}$, the neutral current $\sim 5 \times 10^3/\text{y}$, and the electron scattering rate $\sim 10^3/\text{y}$. The great depth of the laboratory results in a very low cosmic ray background (200 times down on the Kamiokande II experiment).

15.4.6 *Direct neutrino mass measurement*

Experiments to set limits on the neutrino mass have been going on for more than forty years. In more recent times the best precision for the electron neutrino has been achieved by studies of the β-decay spectrum near its end point using high precision magnetic and electric spectrometry.

The β-decay spectrum was discussed in section 7.3. If we take the result of **7.2** with only V and A couplings we have

$$\frac{\mathrm{d}n}{\mathrm{d}E_e} = p_e E_e (E_0 - E_e)^2 F$$

where we have written F, the 'Fermi factor', for the bracket containing matrix elements and coupling constants and where p_e and E_e are the

Fig. 15.15. Distribution of the cosine of the angle between the electron trajectory and the direction of the Sun for data from the Kamiokande II detector for thresholds of (a) 9.3 MeV and (b) 10.1 MeV.

momentum and energy of the electron and E_0 is the maximum electron energy achieved when the neutrino takes zero energy, i.e. E_0 is the spectrum end-point. This expression assumes zero mass for the neutrino. If the neutrino mass, m_ν, is not zero then the phase space part of the expression is modified and we have

$$\frac{dn}{dE_e} = p_e E_e [(E_0 - E_e) + m_\nu] [((E_0 - E_e) + m_\nu)^2 - m_\nu^2]^{\frac{1}{2}} F$$

The spectrum shape is affected mainly by the second bracket which modifies the shape near the end-point to the extent that the spectrum slope, which is zero for $m_\nu = 0$, becomes infinite for finite mass, while the maximum electron energy is reduced by the amount of the neutrino rest mass.

Tritium is a particularly attractive β-emitter for such studies. Its β-decay

$$^3H \rightarrow {}^3He + e^- + \bar{\nu}_e$$

is an allowed ground to ground state transition of low and well known end-point energy (18.59 keV) and convenient half life (12.26 y). In practice the detail of the spectrum shape near the end point depends on the resolution and energy response of the spectrometer and also on subtle effects associated with the molecular environment in the source when the tritium is implanted in a solid. In consequence these measurements are difficult and delicate.

After a series of experiments progressively reducing the upper mass limit for ν_e, much excitement was generated by a finite mass result for m_ν from a group at ITEP in the USSR. These workers used tritium implanted in a complex organic compound known as valine. The spectrum was measured using an iron-free toroidal magnetic spectrometer. In a series of papers the mass limits were reduced to

$$17 < m_\nu < 40 \text{ eV}/c^2$$

(Boris *et al.* 1987). Apart from its intrinsic interest such a neutrino mass could have important implications for the 'dark matter problem' (see section 15.2) and closure of the universe (see section 15.2.2).

Other experiments with better resolution and with other kinds of tritium sources have not confirmed this result. Fritschi *et al.* (1986) using tritium absorbed in carbon find a best fit for $m_\nu = 0$ and set a limit of $m_\nu < 18 \text{ eV}/c^2$ with 90 % confidence level. Recent data from the Tokyo INS group using tritium absorbed in an organic molecule source find $m_\nu < 11 \text{ eV}/c^2$ while another recent result from Los Alamos using tritium gas finds $m_\nu < 13.4 \text{ eV}/c^2$ (both these limits at 95 % c.l.).

As we shall see in the following section, data from the 1987 supernova support these latter limits and are consistent with a very small or zero mass electron neutrino. We have already seen that the oscillation/MSW effect hypothesis concerning the solar neutrino problem suggests a mass many orders of magnitude below the sensitivity of these direct measurements.

Limits on the masses of the muon and tau neutrinos are much higher.

For v_μ a limit can be set from study of the muon momentum in the pion decay

$$\pi^+ \to \mu^+ + v_\mu$$

It is a simple kinematic exercise to show that for π^+ decay at rest

$$m_{v_\mu}^2 = m_\pi^2 + m_\mu^2 - 2m_\pi (m_\mu^2 + p_\mu^2)^{\frac{1}{2}}$$

The limit on m_v depends on the measurement accuracy of p_μ and on the precision to which the pion and muon masses are known. Experiments at the Swiss laboratory SIN (Abela *et al.* 1984) used a π^+ beam stopped in a scintillator followed by magnetic analysis of the muons which were detected in a silicon surface barrier detector. Using the SIN results and the best values for m_π and m_μ Jeckelman (1986) has obtained a limit

$$m_{v_\mu} < 270 \text{ keV}/c^2$$

For v_τ the best limit to date comes from studies using the ARGUS spectrometer at the DESY laboratory at Hamburg. τs were produced by the e^+e^- collider

$$e^+ + e^- \to \tau^+ + \tau^-$$

with each τ having half the total c.m. energy. The decays studied were

$$\tau^\pm \to \pi^\pm \pi^+ \pi^- v_\tau$$

and

$$\tau^\pm \to \pi^\pm \pi^+ \pi^+ \pi^- \pi^- v_\tau$$

The pion momenta were measured by curvature in the magnetic field so the neutrino energy could be determined. In the three pion case the neutrino mass limit can then be obtained from the lower edge of the neutrino spectrum and the three pion effective mass. Including systematic errors the result is

$$m_{v_\tau} < 70 \text{ MeV}/c^2$$

Using five pion data a somewhat different analysis yields (Albrecht *et al.* 1988)

$$m_{v_\tau} < 35 \text{ MeV}/c^2$$

which is the best limit to date.

15.4.7   *SN1987A*

The Sun is not the only source of extraterrestrial neutrinos. On February 23rd 1987 the explosion of a supernova was observed in the Large Magellanic Cloud, a neighbour of our own galaxy. A burst of neutrinos from this explosion which had been travelling through space for 170 000 years was detected on Earth over a period of 13 seconds.

Such a supernova explosion is a rare event which occurs in a galaxy such as ours only about once in every forty years. The position of the supernova SN1987A coincided with that of a previously known blue supergiant star – Sanduleak 69 202, which was found to have disappeared following the explosion. From its spectral and luminosity characteristics the mass of the progenitor Sanduleak was inferred to be $\sim 20\ M_\odot$.

Stellar evolution theory predicts that large stars with mass greater than 8–10 $M_\odot$ will suffer catastrophic collapse to become neutron stars when they have exhausted the burnable nuclear fuel in their cores. When this happens the thermal pressure associated with the release of fusion energy disappears and is no longer available to support the gravitational forces so that the star collapses.

We can estimate the order of magnitude of the neutrino flux in terms of the parameters of the final neutron star. The mass of such residual stars is known from both observation and theory to be $\sim (1\text{–}2)M_\odot$ with gravitational binding energy $\sim 3 \times 10^{46}$ joules most of which will be carried away by neutrinos. During the very early cooling phase neutrinos arise mainly from thermal processes such as

$$e^+ + e^- \rightarrow \bar{\nu} + \nu$$

and each flavour of neutrino and antineutrino will carry away roughly the same amount of energy. In this case where $\bar{\nu}_e$ are present the dominant process for the large water detectors IMB and Kamiokande is

$$\bar{\nu}_e + p \rightarrow n + e^+$$

and the relevant energy for the $\bar{\nu}_e$ is thus $\sim 5 \times 10^{45}$ joules.

The process by which the collapse leads to the final neutron star is complicated. During the infall stage electrons are captured on protons and nuclei, producing electron neutrinos and turning protons into neutrons. The density of the surrounding matter however is such that the mean free path for the neutrinos is small, their diffusion time is much greater than the time for collapse (10 ms) and these neutrinos are largely trapped within the collapsing star. It is generally estimated that only $\sim 1\%$ of the binding energy of the neutron star is emitted as neutrinos during this phase.

The initial collapse is followed by a 'core bounce' in which an outgoing shock wave is produced with an expanding shock front where the wave meets the material still falling in from the outer layers. The temperature is high enough for production of e^+e^- pairs and thus antineutrinos can now also be produced via processes

$$e^+e^- \rightarrow \bar{v}+v \quad e^+ + n \rightarrow \bar{v}_e + p$$

These neutrinos and antineutrinos diffuse out of the collapsed star in times ~ 1–10 s.

The energy spectrum of the emitted neutrinos depends on the details of the supernova collapse model which involves processes not all of which are well understood. Different authors have used Boltzmann and Fermi–Dirac distributions. In fact the effective temperature T is not very sensitive to the detailed shape of the distribution.

The data

The existence of the large-mass real-time water Cerenkov detectors, IMB and Kamiokande II, allowed the observation of neutrino signals from SN1987A. These were the first-ever observations of neutrino signals from outside our solar system. They permit quite new tests of stellar evolution theory as well as providing a new laboratory for the study of neutrino properties. Signals were also recorded in the LSD detector in the Mont Blanc tunnel and in the Baksan laboratory liquid scintillation telescope.

The IMB and Kamiokande II detectors have already been described (sections 14.2.2 and 15.4.5 respectively). Kamiokande recorded a burst of 12 neutrino events within a 13 second interval shortly prior to the optical observation of the supernova. At the same time within an uncertainty of one minute, a burst of 8 events within 6 seconds was observed in the IMB detector. The Baksan liquid scintillation telescope recorded a burst of 5 events at the same time within 30 seconds as the IMB and Kamiokande bursts.

The LSD, a liquid scintillation detector, recorded a burst of 5 events which occurred 4.7 hours *before* the signals in the other detectors. This observation is difficult to understand. Although the events were all of lower energies (5.8–7.8 MeV) where the Kamiokande II detector is less efficient, nevertheless 20 events would have been expected in the Kamiokande detector on the basis of the LSD observation. Although the probability is small, it is possible that the LSD signal is due to random noise.

Except possibly in respect of their angular distribution (see below) all the data are consistent with events of the kind

$$\bar{v}_e + p \rightarrow e^+ + n$$

The Kamiokande II data have energies 6.3–35.4 MeV. The IMB detector is not sensitive to the lower energies of this range and its events lie between 19 and 41 MeV. The Baksan data range from 12–23.3 MeV. Kamiokande and IMB also record the electron angle. All detectors have precise relative timing of the events within their own data.

This statistically rather limited set of data has been subject to intense study. The fluxes and energy distributions observed in IMB and Kamiokande are broadly mutually consistent and also consistent with the expectations from stellar collapse models with a neutrino temperature ~ 4 MeV and a total energy in three neutrino flavours $\sim 3 \times 10^{46}$ joules. This direct observational evidence is of exceptional value in confirming the general correctness of the theoretical models for stellar collapse developed over many years without very much experimental data. The data from SN1987A are insufficient to test the more detailed features of the models.

There are however two possibly troublesome features evident in the present data. The angular distributions of the positrons in both the Kamiokande and IMB data are not isotropic as would be expected if the events are due to the $\bar{v}_e p \rightarrow e^+ n$ process, but peaked forward (15 forward/5 back). This might suggest that some of the events are in fact neutrino–electron scattering despite the fact that the cross section for scattering is 10^{-2} times that for the inverse β process. If this were to be the case it would require an electron neutrino luminosity for the supernova much higher than in the normal models. The early burst observed in the LSD detector, if real, is also not accounted for in the usual models.

Neutrino properties

The SN1987A data has also been used to obtain new limits on neutrino properties.

For the *rest mass* it is clear that if the rest mass is different from zero then the more energetic neutrinos will arrive before those with smaller energies *provided that* they all start at the same time or at least within a time span short compared with the arrival time span. In this case it is a simple exercise to show that, if E_{max} and E_{min} are the maximum and minimum neutrino energies and Δt is the spread of the signal, then ($E \gg m$) the mass is given by

$$m_v = E_{min} \left(\frac{2\Delta t}{d}\right)^{\frac{1}{2}} \left(\frac{\alpha^2}{\alpha^2 - 1}\right)^{\frac{1}{2}}$$

where d is the distance of the source and $\alpha = E_{\max}/E_{\min}$. Expressed in conventional units this gives

$$m_\nu = 0.612 \, E_{\min} \left(2\Delta t \, \frac{52 \, \text{kpc}}{d} \right)^{\frac{1}{2}} \left(\frac{\alpha^2}{\alpha^2 - 1} \right)^{\frac{1}{2}}$$

where $E_{\min}$ is in MeV, m_ν is in eV/c^2 and 1 kiloparsec (kpc) = 3.086×10^{19} m.

With no model for the time spread in origin of the neutrinos this expression gives an upper limit for the rest mass. This yields for the IMB and Kamiokande data respectively $m_\nu \leqslant 33$ eV/c^2 and $m_\nu \leqslant 15$ eV/c^2. However the data show no sign of the expected correlation between energy and arrival time.

In fact one can not extract a convincing mass limit without a model relating emission time and neutrino energy. For instance the dominant interaction for neutrinos as they diffuse out of the supernova may be scattering off neutrons, protons, alpha particles or heavier nuclei. Such neutral current processes have cross sections proportional to E_ν^2 and in these collisions the neutrino energy is not significantly changed. In this case the average energy of the neutrinos would increase with time – an effect in the opposite direction to that produced by the mass in free flight. On the other hand if neutrino–electron elastic scattering and charged current neutrino absorption and re-emission by nucleons are important then the neutrinos tend to be re-thermalised and the average neutrino energy may remain constant or even decrease with time. Models where E decreases with time give tighter limits on m_ν than those where E increases with time.

Detailed calculations using models where E increases or decreases with time generally agree that $m_\nu \leqslant 16$ eV/c^2 while some studies claim tighter limits. Abbott *et al.* (1988) for instance get good agreement with the data with a maximum likelihood fit using a 'hot ball' model where the *shape* of the energy spectrum turns out to be independent of time with the overall magnitude decreasing exponentially with time. The result of this fit is

$$m_\nu^2 = 4^{+28}_{-63} \, (\text{eV}/c^2)^2$$

The SN1987A data have also been used to place lower limits on the *lifetime* of the (probably stable) neutrino. For a distance of 52 kpc it is easy to calculate that the lifetime at rest $\tau_0 \gtrsim 5 \times 10^5 \, m_\nu$ s where m_ν is in eV/c^2 and we have taken a typical energy as 10 MeV. This argument is not however complete since if the $\bar{\nu}_e$ is a mixture of two mass eigenstates then it would be possible for all of the heavier state to decay and for the lighter state to be completely stable.

Deductions concerning the neutrino *magnetic moment* are presently too model dependent to warrant discussion here.

16
Epilogue

Particle physicists sometimes complain that there has been 'no exciting new development since last year's conference'. The complaint is justified only because the overall pace of discovery in this field has been so great. In the past decade the pace has been maintained with the discovery of the Z and W bosons and the establishment of three as the number of light neutrinos thus most probably fixing the number of the lepton and quark generations in the Universe.

Technical achievements have included the bringing into successful operation of LEP at CERN and the Tevatron at Fermilab and the collision of microscopically tightly focussed beams at SLAC. The detectors at these accelerators are among the most sophisticated pieces of hardware ever built. The great proton decay detectors have detected no decaying protons but have opened a new window on the Universe in neutrino astronomy.

In the 1960s and much of the 1970s particle physics was primarily concerned with getting the data on which QCD and Electroweak Unification have been built. Our understanding of the fundamental structure of matter has been transformed by these achievements. Current and future work is focussed on very fundamental questions which were not even crystallised in the earlier era. There remain many challenging problems:

What is the mass of the top quark – when found this will open up
   a new field of toponium spectroscopy.
Can we measure the proton lifetime?
Are the neutrinos massive or not?
What is the answer to the solar neutrino problem?

Is the Higgs mechanism the source of mass and if so what is the
   mass of the Higgs boson and can we find it in experiment?
Why does nature need three families of quarks and leptons?
What sets the relative mass scales of these families?
Are the quarks and leptons really point particles?
What is the true grand unification theory?

What is the nature of the dark matter?
What is the real origin of *CP* violation?
Is the Universe closed and if so what is the mechanism which has
   set the scale of the baryon and dark matter densities?
Why is the cosmological constant zero or close to zero?
Why does our Universe appear to depend on tiny symmetry
   violations?
Is string theory really a 'theory of everything' and how can it be
   tested?

This list is not exhaustive. I give it in order to indicate the nature and
range of the problems addressed. The renewed marriage between particle
physics and cosmology is mutually stimulating for experiment and theory
in both disciplines. Coupled with the advent of the new accelerators
recently commissioned, presently under construction and funded for
completion in the nineties, the prospects for answers to at least some of
these questions are very good.

On a number of occasions in the past it has seemed that physics has
achieved an almost complete understanding of the basic structure of
matter. Nature has however always turned out to be much richer than
anyone had conceived. I expect it will happen again.

Appendix A
Relativistic kinematics and phase space

A.1   **Lorentz transformations**

In elementary particle physics we frequently have to deal with problems of relativistic particle kinematics. Most frequently the problems we meet are concerned with transformations between the laboratory system, where measurements are made, and the centre-of-momentum system (often misleadingly but conveniently called the centre-of-mass system), or cms, to which the theoretical predictions apply directly.

Lorentz transformations are transformations which satisfy the basic postulate of special relativity theory that the velocity of light is the same for all observers, regardless of their state of relative motion. The particular underlying importance of the proper Lorentz transformations lies in the fact that they correspond to the transformations associated with the operators of quantum mechanics, so that for Lorentz transformations a true physical law will have the same form in systems moving relative to each other.

Consider two systems, (1) and (2), moving with uniform relative translational motion (fig. A.1). In the classical case the positions of A in the two systems are clearly related by

$$\mathbf{r} = \mathbf{r}' + \mathbf{V}t. \qquad \text{A.1}$$

$\ddot{\mathbf{r}} = \ddot{\mathbf{r}}'$, so that Newton's laws hold in both systems. However, the relativity postulate does not hold. Suppose a light pulse is emitted from O which after a time t has a spherical waveform on which lies A. Then differentiating **A.1** with regard to time we have

$$\mathbf{c} = \mathbf{c}' + \mathbf{V}.$$

For a satisfactory relativistic theory we proceed as follows: Suppose that O and O′ are coincident at time $t = 0$ and that at this instant a light pulse is emitted from O, In (1) the wavefront has the equation

$$r^2 = x^2 + y^2 + z^2 = c^2 t^2$$

or

$$r^2 - c^2 t^2 = x^2 + y^2 + z^2 - c^2 t^2 = 0.$$

If c is the same in both systems, then in (2) the wavefront has equation

$$r'^2 - c^2 t'^2 = x'^2 + y'^2 + z'^2 - c^2 t'^2 = 0.$$

Thus

$$r'^2 - c^2 t'^2 = r^2 - c^2 t^2$$

or

$$x'^2 + y'^2 + z'^2 - c^2 t'^2 = x^2 + y^2 + z^2 - c^2 t^2.$$

We can write this as

$$\sum_{i=1}^{4} x_i^2 = \sum_{i=1}^{4} x_i'^2 = \text{constant}, \qquad \textbf{A.2}$$

where we have written

$$x = x_1, \quad y = x_2, \quad z = x_3, \quad ict = x_4.$$

The x_i are the components of a *Lorentz-invariant four-vector*. We shall discuss other such four-vectors below.

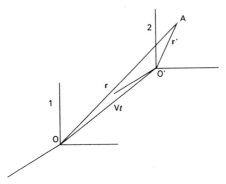

Fig. A.1

Let us examine the *transformation operators*. If we have an observable Q in one system and the value in the second system is Q' then the transformation operator L is such that

$$Q' = LQ. \qquad \textbf{A.3}$$

Since for invariant four-vectors we have the relationship **A.2** the transformations must be equivalent to rotations in a four-dimensional space with orthogonal axes along the x_i directions. First we make a spatial rotation to get V, the relative velocity, along x_3. We now take x_1, x_2, x_3 and x_4 to be the rotated coordinates. x_1 and x_2 are then unaffected by the Lorentz transformation, which is equivalent to a rotation in the x_3, x_4 plane. We now write the transformed quantities x' as

$$x'_j = \sum_{i=1}^{4} a_{ji} x_i, \qquad \textbf{A.4}$$

where the a_{ji} are linked by the orthogonality conditions

$$\sum_{i} a_{ji} a_{ki} = \delta_{jk}. \qquad \textbf{A.5}$$

Since x'_3 and x'_4 cannot depend on x_1 and x_2, L can be shown to be a 4×4 matrix having the form

$$L = \begin{bmatrix} 1 & 0 & 0 & 0 \\ 0 & 1 & 0 & 0 \\ 0 & 0 & a_{33} & a_{34} \\ 0 & 0 & a_{43} & a_{44} \end{bmatrix}. \qquad \textbf{A.6}$$

This transformation operator can be applied to any four-vector. If we take the particular case of the (x, y, z, ict) four-vector, we can write

$$x_3 = Vt = -i\beta x_4, \qquad \textbf{A.7}$$

where $\beta = V/c$.

Using equations **A.3–7** the student should show that

$$a_{33} = a_{44} = \frac{1}{\sqrt{(1-\beta^2)}} = \gamma,$$

$$a_{34} = i\beta\gamma \quad \text{and} \quad a_{43} = -i\beta\gamma.$$

One of the most useful of four-vectors is that formed by the energy and the three components of the momentum. If we write the total energy as ε, the momentum as p and the rest mass as m then we can write

$$\varepsilon^2 - p^2 c^2 = m^2 c^4.$$

We may choose to use units such that $c = 1$ and write this as

$$\varepsilon^2 - p^2 = m^2.$$

The rest mass is invariant, so that $(p_x, p_y, p_z, i\varepsilon) = (\mathbf{p}, i\varepsilon)$ is a Lorentz-invariant four-vector.

Writing

$$\mathbf{p} = m\mathbf{v}\gamma = m\boldsymbol{\beta}\gamma \quad \text{and} \quad \varepsilon = m\gamma,$$

where $\mathbf{v}$ is a particle velocity we have

$$\mathbf{v} = \mathbf{p}/\varepsilon.$$

Also for a system of particles with total energy E and total momentum $\mathbf{P}$ the velocity $\mathbf{V}$ is given by $\mathbf{P}/E$. Thus for two particles

$$\mathbf{V}_{cms} = \frac{\mathbf{P}}{E} = \frac{\mathbf{p}_1 + \mathbf{p}_2}{\varepsilon_1 + \varepsilon_2}.$$

Suppose we now have a particle moving with momentum p at an angle θ to the x_3 axis, which is the direction of motion of the centre of momentum. Then if V is the cms velocity we can apply the operator L to derive the following transformation equations, linking laboratory scalar momentum p, the angle θ and the energy ε to the cms quantities. We find

$$\varepsilon' = \gamma(\varepsilon - Vp\cos\theta),$$
$$p'\cos\theta' = \gamma(p\cos\theta - V\varepsilon),$$
$$p'\sin\theta' = p\sin\theta,$$
$$\varepsilon = \gamma(\varepsilon' + Vp'\cos\theta'),$$
$$p\cos\theta = \gamma(p'\cos\theta' + V\varepsilon').$$
$$(\mathbf{V}\cdot\mathbf{p} = Vp\cos\theta \text{ and } \mathbf{V}\cdot\mathbf{p}' = Vp'\cos\theta')$$

The student should check these relations as an exercise.

A.2   The centre-of-mass system

In the cms the total momentum is zero; for a two-particle collision the particles approach each other with equal and opposite momenta. This system is unique, and calculations which depend on energy and momentum are considerably simplified by its use. We use the following notation:

E, ε: total energy of system or particle
T, t: kinetic energy of system or particle
$\mathbf{P}, \mathbf{p}$: momentum of system or particle
M, m: mass of system or particle
$\mathbf{V}, \mathbf{v}$: velocity of cms, velocity of particle.

As before, we write $c = 1$ and used dashed quantities for the cms and undashed quantities for the laboratory system.

Then

$$\varepsilon^2 = p^2 + m^2, \quad \varepsilon = t + m, \quad p^2 = t^2 + 2mt.$$

Using the energy momentum four-vector invariance we have

$$E^2 - P^2 = E'^2 - P'^2.$$

But, by definition, $P' = 0$ so that E', the energy available in the cms, is given by

$$E'^2 = E^2 - P^2.$$

For example, consider a collision between a particle m_1, v_1, ε_1, $\mathbf{p}_1$ in the laboratory and a stationary particle m_2, $v_2 = 0$, $\varepsilon_2 = m_2$, $\mathbf{p}_2 = 0$. Then

$$\begin{aligned} E'^2 &= (\textstyle\sum \varepsilon)^2 - (\textstyle\sum \mathbf{p})^2 \\ &= (\varepsilon_1 + m_2)^2 - p_1^2. \end{aligned}$$

Note that in the system of units with $c = 1$ we express momenta in units of MeV/c and masses in units of MeV/c^2. Suppose we wish to determine the threshold momentum for the process

$$\pi^- p \to K^0 \Lambda^0.$$

The reaction will become possible when E' is just adequate to provide the masses of the secondary particles.

$$E' = m_\Lambda + m_K = 1.613.$$

Writing

$$E'^2 = \varepsilon_1^2 - p_1^2 + m_2^2 + 2\varepsilon_1 m_2$$

we have

$$\begin{aligned} \varepsilon_1 &= \frac{1}{2m_2}(E'^2 - m_1^2 - m_2^2) \\ &= 0.904 \text{ GeV}. \end{aligned}$$

Thus

$$t_1 = 0.764 \text{ GeV}, \quad p = 0.89 \text{ GeV}/c.$$

The following easily remembered formula for the threshold kinetic energy for a process may be derived as an exercise. If the total initial and final masses are M_i and M_f, respectively, while the mass of the target particle is m_s, then

$$T_{\text{threshold}} = \frac{(M_f - M_i)(M_f + M_i)}{2m_s}.$$

The slow increase of E' with bombarding energy, for a stationary target, illustrates the advantage of colliding-beam experiments possible with storage rings, where E' is simply the sum of the energies of the colliding particles (if they are of equal mass).

A.3   **Geometrical picture of transformations**

A method of gaining a picture of the effect of the laboratory–cms transformation, as well as a simple way of deriving certain formulae, is afforded by the momentum ellipsoid proposed by Blaton (1950).

We first represent the non-relativistic situation in fig. A.2.

We draw a velocity-vector diagram where OO′ represents the velocity of the cms and OM the velocity of a particle in the laboratory. O′M then represents the velocity in the cms. Multiplying these vectors by m_1, the rest mass of the particle, OM now represents the laboratory momentum and O′M the cms momentum. In a two-particle process the momentum of the other particle must be given by O′N in the cms. As θ and ϕ, the azimuthal angle, vary, M and N move on the surface of the sphere centred on O′ with radius p'.

We can readily generalize to the relativistic case. With V along the x-axis we have the relations

$$p'_x = \frac{1}{\gamma}(p_x - V\gamma\varepsilon'), \quad p'_y = p_y, \quad p'_z = p_z$$

so that

$$p'^2 = p'^2_x + p'^2_y + p'^2_z$$
$$= p^2_z + p^2_y + \frac{1}{\gamma^2}(p_x - V\gamma\varepsilon')^2$$

or

$$\frac{p^2_z + p^2_y}{p'^2} + \frac{(p_x - V\gamma\varepsilon')^2}{p'^2\gamma^2} = 1,$$

which is the equation of a prolate ellipsoid of revolution with axis along the x-axis and its centre displaced a distance $V\gamma\varepsilon'$ from the origin of the

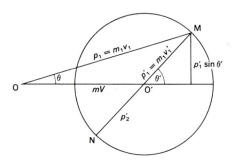

Fig. A.2.

p_x vector. The section in the xy plane is shown in fig. A.3. The size and shape of the ellipsoid are determined by the masses and energies available in the reaction. When two particles emerge one of which has cms momentum $\mathbf{p}_1'$, the other must have momentum $\mathbf{p}_2'$. These can be seen at once to transform to the laboratory momenta $\mathbf{p}_1 = \text{OM}$ and $\mathbf{p}_2 = \text{ON}$. It is also clear that the maximum and minimum laboratory momenta are OB and OA, and the maximum angle of emission of a particle in the laboratory corresponds to the tangent OL and is thus $\hat{\text{L}}\text{OA}$.

The student can show that the eccentricity of the ellipse is V and that the ratio of the semi-major to the semi-minor axis, $a/b = \gamma$.

It is useful to distinguish two different geometrical configurations corresponding to different physical situations. If we write $\alpha = V\gamma\varepsilon'$, then if $\alpha > a$, O is outside the ellipse as in fig. A.3. Since $a = p'\gamma$

$$\alpha > a \quad \text{corresponds to} \quad V\varepsilon' > p',$$
$$\text{i.e.} \quad V > v',$$

the velocity of the centre-of-momentum is greater than the velocity of the particle in the cms. In this case all particles go forward in the laboratory and there exists a maximum angle of emission. If $\alpha < a$ then O is inside the ellipsoid, the velocity of the cms is less than the particle velocity in the cms, and θ can vary from 0 to π.

From the geometry of the ellipse (or by differentiating the expression for $\tan \theta$) it can be shown that

$$\sin \theta_{\max} = \sqrt{\left(\frac{1 - V^2}{\gamma_n^2 - V^2}\right)} = \frac{b^2}{fm}, \qquad \textbf{A.8}$$

where we have written γ_n for V/v_n', where v_n' is the cms velocity, m the mass of the particle and f is the focal distance of the ellipse.

A.4   Decay length

As a final application of the momentum ellipsoid we may use it to visualise the variation of the decay length with angle for unstable particles emitted from a reaction and decaying in flight.

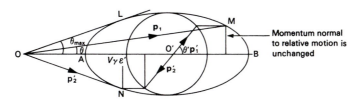

Fig. A.3

We define the *decay length* l in the laboratory system as the distance travelled in one mean life. Then

$$l_{\text{lab}} = v_{\text{lab}}\tau.$$

As is well known, a moving clock appears to the 'stationary' observer to run slow. If τ_0 is the mean life of the particle at rest then

$$\tau = \gamma\tau_0$$

and

$$l = v\gamma\tau_0 = \frac{p}{m}\tau_0 = p\frac{\tau_0}{m},$$

where p is the laboratory momentum. Thus

$$l \propto p$$

and if we look on the momentum vector diagram as a space decay diagram then the ellipsoid is the locus of the decay points of all particles which live exactly one mean life. For instance, consider the decay in flight of Σ-hyperons produced in the reactions

$$\text{K}^-\text{p} \begin{array}{l} \to \Sigma^+\pi^- \\ \to \Sigma^-\pi^+, \end{array}$$

by K^- mesons of a given momentum. A simple calculation gives us p', the Σ momentum in the cms. We can now draw the ellipse with semi-minor axis scaled to equal

$$v'\tau_0 = \frac{p'}{m}\tau_0.$$

The ellipse will then represent the locus of decay points for Σs living for one mean life.

A.5   Limiting kinematic relationships for many-particle systems

The relationships discussed above are general for all laboratory–cms transformations. For reactions from which only two particles emerge they have a direct application since the cms momenta of the secondary particles are equal and opposite. Thus in a given reaction between two particles of fixed energy there is a *unique relationship* between the angles and momenta of the emerging particles given by the foregoing equations or the ellipse diagrams.

If three or more particles emerge there are no longer unique relationships of this kind but only limiting relationships. As an example we calculate the *maximum momentum* of the $(j+1)$th particle in the cms for a reaction from which $(j+1)$ particles emerge. The basis of this calculation is that any particle will attain its maximum momentum for a fixed total energy E' of all particles in the cms, when all the other particles go off *together* in the opposite direction. Thus

$$v_1' = v_2' = v_3' = \cdots = v_j'$$

so that

$$\frac{p_1'}{\varepsilon_1'} = \frac{p_2'}{\varepsilon_2'} = \cdots = \frac{p_j'}{\varepsilon_j'}$$

and

$$\frac{p_1'}{m_1} = \frac{p_2'}{m_2} = \cdots = \frac{p_j'}{m_j}.$$

We can thus write all the 1 to j momenta in terms of p_1' and the masses

$$p_j' = \frac{m_j}{m_1}p_1', \text{ etc.}$$

Then for maximum $p_{j+1}' = p_{max}'$

$$p_{max}' = p_1' + p_2' \cdots + p_j'$$
$$= p_1'\left(1 + \sum_{i=2}^{j}\frac{m_i}{m_1}\right). \qquad \textbf{A.9}$$

Also

$$E' = \varepsilon_1' + \varepsilon_2' + \cdots + \varepsilon_j' + \varepsilon_{j+1}'$$
$$= (p_1'^2 + m_1^2)^{\frac{1}{2}} + \sum_{i=2}^{j}\left(\frac{m_i^2}{m_1^2}p_1'^2 + m_i^2\right)^{\frac{1}{2}} + \left[\left(1 + \sum_{i=2}^{j}\frac{m_i}{m_1}\right)^2 p_1'^2 + m_{j+1}^2\right]^{\frac{1}{2}}. \qquad \textbf{A.10}$$

We can eliminate p_1' from equations **A.9** and **A.10** to solve for p_{max}' as

$$p_{max}'^2 = \frac{\left[E'^2 - \left(m + \sum_{i=1}^{j} m_i\right)^2\right]\left[E'^2 - \left(m - \sum_{i=1}^{j} m_i\right)^2\right]}{4E'^2}$$

or

$$\varepsilon_{max}' = \frac{E'^2 + m^2 - \left(\sum_{i=1}^{j} m_i\right)^2}{2E'}.$$

where we have written $m = m_{j+1}$. In this limiting situation we have effectively a two-body process so that the maximum of all $\sin\theta_{max}$ is given by substituting the velocity corresponding to the above ε_{max}' into equation **A.8**.

As an example we show in fig. A.4 the accessible kinematic regions for particles from the reaction pp → pnπ⁺, where the incident energy has been chosen such that for the nucleon $V > v_{n(max)}$. Thus at any angle the limiting value of ε_n is double valued as one would expect from the ellipsoid. For the pion $V < v_{max}$, ε_π is thus single-valued and there is no θ_{max}.

A.6   Processes with three particles in the final state: the Dalitz plot

We have seen that, for three or more particles in the final state, only limiting kinematic relationships may be given. A particularly useful way of treating the three-particle case was proposed by Dalitz. The particular merit of this kind of plot arises from the uniform distribution of events within it if the reaction proceeds according to the available density of states. We shall return to this topic in the next section. Here, we deal only with the plot limits.

We first consider the case in which all three outgoing particles have the same mass, such as, for instance, the τ-decay of the K-meson into three pions. If the total energy in the cms is $E' (= m_\tau + T_\tau)$ then

$$E' = \varepsilon_1' + \varepsilon_2' + \varepsilon_3'$$

and

$$E' - \sum_{i=1}^{3} m_i = T' = t_1' + t_2' + t_3'.$$

T' is a constant and so the three-particle kinetic energies in the cms can be

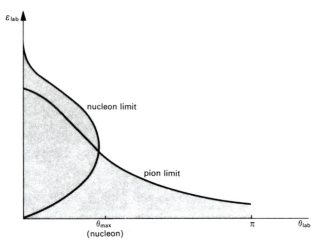

Fig. A.4

represented as perpendicular distances from a point within a triangle to the three sides. With the x- and y-axes drawn as in fig. A.5. we see that $y = t_1'$ and

$$x = \frac{1}{\sqrt{3}}(t_3' - t_2').$$

The conservation of energy required can then be satisifed for all points within the triangle. However, the conservation of momentum imposes a tighter limit on the allowed region. In the classical case this limit is the inscribed circle, as can be seen by noting that the momentum limit is set by the condition

$$\mathbf{p}_1 + \mathbf{p}_2 + \mathbf{p}_3 = \mathbf{0}, \qquad\qquad \textbf{A.11}$$

so that in the classical case

$$(m\mathbf{v}_1 + m\mathbf{v}_2)^2 = m^2\mathbf{v}_3^2$$

from which

$$t_1^2 + t_2^2 + t_3^2 - 2(t_1 t_2 + t_2 t_3 + t_3 t_1) = 0,$$

which is the inscribed circle

$$x^2 + \left(y - \frac{t_1 + t_2 + t_3}{3}\right)^2 = \left(\frac{t_1 + t_2 + t_3}{3}\right)^2.$$

In the extreme relativistic case we get a triangle, as shown in the figure, while in the usual intermediate situation we have a shape between these extremes, the equation of which may again be calculated by applying equation **A.11**.

Before leaving the symmetrical case we note some other features of the plot. If in the cms we group two particles which then have relative momentum $\mathbf{q}$, and with respect to which the third particle has momentum

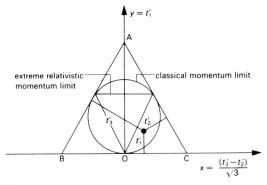

Fig. A.5

p, we have the configuration shown in fig. A.6. $\mathbf{p}_1$, $\mathbf{p}_2$ and $-\mathbf{p}$ are then the momenta in the overall cms. The student may check that

$$p^2 \propto PN,$$
$$q^2 \propto PQ,$$
$$\cos\theta = \frac{GP}{GH}.$$

A.12

where P, N, Q, G, H are shown in fig. A.7. The dependences of **A.12** mean that if, for instance, the cross-section for a process is a function of $\cos\theta$, then this will be reflected as a variation in the density of the population across the circle.

If the masses of the three particles involved are not equal, the Dalitz plot is no longer symmetrical and there is then little merit in plotting

$$\frac{1}{\sqrt{3}}(t_2' - t_3')$$

along the x-axis rather than simply t_2', say. Fig. A.8 shows the Dalitz plot for a process such as $K^-p \rightarrow \Lambda\pi^+\pi^-$. In this case we have plotted t_{π^+}' *against* t_{π^-}' so that the plot is symmetrical about the 45° line. As before, the boundary corresponds to collinear particles. At A, $t_{\pi^-}' = t_{\pi^+}'$ and t_Λ' is a maximum; at B, $t_\Lambda' = 0$. Along the tangent at B,

$$t_{\pi^+}' + t_{\pi^-}' = \text{constant} = E' - t_\Lambda',$$

so t_Λ' is constant along this and parallel lines, increasing from zero at B to a maximum at A.

Fig. A.6

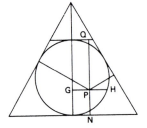

Fig. A.7

A.7 Phase space

Having examined the kinematic limits we now turn to the distribution within the limits, that is, to the probability of the occurrence of a given set of particle momenta within the allowed range of momenta.

We start from the formula for the transition rate from the initial to the final state

$$T = \frac{2\pi}{\hbar}|M|^2\frac{dN}{dE}. \qquad \textbf{A.13}$$

$|M|^2$ is the square of the modulus of the matrix element linking the states. dN/dE is the density of final states or 'phase-space' factor. If M is not a function of the final-state momenta, then such a dependence will be contained only in the phase-space factor.

It is convenient to work in a phase space where the coordinates represent the momentum components of a given particle. If there are n particles in the final state, then it can be represented by a vector in a $3n$-dimensional phase space. Imposition of energy conservation limits this vector, so that all possible states are seen to lie within the volume swept out by it in the $3n$-dimensional space. The number of states is then obtained by dividing this volume by the volume per state, and the density of states is obtained by differentiating this number with respect to E. Thus if $Q(E)$ is the volume in the $3n$-dimensional space we write

$$Q(E) = \prod_{i=1}^{n}\int d\mathbf{p}_i,$$

where there are implicit limits set by momentum and energy conservation. We shall make these limits explicit, and carry out the differentiation with

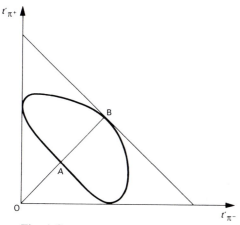

Fig. A.8

respect to energy, below, but shall first divide by the volume per state, which is obtained by the standard method of statistical mechanics.

The number of plane waves in a frequency interval $v - v + \delta v$ within a volume V is given by

$$\delta N(v) = N(v)\,\delta v = 4\pi V \frac{v^2\,\delta v}{c^3}.$$

In terms of momentum this translates into a number of de Broglie waves

$$N(p)\,\delta p = \frac{4\pi V}{h^3} p^2\,\delta p = \frac{V}{(2\pi\hbar)^3} 4\pi p^2\,\delta p.$$

These are contained in a phase-space element which is a shell of thickness δp, radius p and volume $4\pi p^2\,\delta p$. Each possible state corresponds to a discrete point in phase space, so that the discrete quantized volume occupied by each is $(2\pi\hbar)^3/V$. For the n-particle case, the unit cell volume is $(2\pi\hbar/V^{\frac{1}{3}})^{3n}$, and the number of states is

$$N = \left(\frac{V^{\frac{1}{3}}}{2\pi\hbar}\right)^{3n} Q(E).$$

We may impose the momentum and energy conservation explicitly by considering first one integral

$$G(E) = \int d^3\mathbf{p}_1\,\theta(x),$$

where $\theta(x)$ is a step function inserted to limit the states to the allowed region, and is illustrated in fig. A.9. Only at A is $d\theta/dx$ different from zero, and at A

$$\frac{d\theta}{dx} = \delta(x_A).$$

Thus

$$\frac{dG(E)}{dE} = \int d^3\mathbf{p}_1\,\frac{d\theta(E'-\varepsilon_1)}{dE} = \int d^3\mathbf{p}_1\,\delta(E'-\varepsilon_1).$$

We can ensure momentum conservation by including another δ-function. Returning also to the multiparticle equation we have

$$\frac{dN}{dE} = \left(\frac{V^{\frac{1}{3}}}{2\pi\hbar}\right)^{3n} \prod_{i=1}^{n} \int d^3\mathbf{p}_i\,\delta\left(E' - \sum_{i=1}^{n}\varepsilon_i\right) \delta^3\left(\sum_{i=1}^{n}\mathbf{p}_i\right). \qquad \textbf{A.14}$$

The expression for the density of states which follows from **A.14** is not itself Lorentz invariant, which means that, since the transition rate *is* Lorentz invariant, the matrix element itself must still contain kinematic factors.

It is easy to see qualitatively how the result must be modified to render the phase space or density of states factor invariant and thus extract such kinematic factors from the matrix element. The normal probability density is expressed in quantum mechanics as the square of the modulus of the wave function for each particle $|\psi|^2$, with normalisation $\int |\psi|^2 \, \mathrm{d}x \, \mathrm{d}y \, \mathrm{d}z = 1$. Due to the relativistic contraction of a moving system, this density increases by a factor γ for a moving system so that the density is not invariant. However, since the total energy ε also varies as γ, we can write a density $|\sqrt{\varepsilon}\psi|^2$ which is invariant. In fact, it is convenient to use $|\sqrt{(2\varepsilon)}\psi|^2$. Then, introducing $\sqrt{(2\varepsilon_i)}$ to the matrix element for each particle, we have a Lorentz-invariant element M' and we write instead of **A.13** and **A.14**

$$T = \frac{2\pi}{h} |M'|^2 \left(\frac{V^{\frac{1}{3}}}{2\pi h} \right)^{3n} \prod_{i=1}^{n} \int \frac{\mathrm{d}^3 \mathbf{p}_i}{2\varepsilon_i} \delta \left(E' - \sum_{i=1}^{n} \varepsilon_i \right) \delta^3 \left(\sum_{i=1}^{n} \mathbf{p}_i \right). \qquad \textbf{A.15}$$

We may also approach this equation by working directly in terms of the energy momentum four-vectors $\tilde{p}_i$ and writing the density of states factors as

$$\frac{\mathrm{d}N}{\mathrm{d}E} \propto \prod_{i=1}^{n-1} \int \mathrm{d}^4 \tilde{p}_i \, \delta(p_i^2 - m_i^2)$$

$$= \prod_{i=1}^{n-1} \int \mathrm{d}^3 \mathbf{p}_i \, \mathrm{d}\varepsilon_i \, \delta(\varepsilon_i^2 - p_i^2 - m_i^2)$$

$$= \prod_{i=1}^{n-1} \int \mathrm{d}^3 \mathbf{p}_i \, \delta(\varepsilon_i^2 - p_i^2 - m_i^2) \, \mathrm{d}(\varepsilon_i^2 - p_i^2 - m_i^2) \frac{\mathrm{d}\varepsilon_i}{\mathrm{d}(\varepsilon_i^2 - p_i^2 - m_i^2)}$$

$$= \prod_{i=1}^{n-1} \int \frac{\mathrm{d}^3 \mathbf{p}_i}{2\varepsilon_i} \, \mathrm{d}(\varepsilon_i^2 - p_i^2 - m_i^2) \, \delta(\varepsilon_i^2 - p_i^2 - m_i^2).$$

But the integration over $(\varepsilon_i^2 - p_i^2 - m_i^2)$ is cancelled by the δ-function, so that the expression becomes

$$\prod_{i=1}^{n-1} \int \frac{\mathrm{d}^3 \mathbf{p}_i}{2\varepsilon_i} \, \delta(\varepsilon_i^2 - p_i^2 - m_i^2),$$

which is directly related to equation **A.15**.

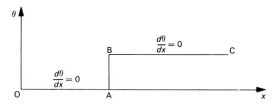

Fig. A.9

We now specialise to the three-particle case and integrate over the uninteresting variables.

$$\frac{\mathrm{d}N}{\mathrm{d}E} = \frac{V^3}{(2\pi\hbar)^9} \int \frac{\mathrm{d}\mathbf{p}_1}{2\varepsilon_1} \int \frac{\mathrm{d}\mathbf{p}_2}{2\varepsilon_2} \int \frac{\mathrm{d}\mathbf{p}_3}{2\varepsilon_3} \delta\left(\sum_{i=1}^{3} \mathbf{p}_i\right) \delta\left(E' - \sum_{i=1}^{3} \varepsilon_i\right).$$

The momenta are related as in fig. A.10 and

$$\delta\mathbf{p} = \delta p + p\, \delta\theta + p \sin\theta\, \delta\phi$$

so that

$$\int \mathrm{d}\mathbf{p} = \int \mathrm{d}p \int p\, \mathrm{d}\theta \int p \sin\theta\, \mathrm{d}\phi$$

$$= \int p^2\, \mathrm{d}p \int \mathrm{d}\Omega.$$

Thus

$$\iint \frac{\mathrm{d}\mathbf{p}_1\, \mathrm{d}\mathbf{p}_2}{\varepsilon_1\, \varepsilon_2} = \int \frac{4\pi p_1^2}{\varepsilon_1}\, \mathrm{d}p_1 \int p_2^2 \frac{\mathrm{d}p_2}{\varepsilon_2} \int \mathrm{d}\phi\, \mathrm{d}(\cos\theta)$$

$$= \int \frac{4\pi p_1^2}{\varepsilon_1}\, \mathrm{d}p_1 \int 2\pi \frac{p_2^2}{\varepsilon_2}\, \mathrm{d}p_2 \int \mathrm{d}(\cos\theta),$$

while the integration over $\mathbf{p}_3$ is now redundant since it is fixed.

To express the integration over $\cos\theta$ in terms of the other quantities we write

$$\varepsilon_3^2 = p_3^2 + m_3^2$$
$$= m_3^2 + p_1^2 + p_2^2 + 2p_1 p_2 \cos\theta.$$

Then, taking differentials with respect to $\cos\theta$ and holding p_1 and p_2 fixed, we have

$$2p_1 p_2\, \delta(\cos\theta) = 2\varepsilon_3\, \delta\varepsilon_3.$$

Using $p\, \delta p = \varepsilon\, \delta\varepsilon$ (since $\varepsilon^2 = p^2 + m^2$), we have

$$\frac{\mathrm{d}N}{\mathrm{d}E} = \frac{V^3}{(2\pi\hbar)^9} \frac{8\pi^2}{8} \int \mathrm{d}\varepsilon_1 \int \mathrm{d}\varepsilon_2 \int \mathrm{d}\varepsilon_3\, \delta\left(E' - \sum_{i=1}^{3} \varepsilon_i\right)$$

and finally dropping the redundant integration

$$T = \frac{2\pi}{\hbar} |M'|^2 \frac{V^3}{(2\pi\hbar)^9\, \pi^2} \iint \mathrm{d}\varepsilon_1\, \mathrm{d}\varepsilon_2,$$

Fig. A.10

i.e.

$$T \propto \iint d\varepsilon_1 \, d\varepsilon_2$$

$$= \iint dt_1 \, dt_2.$$

But $\delta t_1 \, \delta t_2$ is the element of area in the Dalitz plot so that:

If a process is governed solely by the density of states factor the density of the population in the Dalitz plot will be uniform. Any deviation from uniformity is indicative of a special interaction depending on the relative momenta of the particles, such as, for instance, a strong force binding two particles together.

It is easy to show that in the symmetrical Dalitz plot for three particles of equal mass, where $x \propto (t_3 - t_2)$ and $y \propto t_1$, then $\delta y \propto \delta t_1$ and $\delta x \propto \delta t_2$. The Dalitz plot is frequently made in terms of the effective-mass squared of particle pairs. Such a plot still has the property that the elementary area $\delta m_{ij}^2 \, \delta m_{jk}^2$ is proportional to the element of phase space, since

$$m_{ij}^2 = (\varepsilon_i + \varepsilon_j)^2 - (\mathbf{p}_i + \mathbf{p}_j)^2$$
$$= (E - \varepsilon_k)^2 - p_k^2,$$
$$\delta m_{ij}^2 = -2(E - \varepsilon_k)\delta\varepsilon_k - 2p_k\,\delta p_k$$
$$= -2E\,\delta\varepsilon_k.$$

So that

$$\delta m_{ij}^2 \, \delta m_{jk}^2 \propto \delta\varepsilon_k \, \delta\varepsilon_i$$

and thus the effective-mass squared Dalitz plot will also be uniformly populated if the reaction is regulated by phase space.

A number of examples of the Dalitz plot are shown in chapter 9.

Appendix B

Clebsch–Gordan coefficients and particle properties

B.1 Clebsch–Gordan coefficients

$$\psi(J,M) = \sum_{m_1, m_2} C(J,M,j_1,m_1,j_2,m_2)\phi(j_1,m_1)\chi(j_2,m_2)$$

For each pair of values of j_1 and j_2 the tables give:

$$
\begin{array}{cc}
J & J & \dots \\
M & M & \dots
\end{array}
$$

$$
\begin{array}{cc}
m_1 & m_2 \\
m_1 & m_2 \\
\vdots & \vdots
\end{array}
$$

In all cases the *squares of the coefficients are given.*

(a) $j_1 = \tfrac{1}{2}$, $j_2 = \tfrac{1}{2}$

		1	1	0	1
		+1	0	0	−1
$+\tfrac{1}{2}$	$+\tfrac{1}{2}$	1			
$+\tfrac{1}{2}$	$-\tfrac{1}{2}$		$\tfrac{1}{2}$	$\tfrac{1}{2}$	
$-\tfrac{1}{2}$	$+\tfrac{1}{2}$		$\tfrac{1}{2}$	$-\tfrac{1}{2}$	
$-\tfrac{1}{2}$	$-\tfrac{1}{2}$				1

(b) $j_1 = 1$, $\quad j_2 = \frac{1}{2}$

		$\frac{3}{2}$ $+\frac{3}{2}$	$\frac{3}{2}$ $+\frac{1}{2}$	$\frac{1}{2}$ $+\frac{1}{2}$	$\frac{3}{2}$ $-\frac{1}{2}$	$\frac{1}{2}$ $-\frac{1}{2}$	$\frac{3}{2}$ $-\frac{3}{2}$
$+1$	$+\frac{1}{2}$	1					
$+1$	$-\frac{1}{2}$		$\frac{1}{3}$	$\frac{2}{3}$			
0	$+\frac{1}{2}$		$\frac{2}{3}$	$-\frac{1}{3}$			
0	$-\frac{1}{2}$				$\frac{2}{3}$	$\frac{1}{3}$	
-1	$+\frac{1}{2}$				$\frac{1}{3}$	$-\frac{2}{3}$	
-1	$-\frac{1}{2}$						1

(c) $j_1 = 1$, $\quad j_2 = 1$

		2 $+2$	2 $+1$	1 $+1$	2 0	1 0	0 0	2 -1	1 -1	2 -2
$+1$	$+1$	1								
$+1$	0		$\frac{1}{2}$	$\frac{1}{2}$						
0	$+1$		$\frac{1}{2}$	$-\frac{1}{2}$						
$+1$	-1				$\frac{1}{6}$	$\frac{1}{2}$	$\frac{1}{3}$			
0	0				$\frac{2}{3}$	0	$-\frac{1}{3}$			
-1	$+1$				$\frac{1}{6}$	$-\frac{1}{2}$	$\frac{1}{3}$			
0	-1							$\frac{1}{2}$	$\frac{1}{2}$	
-1	0							$\frac{1}{2}$	$-\frac{1}{2}$	
-1	-1									1

(d) $j_1 = \frac{3}{2}$, $\quad j_2 = \frac{1}{2}$

		2 $+2$	2 $+1$	1 $+1$	2 0	1 0	2 -1	1 -1	2 -2
$+\frac{3}{2}$	$+\frac{1}{2}$	1							
$+\frac{3}{2}$	$-\frac{1}{2}$		$\frac{1}{4}$	$\frac{3}{4}$					
$+\frac{1}{2}$	$+\frac{1}{2}$		$\frac{3}{4}$	$-\frac{1}{4}$					
$+\frac{1}{2}$	$-\frac{1}{2}$				$\frac{1}{2}$	$\frac{1}{2}$			
$-\frac{1}{2}$	$+\frac{1}{2}$				$\frac{1}{2}$	$-\frac{1}{2}$			
$-\frac{1}{2}$	$-\frac{1}{2}$						$\frac{3}{4}$	$\frac{1}{4}$	
$-\frac{3}{2}$	$+\frac{1}{2}$						$\frac{1}{4}$	$-\frac{3}{4}$	
$-\frac{3}{2}$	$-\frac{1}{2}$								1

B.2   Particle properties

(Adapted from 'Review of Particle Properties', G. P. Yost *et al.*, *Rev. Mod. Phys.* **204**, 1988.)

Where properties are unmeasured or very uncertain the entry is omitted except in a few cases where a strong theoretical expectation is indicated.

Small fraction ($< 1\%$) or poorly-determined decay modes are also omitted.

B.2.1. 'Stable' particles (stable or decaying by weak or electromagnetic transitions)

	J^P	I^G	C	Mass (MeV/c^2)	Mean life (s)	Decays	Fraction (%)
2.1.1. Gauge bosons							
γ	1^-	0	$-$	$< 3.10^{-33}$	$-$	stable	$-$
W				81000 ± 1300	$\Gamma < 6.5$ GeV	$e\nu$	10.0 ± 0.3
						$\mu\nu$	12.0 ± 7
						$\tau\nu$	10.2 ± 4
Z†				91182	$\Gamma = 2.541 \pm 0.056$ GeV	e^+e^-	3.23 ± 0.21
					± 26(stat.) ± 30(syst.)	$\mu^+\mu^-$	2.98 ± 0.46
						$\tau^+\tau^-$	3.31 ± 0.41
						hadrons	68.9 ± 3.0
						invisible	21.4 ± 2.9
2.1.2. Leptons							
ν_e	$\frac{1}{2}$			< 0.000018	$-$	stable	
e	$\frac{1}{2}$			0.51099906 ± 0.00000015	$-$	stable	
ν_μ	$\frac{1}{2}$			< 0.25	$-$	stable	
μ	$\frac{1}{2}$			105.65839 ± 0.00006	2.19703×10^{-6} ± 0.00004	$e\bar{\nu}\nu$	100
ν_τ	$\frac{1}{2}$			< 35			
τ	$\frac{1}{2}$			1784.1 ± 3.2	$(3.04 \pm 0.09) \times 10^{-13}$	$\mu\bar{\nu}\nu$	17.8 ± 0.4
						$e\bar{\nu}\nu$	17.5 ± 0.4
						charged hadron + neutrals	50.3 ± 0.6
						3 charged hadrons + neutrals	17.0 ± 1.3
2.1.3. Non-strange mesons							
$\pi^\pm$	0^-	1^-		139.56755 ± 0.00033	2.6029×10^{-8} ± 0.0023	$\mu\nu$	100
π^0	0^-	1^-		134.9734 ± 0.0025	0.84×10^{-16} ± 0.06	$\gamma\gamma$	98.8
						γe^+e^-	1.2
η	0^-	0^+	$+$	548.8 ± 0.6	$\Gamma = (1.008 \pm 0.19)$ keV	$\gamma\gamma$	38.9 ± 0.4
						$3\pi^0$	31.9 ± 0.34
						$\pi^+\pi^-\pi^0$	23.7 ± 0.5
						$\pi^+\pi^-\gamma$	4.9 ± 0.1
2.1.4. Strange mesons							
$K^\pm$	0^-	$\frac{1}{2}$		493.646 ± 0.009	1.2371×10^{-8} ± 0.0026	$\mu\nu$	63.5 ± 0.2
						$\pi^\pm\pi^0$	21.2 ± 0.2
						$\pi^\pm\pi^+\pi^-$	5.59 ± 0.03
						$\pi^\pm\pi^0\pi^0$	1.73 ± 0.05
						$\pi^0\mu\nu$	3.2 ± 0.1
						$\pi^0 e\nu$	4.82 ± 0.05
$K^0, \bar{K}^0$	0^-	$\frac{1}{2}$		497.671 ± 0.030		50 % K_S^0	
						50 % K_L^0	
K_S^0	0^-	$\frac{1}{2}$			0.8923×10^{-10} ± 0.0022	$\pi^+\pi^-$	68.6 ± 0.26
						$\pi^0\pi^0$	31.4 ± 0.26
K_L^0	0^-	$\frac{1}{2}$			5.18×10^{-8} ± 0.04	$\pi^0\pi^0\pi^0$	21.7 ± 0.7
						$\pi^+\pi^-\pi^0$	12.4 ± 0.2
						$\pi^\pm\mu^\mp\nu$	27.1 ± 0.4
						$\pi^\pm e^\mp\nu$	38.6 ± 0.4

† Data from ALEPH; Decamp *et al. Phys. Lett.*, **B235**, 399 (1990); **B234**, 399 (1990)

	J^P	I^G	C	Mass (MeV/c^2)	Mean life (s)	Decays	Fraction (%)
2.1.5. Charmed non-strange mesons (see chapter 13 for discussion of decays)							
$D^\pm$	0^-	$\frac{1}{2}$		1869.4 ± 0.6	$10.7 \pm 0.3 \times 10^{-13}$	e ...	19 ± 2
						charged K ...	22.8 ± 4
						K^0 or $\bar{K}^0$ ...	48 ± 15
$D^0, \bar{D}^0$	0^-	$\frac{1}{2}$		1864.7 ± 0.6	$4.28 \pm 0.11 \times 10^{-13}$	e ...	7.7 ± 1.1
						charged K ...	49 ± 6
						K^0 or $\bar{K}^0$ ...	33 ± 10
2.1.6. Charmed strange mesons							
$D_s^\pm$ was $F^\pm$	0^-	0		1969.3 ± 1.1	$4.36 {+0.38 \atop -0.32} \times 10^{-13}$	$\phi\pi^\pm$	8 ± 5
						$\phi\pi^+\pi^-$	4 ± 3
						$\bar{K}^*(892)^\circ K^\circ$	8 ± 5
$D_s^{*\pm}$ was $F^{*\pm}$				2112.7 ± 2.3		$D_s^+\gamma$	dominant
2.1.7. Bottom mesons							
$B^\pm$	0^-	$\frac{1}{2}$		5277.6 ± 1.4		$\bar{D}^0\pi$	0.5 ± 0.2
						$D^*\pi\pi$	0.25 ± 0.15
						$D^*\pi\pi\pi$	4.3 ± 2.9
$B^0, \bar{B}^0$	0^-	$\frac{1}{2}$		5279.4		$D^*\rho$	8 ± 7
	(predictions)			± 1.5			
$B^\pm, B^0, \bar{B}^0$ not separated					$(13.1 \pm 1.4) \times 10^{-13}$	$e\nu$ hadrons	12.3 ± 0.8
						$\mu\nu$ hadrons	11 ± 0.9
						D^0 ...	39 ± 6
2.1.8 Non-strange baryons							
p	$\frac{1}{2}^+$	$\frac{1}{2}$		938.2231 ± 0.00028	$> 10^{32}$ y		
n	$\frac{1}{2}^+$	$\frac{1}{2}$		939.56563 ± 0.00028	896 ± 10	$pe^-\bar{\nu}$	100
2.1.9. Strangeness – 1 baryons							
Λ	$\frac{1}{2}^+$	0		1115.60 ± 0.05	2.632×10^{-10} ± 0.020	$p\pi^-$	64.2
						$n\pi^0$	35.8
Σ^+	$\frac{1}{2}^+$	1		1189.36 ± 0.06	0.800×10^{-10} ± 0.004	$p\pi^0$	51.6
						$n\pi^+$	48.4
Σ^0	$\frac{1}{2}^+$	1		1192.46 ± 0.08	7.4×10^{-20} ± 0.7	$\Lambda\gamma$	100
Σ^-	$\frac{1}{2}^+$	1		1197.43 ± 0.05	1.48×10^{-10} ± 0.01	$n\pi^-$	100
2.1.10. Strangeness – 2 baryons							
Ξ^0	$\frac{1}{2}^+$	$\frac{1}{2}$		1314.9 ± 0.6	2.90×10^{-10} ± 0.10	$\Lambda\pi^0$	100
Ξ^-	$\frac{1}{2}^+$	$\frac{1}{2}$		1321.32 ± 0.13	1.64×10^{-10} ± 0.02	$\Lambda\pi^-$	100
2.1.11. Strangeness – 3 baryons							
Ω^-	$\frac{3}{2}^+$	0		1672.45 ± 0.32	0.822×10^{-10} ± 0.012	ΛK^-	67.8 ± 0.7
						$\Xi^0\pi^-$	23.6 ± 0.7
						$\Xi^-\pi^0$	8.6 ± 0.4

	J^P	I^G	C	Mass (MeV/c^2)	Mean life (s)	Decays	Fraction (%)
2.1.12. Charmed baryons							
Λ_c^+	$\frac{1}{2}^+$		0	2284.9 ± 1.5	$1.79 \pm 0.2 \times 10^{-13}$	$pK^-\pi^+$	2.6 ± 0.9
						$p\bar{K}^0$	1.5 ± 0.6
						$\Lambda \ldots$	27 ± 9
						$e^+ \ldots$	4.5 ± 1.7
Ξ_c^+	$\frac{1}{2}^+$	$\frac{1}{2}$		2460 ± 19	$4.3 \pm 1.7 \times 10^{-13}$	$\Lambda K^-\pi^+\pi^+$	
	predictions					$\Sigma^0 K^-\pi^+\pi^+$	

B.2.2. Strongly-decaying particles

	J^P	I^G	C	Mass (MeV/c^2)	Full width, Γ (MeV/c^2)	Decays	Fraction (%)
2.2.1. Non-strange mesons							
$\rho(770)$	1^-	1^+	$-$	770 ± 3	153 ± 2	$\pi\pi$	100
$\omega(783)$	1^-	0^-	$-$	782.0 ± 0.1	8.5 ± 0.1	$\pi^+\pi^-\pi^0$	89.3 ± 0.6
						$\pi^0\gamma$	8.0 ± 0.9
						$\pi^+\pi^-$	1.7 ± 0.3
$\eta'(958)$	0^-	0^+	$+$	957.50 ± 0.24	0.21 ± 0.02	$\eta\pi\pi$	64.6 ± 2.0
						$\rho^0\gamma$	30.1 ± 1.4
						$\omega\gamma$	3.0 ± 0.3
						$\gamma\gamma$	2.2 ± 0.2
$f_0(975)$ was S(975)	0^+	0^+	$+$	976 ± 3	34 ± 6	$\pi\pi$	78 ± 3
						$K\bar{K}$	22 ± 3
$a_0(980)$ was $\delta(980)$	0^+	1^-	$+$	983 ± 3	57 ± 11	$\eta\pi$	
						$K\bar{K}$	
$\phi(1020)$	1^-	0^-	$-$	1019.41 ± 0.01	4.41 ± 0.05	K^+K^-	49.3 ± 1.0
						$K^0_L K^0_S$	34.7 ± 1.0
						$\pi^+\pi^-\pi^0$	14.8 ± 0.7
						$\eta\gamma$	1.2 ± 0.2
$h_1(1170)$	1^+	0^-	$-$	1170 ± 40	335 ± 26	$\rho\pi$	
$b_1(1235)$ was B(1235)	1^+	1^+	$-$	1233 ± 10	150 ± 10	$\omega\pi$	~ 100
$a_1(1260)$ was A$_1$(1270)	1^+	1^-	$+$	1260 ± 30	300–600	$\rho\pi$	~ 100
$f_2(1270)$	2^+	0^+	$+$	1274 ± 5	185 ± 20	$\pi\pi$	86 ± 2
						$2\pi^+2\pi^-$	2.9 ± 0.4
						$\pi^+\pi^-2\pi^0$	6.4 ± 2.5
						$K\bar{K}$	4.2 ± 1.0
$\eta(1280)$	0^-	0^+	$+$	1279 ± 5	32 ± 10	$\eta\pi\pi$	
$f_1(1285)$ was D(1285)	1^+	0^+	$+$	1283 ± 5	25 ± 3	$K\bar{K}\pi$	11 ± 3
						$\eta\pi\pi$	49 ± 6
						4π	40 ± 7
$a_2(1320)$ was A$_2$(1320)	2^+	1^-	$+$	1318 ± 5	110 ± 5	$\rho\pi$	70.1 ± 2.2
						$\eta\pi$	14.5 ± 1.1
						$\omega\pi\pi$	10.6 ± 2.5
						$K\bar{K}$	4.9 ± 0.8
$f_0(1400)$ was $\varepsilon(1300)$	0^+	0^+	$+$	~ 1400	150–400	$\pi\pi$	93 ± 2
						$K\bar{K}$	7 ± 2

	J^P	I^G	C	Mass (MeV/c^2)	Full width, Γ (MeV/c^2)	Decays	Fraction (%)
$f_1(1420)$ was E(1420)	1^+	0^+	$+$	1422 ± 10	55 ± 3	$K\bar{K}\pi$	~ 100
$\eta(1430)$ was i(1440)	0^-	0^+	$+$	1440 ± 20	60 ± 30	$K\bar{K}\pi$ $\eta\pi\pi$	
$f_2'(1525)$ was f'(1525)	2^+	0^+	$+$	1525 ± 5	76 ± 10	$K\bar{K}$	~ 100
$f_1(1530)$	1^+	0^+	$+$	1527 ± 5	106 ± 14	$K^*\bar{K} + \bar{K}^*K$	
$f_0(1590)$	0^+	0^+	$+$	1587 ± 11	$\sim 175 \pm 19$	$\eta'(958)\eta$	~ 100
$\omega_3(1670)$ was $\omega(1670)$	3^-	0^-	$-$	1668 ± 15	166 ± 15	3π 5π	
$\pi_2(1670)$ was $A_3(1680)$	2^-	1^-	$+$	1665 ± 20	250 ± 20	$f_2(1270)\pi$ $\rho\pi$ $\pi\pi\pi$ $K^*\bar{K} + \bar{K}^*K$	53 ± 5 34 ± 6 9 ± 5 4 ± 1
$\rho_3(1690)$ was g(1690)	3^-	1^+	$-$	1691 ± 5	215 ± 20	2π 4π $K\bar{K}\pi$ $K\bar{K}$	23.8 ± 1.3 70.9 ± 1.9 3.8 ± 1.2 1.5 ± 0.3
$\rho(1700)$	1^-	1^+	$-$	1700 ± 20	235 ± 50	4π 2π $K^*\bar{K} + \bar{K}^*K$ $\eta\pi\pi$ $\bar{K}K$ e^+e^-	
$f_2(1720)$ was $\theta(1690)$	2^+	0^+	$+$	1721 ± 4	138 ± 11	$\eta\eta$ $K\bar{K}$ $\pi\pi$	18 ± 13 38 ± 19 4 ± 2
$f_2(2010)$	2^+	0^+	$+$	2011 ± 70	202 ± 65	$\phi\phi$	
$f_4(2050)$ was h(2030)	4^+	0^+	$+$	2047 ± 11	204 ± 13	$\pi\pi$	17 ± 2
$f_2(2300)$	2^+	0^+	$+$	2297 ± 28	149 ± 41	$\phi\phi$	
$f_2(2340)$	2^+	0^+	$+$	2339 ± 55	319 ± 75	$\phi\phi$	

2.2.2. $c\bar{c}$ Mesons

	J^P	I^G	C	Mass (MeV/c^2)	Full width, Γ (MeV/c^2)	Decays	Fraction (%)
$\eta_c(1S)$ or $\eta_c(2980)$	0^-	0^+	$+$	2979.6 ± 1.7	$10.3 + 3.8 - 3.4$	$K\bar{K}\pi$ $\eta\pi\pi$ $K^+K^-\pi^+\pi^-$ $2(\pi^+\pi^-)$ $\eta'(958)\pi\pi$ $K^{*0}K^-\pi^+ + $c.c.	5.5 ± 0.8 5.0 ± 1.1 2.0 ± 0.3 1.2 ± 0.3 4.1 ± 1.7 2.0 ± 0.7
$J/\psi(1S)$ or $J/\psi(3097)$	1^-	0^-	$-$	3096.9 ± 0.1	0.068 ± 0.010	e^+e^- $\mu^+\mu^-$ hadrons $+$radiative	6.9 ± 0.9 6.9 ± 0.9 86.2 ± 2.0
$\chi_{c0}(1P)$ was $\chi(3415)$	0^+	0^+	$+$	3415.0 ± 1.0	13.5 ± 5.3	$2(\pi^+\pi^-)$ $\pi^+\pi^-K^+K^-$ $3(\pi^+\pi^-)$	3.7 ± 0.7 3.0 ± 0.7 1.5 ± 0.5
$\chi_{c1}(1P)$ was $\chi(3510)$	1^+	0^+	$+$	3510.6 ± 0.5		$\gamma J/\psi$ $3(\pi^+\pi^-)$ $2(\pi^+\pi^-)$	27.3 ± 1.6 2.2 ± 0.8 1.6 ± 0.5
$\chi_{c2}(1P)$ was $\chi(3555)$	2^+	0^+	$+$	3556.3 ± 0.4		$\gamma J/\psi$ $2(\pi^+\pi^-)$ $\pi^+\pi^-K^+K^-$ $3(\pi^+\pi^-)$	13.5 ± 1.1 2.3 ± 0.5 2.0 ± 0.5 1.2 ± 0.8

	J^P	I^G	C	Mass (MeV/c^2)	Full width, Γ (MeV/c^2)	Decays	Fraction (%)
ψ (2S)	1^-	0^-	$-$	3686.0 ±0.1	0.243 ±0.043	e^+e^- $\mu^+\mu^-$ hadrons + radiative	0.9±0.1 0.8±0.2 98.1±0.3
ψ (3770)	1^-		$-$	3770±3	25±3	$D\bar{D}$	dominant
ψ (4040)	1^-		$-$	4040±10	52±10	hadrons	dominant
ψ (4160)	1^-		$-$	4159±20	78±20	hadrons	dominant
ψ (4415)	1^-		$-$	4415±6	43±20	hadrons	dominant

2.2.3. $b\bar{b}$ Mesons

	J^P	I^G	C	Mass (MeV/c^2)	Full width, Γ (MeV/c^2)	Decays	Fraction (%)
Υ(1S)	1^-		$-$	9460.0 ±0.2	0.052 ±0.003	$\mu^+\mu^-$ e^+e^- $\tau^+\tau^-$	2.6±0.2 2.5±0.2 3.0±0.4
χ_{b0}(1P)			$+$	9859.8		$\gamma\Upsilon$(1S)	
χ_{b1}(1P)	1^+		$+$	9891.9		$\gamma\Upsilon$(1S)	35±8
χ_{b2}(1P)	2^+		$+$	9913.2		$\gamma\Upsilon$(1S)	22±4
Υ(2S)	1^-		$-$	10023.3 ±0.3	0.044 ±0.009	$\tau^+\tau^-$ $\mu^+\mu^-$ Υ(1S)$\pi\pi$	1.7±1.6 1.4±0.3 27.3±1.5
χ_{b0}(2P)			$+$	10235.3 ±1.1		$\chi_b\gamma$ Υ(2S)γ	17.6±1.7 6.9±4.1
χ_{b1}(2P)			$+$	10255.2 ±0.4		Υ(1S)γ Υ(2S)γ	1.4±1.0 25±8
χ_{b2}(2P)			$+$	10269.0 ±0.7		Υ(1S)γ Υ(2S)γ	6.1±1.7 19±7
Υ(3S)	1^-		$-$	10355.3 ±0.5	0.026 ±0.006	Υ(1S)γ $\mu^+\mu^-$ χ_b(2P)γ Υ(2S)...	6.3±1.8 1.6±0.4 30±6 10.1±1.7
Υ(4S)	1^-		$-$	10580 ±3.5	24±2	Υ(1S)$\pi^+\pi^-$	3.6±0.3
Υ(10860)	1^-		$-$	10865 ±8	110±13		
Υ(11020)	1^-		$-$	11019 ±8	79±16		

2.2.4. Strange mesons

	J^P	I^G	C	Mass (MeV/c^2)	Full width, Γ (MeV/c^2)	Decays	Fraction (%)
K*(892)	1^-	$\frac{1}{2}$		892.1 ±0.3	51.3±0.8	$K\pi$	100
K$_1$(1270) was Q$_1$(1280)	1^+	$\frac{1}{2}$		1270±10	90±20	$K\rho$ $K\pi$ K*(892)π $K\omega$ Kf$_0$(1400)	42±6 28±4 16±5 11±2 3±2
K$_1$(1400) was Q$_2$(1400)	1^+	$\frac{1}{2}$		1401±10	184±9	K*(892)π $K\rho$ Kf$_0$(1400) $K\omega$	94±6 3±3 2±2 1±1
K*(1415)	1^-	$\frac{1}{2}$		1415±15	240±20	K*(892)π $K\pi$	dominant 6.6±1.3
K$_0^*$(1430)	0^+	$\frac{1}{2}$		1429±7	287±23	$K\pi$	
K$_2^*$(1430) was K*(1430)	2^+	$\frac{1}{2}$		1426±2	99±3	$K\pi$ K*(892)π K*(892)$\pi\pi$ $K\rho$ $K\omega$	49.7±1.2 25.2±1.7 13.0±2.3 8.8±0.8 2.9±0.8

	J^P	I^G	C	Mass (MeV/c^2)	Full width, Γ (MeV/c^2)	Decays	Fraction (%)
K*(1715)	1^-	$\frac{1}{2}$		1717 ± 27	200–400	Kπ	39 ± 3
						Kρ	$31 + 7$ -3
						K*(892)π	$30 + 3$ -7
K$_2$(1770) was L	2^-	$\frac{1}{2}$		~ 1770	136 ± 18	K$_2^*$(1430)π	dominant
K$_3^*$(1780) was K*(1780)	3^-	$\frac{1}{2}$		1776 ± 4	160 ± 15	Kρ	44 ± 4
						K*(892)π	27 ± 3
						Kπ	19 ± 1
						Kη	11 ± 5
K$_4^*$(2075)	4^+	$\frac{1}{2}$		2074 ± 14	210 ± 50	Kπ	10 ± 1
						K*(892)$\pi\pi$	9 ± 5
						K*(892)$\pi\pi\pi$	7 ± 5
						ρKπ	6 ± 3
						ωKπ	5 ± 3
						ϕKπ	3 ± 1
						ϕK*(892)	1.4 ± 0.7

2.2.5. Charmed non-strange mesons

	J^P	I^G	C	Mass (MeV/c^2)	Full width, Γ (MeV/c^2)	Decays	Fraction (%)
D*(2010)$^+$	1^-	$\frac{1}{2}$		2010.1 ± 0.6	< 2.0	D$^0\pi^+$	49 ± 8
						D$^+\pi^0$	34 ± 7
						D$^+\gamma$	17 ± 11
D*(2010)^0	1^-	$\frac{1}{2}$		2007.1 ± 1.4	< 5.0	D$^0\pi^0$	52 ± 7
						D$^0\gamma$	48 ± 7
D$_J$(2420)^0		$\frac{1}{2}$		2422 ± 4	60 ± 13	D*(2010)$^+\pi^-$	

2.2.6. Charmed strange mesons

				Mass (MeV/c^2)	Full width, Γ (MeV/c^2)	Decays	Fraction (%)
D$_s^{*\pm}$ was F*				2112.7 ± 2.3	< 22	D$_s^+\gamma$	

	J^P	I	$L_{2I,2J}$	Mass (MeV/c^2)	Full width, Γ (MeV/c^2)	Decays	Fraction (%)

2.2.7. Nucleon baryon resonances (S = 0, $I = \frac{1}{2}$) N

	J^P	I	$L_{2I,2J}$	Mass (MeV/c^2)	Full width, Γ (MeV/c^2)	Decays	Fraction (%)
N(1440)	$\frac{1}{2}^+$	$\frac{1}{2}$	P$_{11}$	1400–1480	120–350	Nπ	50–70
						N$\pi\pi$	30–50
						$\Delta\pi$	10–20
						Nρ	10–15
N(1520)	$\frac{3}{2}^-$	$\frac{1}{2}$	D$_{13}$	1510–1530	100–140	Nπ	50–60
						N$\pi\pi$	40–50
						$\Delta\pi$	20–30
						Nρ	15–25
N(1535)	$\frac{1}{2}^-$	$\frac{1}{2}$	S$_{11}$	1520–1560	100–250	Nπ	35–50
						Nη	45–55
						N$\pi\pi$	~ 10
N(1650)	$\frac{1}{2}^-$	$\frac{1}{2}$	S$_{11}$	1620–1680	100–200	Nπ	55–65
						N$\pi\pi$	20–35
						Nρ	5–30
						ΛK	~ 8
N(1675)	$\frac{5}{2}^-$	$\frac{1}{2}$	D$_{15}$	1660–1690	120–180	Nπ	35–40
						N$\pi\pi$	60–65
						$\Delta\pi$	55–60

	J^P	I	$L_{2I,2J}$	Mass (MeV/c^2)	Full width, Γ (MeV/c^2)	Decays	Fraction (%)
N (1680)	$\frac{5}{2}^+$	$\frac{1}{2}$	F_{15}	1670–1690	110–140	Nπ	55–65
						N$\pi\pi$	35–45
						$\Delta\pi$	10–15
						Nρ	10–20
N (1700)	$\frac{3}{2}^-$	$\frac{1}{2}$	D_{13}	1670–1730	70–120	Nπ	5–15
						N$\pi\pi$	80–90
						$\Delta\pi$	15–70
						Nη	~ 4
N (1710)	$\frac{1}{2}^+$	$\frac{1}{2}$	P_{11}	1680–1740	90–130	Nπ	10–20
						N$\pi\pi$	< 50
						$\Delta\pi$	10–20
						Nη	~ 25
						ΛK	~ 15
						ΣK	2–10
N (1720)	$\frac{3}{2}^+$	$\frac{1}{2}$	P_{13}	1690–1800	125–250	Nπ	10–20
						N$\pi\pi$	< 75
N (2190)	$\frac{7}{2}^-$	$\frac{1}{2}$	G_{17}	2120–2230	200–500	Nπ	~ 14
						Nη	~ 3
N (2220)	$\frac{9}{2}^+$	$\frac{1}{2}$	H_{19}	2150–2300	300–500	Nπ	~ 18
N (2250)	$\frac{9}{2}^-$	$\frac{1}{2}$	G_{19}	2130–2270	200–500	Nπ	~ 10
N (2600)	$\frac{11}{2}^-$	$\frac{1}{2}$	I_{111}	2580–2700	> 300	Nπ	~ 5

2.2.8. Delta baryon resonances ($S = 0$, $I = \frac{3}{2}$) Δ

	J^P	I	$L_{2I,2J}$	Mass (MeV/c^2)	Full width, Γ (MeV/c^2)	Decays	Fraction (%)
Δ (1232)	$\frac{3}{2}^+$	$\frac{3}{2}$	P_{33}	1230–1234	110–120	Nπ	100
Δ (1620)	$\frac{1}{2}^-$	$\frac{3}{2}$	S_{31}	1600–1650	120–160	Nπ	25–35
						N$\pi\pi$	65–75
						$\Delta\pi$	60–70
						Nρ	10–20
Δ (1700)	$\frac{3}{2}^-$	$\frac{3}{2}$	D_{33}	1630–1740	190–300	Nπ	10–20
						N$\pi\pi$	80–90
						$\Delta\pi$	50–90
						Nρ	< 35
Δ (1900)	$\frac{1}{2}^-$	$\frac{3}{2}$	S_{31}	1850–2000	130–300	Nπ	5–15
Δ (1905)	$\frac{5}{2}^+$	$\frac{3}{2}$	F_{35}	1890–1920	250–400	Nπ	5–15
						N$\pi\pi$	< 75
						$\Delta\pi$	~ 25
						Nρ	< 50
Δ (1910)	$\frac{1}{2}^+$	$\frac{3}{2}$	P_{31}	1850–1950	200–330	Nπ	15–25
						N$\pi\pi$	< 75
Δ (1920)	$\frac{3}{2}^+$	$\frac{3}{2}$	P_{33}	1860–2160	190–300	Nπ	15–20
						ΣK	~ 5
Δ (1930)	$\frac{5}{2}^-$	$\frac{3}{2}$	D_{35}	1890–1960	150–350	Nπ	5–15
Δ (1950)	$\frac{7}{2}^+$	$\frac{3}{2}$	F_{37}	1910–1960	200–340	Nπ	35–45
						N$\pi\pi$	< 40
						$\Delta\pi$	~ 30
						Nρ	< 10
Δ (2420)	$\frac{11}{2}^+$	$\frac{3}{2}$	H_{311}	2380–2450	300–500	Nπ	5–15

2.2.9. Lambda baryon resonances ($S = -1$, $I = 0$) Λ

	J^P	I	$L_{2I,2J}$	Mass (MeV/c^2)	Full width, Γ (MeV/c^2)	Decays	Fraction (%)
Λ (1405)	$\frac{1}{2}^-$	0	S_{01}	1405 ± 5	40 ± 10	$\Sigma\pi$	100
Λ (1520)	$\frac{3}{2}^-$	0	D_{03}	1519.5 ± 1.0	15.6 ± 1.0	N$\overline{\text{K}}$	45 ± 1
						$\Sigma\pi$	42 ± 1
						$\Lambda\pi\pi$	10 ± 1
Λ (1600)	$\frac{1}{2}^+$	0	P_{01}	1560–1700	50–250	N$\overline{\text{K}}$	15–30
						$\Sigma\pi$	10–60
Λ (1670)	$\frac{1}{2}^-$	0	S_{01}	1660–1680	25–50	N$\overline{\text{K}}$	15–25
						$\Sigma\pi$	20–60
						$\Lambda\eta$	15–35

	J^P	I	$L_{2I,2J}$	Mass (MeV/c^2)	Full width, Γ (MeV/c^2)	Decays	Fraction (%)
$\Lambda(1690)$	$\frac{3}{2}^-$	0	D_{03}	1685–1695	50–70	$N\bar{K}$	20–30
						$\Sigma\pi$	20–40
						$\Lambda\pi\pi$	~ 25
						$\Sigma\pi\pi$	~ 20
$\Lambda(1800)$	$\frac{1}{2}^-$	0	S_{01}	1720–1850	200–400	$N\bar{K}$	25–40
$\Lambda(1810)$	$\frac{1}{2}^+$	0	P_{01}	1750–1850	50–250	$N\bar{K}$	20–50
						$\Sigma\pi$	10–40
						$N\bar{K}^*(892)$	30–60
$\Lambda(1820)$	$\frac{5}{2}^+$	0	F_{05}	1815–1825	70–90	$N\bar{K}$	55–65
						$\Sigma\pi$	8–14
						$\Sigma(1385)\pi$	5–10
$\Lambda(1830)$	$\frac{5}{2}^-$	0	D_{05}	1810–1830	60–110	$N\bar{K}$	3–10
						$\Sigma\pi$	35–75
						$\Sigma(1385)\pi$	> 15
$\Lambda(1890)$	$\frac{3}{2}^+$	0	P_{03}	1850–1910	60–200	$N\bar{K}$	20–35
						$\Sigma\pi$	3–10
$\Lambda(2100)$	$\frac{7}{2}^-$	0	G_{07}	2090–2110	100–250	$N\bar{K}$	25–35
						$\Sigma\pi$	~ 5
						$N\bar{K}^*(892)$	10–20
$\Lambda(2110)$	$\frac{5}{2}^+$	0	F_{05}	2090–2140	150–250	$N\bar{K}$	5–25
						$\Sigma\pi$	10–40
						$N\bar{K}^*(892)$	10–60
$\Lambda(2350)$	$\frac{9}{2}^+$	0	H_{09}	2340–2370	100–250	$N\bar{K}$	~ 12
						$\Sigma\pi$	~ 10

2.2.10. Sigma baryon resonances $(S = -1,\ I = 1)\ \Sigma$

	J^P	I	$L_{2I,2J}$	Mass (MeV/c^2)	Full width, Γ (MeV/c^2)	Decays	Fraction (%)
$\Sigma(1385)$	$\frac{3}{2}^+$	1	P_{13}	$(+)\,1382.8 \pm 0.4$	36 ± 1	$\Lambda\pi$	88 ± 2
				$(0)\,1383.7 \pm 1.0$	36 ± 5	$\Sigma\pi$	12 ± 2
				$(-)\,1387.2 \pm 0.6$	39 ± 2		
$\Sigma(1660)$	$\frac{1}{2}^+$	1	P_{11}	1630–1690	40–200	$N\bar{K}$	10–30
$\Sigma(1670)$	$\frac{3}{2}^-$	1	D_{13}	1665–1685	40–80	$N\bar{K}$	7–13
						$\Lambda\pi$	5–15
						$\Sigma\pi$	30–60
$\Sigma(1750)$	$\frac{1}{2}^-$	1	S_{11}	1730–1800	60–160	$N\bar{K}$	10–40
						$\Sigma\eta$	15–55
$\Sigma(1775)$	$\frac{5}{2}^-$	1	D_{15}	1770–1780	105–135	$N\bar{K}$	37–43
						$\Lambda\pi$	14–20
						$\Sigma\pi$	2–5
						$\Sigma(1385)\pi$	8–12
						$\Lambda(1520)\pi$	17–23
$\Sigma(1915)$	$\frac{5}{2}^+$	1	F_{15}	1900–1935	80–160	$N\bar{K}$	5–15
$\Sigma(1940)$	$\frac{3}{2}^-$	1	D_{13}	1900–1950	150–300	$N\bar{K}$	< 20
$\Sigma(2030)$	$\frac{7}{2}^+$	1	F_{17}	2025–2040	150–200	$N\bar{K}$	17–23
						$\Lambda\pi$	17–23
						$\Sigma\pi$	5–10
						$\Sigma(1385)\pi$	5–15
						$\Lambda(1520)\pi$	10–20
						$\Delta(1232)K$	10–20
$\Sigma(2250)$		1		2210–2280	60–150	$N\bar{K}$	< 10

2.2.11. Ξ Baryon resonances $(S = -2,\ I = \frac{1}{2})\ \Xi$

	J^P	I	$L_{2I,2J}$	Mass (MeV/c^2)	Full width, Γ (MeV/c^2)	Decays	Fraction (%)
$\Xi(1530)$	$\frac{3}{2}^+$	$\frac{1}{2}$	P_{13}	$(0)\,1531.8 \pm 0.3$	9.1 ± 0.5	$\Xi\pi$	100
				$(-)\,1535.0 \pm 0.6$	10.1 ± 1.9		

	J^P	I	$L_{2I,\,2J}$	Mass (MeV/c^2)	Full width, Γ (MeV/c^2)	Decays	Fraction (%)
$\Xi(1820)$	$\frac{3}{2}^-$	$\frac{1}{2}$	D_{13}	1823 ± 6	$20\,^{+15}_{-10}$	$\Lambda\bar{K}$ $\Sigma\bar{K}$ $\Xi(1530)\pi$	large small large
$\Xi(2030)$		$\frac{1}{2}$		2025 ± 5	$20\,^{+15}_{-5}$	$\Lambda\bar{K}$ $\Sigma\bar{K}$	~ 20 ~ 80

2.2.12. Ω Baryon resonances $(S = -3,\ I = 0)\,\Omega$

	J^P	I	$L_{2I,\,2J}$	Mass (MeV/c^2)	Full width, Γ (MeV/c^2)	Decays	Fraction (%)
$\Omega(2250)$		0		2252 ± 9	55 ± 18	$\Xi^-\pi^+K^-$ $\Xi(1530)^0K^-$	seen seen

2.2.13. Charmed baryon resonances

	J^P	I	$L_{2I,\,2J}$	Mass (MeV/c^2)	Full width, Γ (MeV/c^2)	Decays	Fraction (%)
$\Sigma_c(2455)^{++}$	$\frac{1}{2}^+$			2452 ± 1.7		$\Lambda_c^+\pi$	100
$\Sigma_c(2455)^+$				2452 ± 3.4			
$\Sigma_c(2455)^0$							

References

Abbott, L. F., *et al.*, *Nuc. Phys.* **B299**, 734 (1988)
Abela, R., *et al.*, *Phys. Lett.* **146B**, 431 (1984)
Albrecht, A., Steinhardt, P. J., *Phys. Rev. Lett.* **48**, 1220 (1982)
Albrecht, H., *et al.*, *Phys. Lett.* **B202**, 149 (1988)
ALEPH collaboration, Decamp, D. *et al.*, *Phys. Lett.* **B253**, 399 (1990)
Anderson, C. D., *Science* **76**, 238 (1932)
Anderson, C. D., Neddermeyer, S. H., *Phys. Rev.* **50**, 263 (1936)
Aubert, J. J., *et al.*, *Phys. Rev. Lett.* **33**, 1404 (1974)
Audouze, J., *Cosmology and Particle Physics*, World Scientific, 2 (1987)
Augustin, J. E., *et al.*, *Phys. Rev. Lett.* **33**, 1406 (1974)
Bahcall, J. N., Ulrich, R. K., *Rev. Mod. Phys.* **60**, 297 (1988)
Bahcall, J. N., *Neutrino Astrophysics*, Cambridge University Press, (1989)
Baldo-Ceolin, M., *et al.*, *Nuovo Cim.* **6**, 84 (1957)
Barnes, V. E., *et al.*, *Phys. Rev. Lett.* **15**, 322 (1965)
Behrends, S., *et al.*, *Phys. Rev. Lett.* **50**, 881 (1983)
Bellettini, G., *et al.*, *Nuovo Cim.* **40A**, 1139 (1965)
Bertanza, L., *et al.*, *Phys. Rev. Lett.* **9**, 180 (1962)
Bethe, H. A., *Phys. Rev.* **55**, 434 (1939)
Bethe, H. A., *Phys. Rev. Lett.* **56**, 1305 (1986)
Bethe, H. A., Marshak, R. E., *Phys. Rev.* **72**, 506 (1947)
Bionta, R. M., *et al.*, *Phys. Rev. Lett.* **51**, 27 (1983)
Bjorken, J. D., *Phys. Rev.* **163**, 1767 (1967)
Bjorklund, R., *et al.*, *Phys. Rev.* **77**, 213 (1950)
Bland, R., *et al.*, *Nuc. Phys.* **B13**, 595 (1969)

Blaton, J., *Konelige Danske Videnskabernes Selskab, Mathematisk-Fysike* **24**, 6 (1950)

Blümer, H., Kleinknecht, K., *Phys. Lett.* **B161**, 407 (1985)

Böhm, A., *Proc. Intern. Europhysics Conf. on High Energy Physics*, Brighton (1983)

Bologna, G., Vincelli, M. L., *Data Acquisition in High Energy Physics*, North Holland (1983)

Booth, P. S. L., *et al.*, *Nuc. Phys.* **B242**, 51 (1984)

Boris, S., *et al.*, *Phys. Rev. Lett.* **58**, 2019 (1987)

Brookhaven National Laboratory, *Phys. Rev. Lett.* **12**, 204 (1964)

Burgy, M. T., *et al.*, *Phys Rev.* **110**, 1214 (1958); *Phys. Rev. Lett.* **1**, 324 (1958)

Cabbibo, N., *Phys. Rev. Lett.* **10**, 531 (1963)

Cabrera, B., *Phys. Rev. Lett.* **48**, 1378 (1982)

Callan, C. G., *Phys. Rev.* **D26**, 2058 (1982); *Nuc. Phys.* **B212**, 391 (1983)

Callan, C. G., Gross, D. G., *Phys. Rev. Lett.* **21**, 311 (1968); **22**, 156 (1969)

Carlson, A. G., *et al.*, *Phil. Mag.* **41**, 701 (1950)

Cartwright, W. F., *et al.*, *Phys. Rev.* **91**, 677 (1953)

Chamberlain, O., *et al.*, *Phys. Rev.* **100**, 947 (1955)

Charpak, G., *et al.*, *Nuc. Inst. Meth.* **62**, 235 (1968)

Charpak, G., *et al.*, *Nuc. Inst. Meth.* **A283**, 596 (1989)

Christenson, J. H., *et al.*, *Phys Rev. Lett.* **13**, 138 (1964)

Close, F. E., *An Introduction to Quarks and Partons*, Academic Press (1979)

Cnops, A. M., *et al.*, *Phys. Lett.* **22**, 546 (1966)

Conversi, M., *et al.*, *Phys. Rev.* **71**, 209 (1947)

Courant, E. D., *et al.*, *Phys. Rev.* **88**, 1190 (1952)

Crennell, D. J., *et al.*, *Phys. Lett.* **28B**, 136 (1968)

Danby, G., *et al.*, *Phys. Rev. Lett.* **9**, 36 (1962)

Davis, R., *Proc. of Seventh Workshop on Grand Unification, ICOBAN '86*, Singapore World Scientific, 237 (1987)

Delbourgo-Salvador, P., *et al.*, *Astron: Astrophys.* **150**, 53 (1985)

De Panfilis, S., *et al.*, *Phys. Rev. Lett.* **59**, 839 (1987)

Dirac, P. A. M., *Camb. Phil. Soc.* **26**, 361 (1930)

Dirac, P. A. M., *Proc. Roy. Soc.* **A133**, 60 (1931)

Durbin, R., *et al.*, *Phys. Rev.* **83**, 646 (1951); **84**, 581 (1951)

Englert, F., Brout, R., *Phys. Rev. Lett.* **13**, 321 (1964)

Feynman, R. P., *Phys. Rev. Lett.* **23**, 1415 (1969)

Fermi, E., *Z. Physik* **88**, 161 (1934)

Fickinger, W. J., *et al.*, *Phys. Rev.* **125**, 2082 (1962)

Friedman, J. I., Kendall, H. W., *Ann. Rev. Nat. Sci.* **22**, 203 (1972)

Fritschi, M., *et al.*, *Phys. Lett.* **B173**, 485 (1986)

Gaillard, M. K., Lee, B. W., *Phys. Rev.* **D10**, 897 (1974)

Gell-Mann, M., *Phys. Rev.* **125**, 1067 (1962); see also *The Eightfold Way*, Benjamin (1964), for a collection of the early papers on SU(3)

Gell-Mann, M., Pais, A., *Proc. of the Glasgow Conference on Nuclear and Meson Physics*, 342 (1954)

Georgi, H., Glashow, S. L., *Phys. Rev. Lett.* **32**, 438 (1974)

Glashow, S. L., *Nuc. Phys.* **22**, 579 (1961)

Glashow, S. L., *et al.*, *Phys. Rev.* **D2**, 1285 (1970)

Goldhaber, M. L., *et al.*, *Phys. Rev.* **109**, 1015 (1958)

Greenberg, O. W., *Phys. Rev. Lett.* **13**, 598 (1964)

Guralnick, G., *et al.*, *Phys. Rev. Lett.* **13**, 585 (1964)

Guth, A. H., *Phys. Rev.* **D23**, 347 (1981)

Halzen, F., Martin, A. D., *Quarks and Leptons*, Wiley (1984)

Hanson, G., *et al.*, *Phys. Rev. Lett.* **35**, 1609 (1975)

Heisenberg, W., *Z. Physik* **77**, 1 (1932)

Herb, S. W., *et al.*, *Phys. Rev. Lett.* **39**, 252 (1977)

Higgs, P. W., *Phys. Lett.* **12**, 132 (1964); *Phys. Rev. Lett.* **13**, 508 (1964); *Phys. Rev.* **145**, 1156 (1966)

Hirata, K. S., *et al.*, *Phys. Rev. Lett.* **63**, 16 (1989)

t'Hooft, G., *Phys. Lett.* **37B**, 195 (1971)

t'Hooft, G., *Nuc. Phys.* **B33**, 173 (1971): **B35**, 167 (1971)

t'Hooft, G., *Nuc. Phys.* **B79**, 276 (1974)

t'Hooft, G., Veltmann, M., *Nuc. Phys.* **B44**, 189 (1972)

Innes, W. R., *et al.*, *Phys. Rev. Lett.* **39**, 1240 (1977)

Jauch, J. M., *CERN Report 59–35*, 19 (1959)

Jeckelman, B., *et al.*, *Phys. Rev. Lett.* **56**, 1444 (1986)

Kazanas, D., *Ap. J.* **241**, L59 (1980)

Kemmer, N., *Proc. Roy. Soc.* **A166**, 127 (1938)

Kibble, T. W. B., *Phys. Rev.* **155**, 1554 (1967)

Kobayashi, M., Maskawa, K., *Prog. Theo. Phys.* **49**, 282 (1972)

Koller, K., Krasemann, H., *Phys. Lett.* **88B**, 119 (1979)

Lande, K., *et al.*, *Phys. Rev.* **103**, 1901 (1956)

Langacker, P., *Phys. Rep.* **72**, 186 (1981)

Larribe, A., *et al.*, *Phys. Lett.* **23**, 600 (1966)

LaRue, G. S., *et al.*, *Phys. Rev. Lett.* **38**, 1011 (1977); **46**, 967 (1981)

Lattes, C. M., *et al.*, *Nature* **160**, 453 (1947)

Lee, B. W., *Phys. Rev.* **D5**, 823 (1972)

Lee, T. D., Yang, C. N., *Phys. Rev.* **104**, 254 (1956)

Leprince-Ringuet, L., Lheritier, M., *Comptes Rendues de l'Academie des Sciences* **219**, 618 (1944)

Lim, C. S., Marciano, W. J., *Phys. Rev.* **D37**, 1368 (1988)

Linde, A., *Phys. Lett.* **B108**, 389 (1982)

Lipkin, H. J., *Lie Groups for Pedestrians*, North Holland (1965)

Luders, G., *Kongelige Danske Videnskabarnes Selskab, Mathematisk-Fysike* **28**, no. 5 (1954)

Marinelli, M., Morpurgo, G., *Phys. Lett.* **94B**, 427 and 433 (1980)

Mather, J., *et al.*, *Astrophys. J. Lett.* to be published (1990)

Michel, L., *Rev. Mod. Phys.* **29**, 223 (1957)

Mikeyev, S. P., Smirnov, A. Y., *Sov. J. Nuc. Phys.* **42**, 913 (1986)

Mikeyev, S. P., Smirnov, A. Y., *Sov. Phys. JETP* **64**, 4 (1986)

Mikeyev, S. P., Smirnov, A. Y., *Nuovo Cim.* **9C**, 17 (1986)

Müller, D., *Phys. Rev.* **D5**, 2677 (1972)

Ne'eman, Y., *Nuc. Phys.* **26**, 222 (1961)

Nishijima, K., *Prog. Theo. Phys.* **13**, 285 (1954)

O'Ceallaigh, C, *Phil. Mag.* **41**, 838 (1950)

Okubo, S., *Phys. Lett.* **5**, 165 (1963)

OPAL collaboration, (Akrawy *et al.*) *Phys. Lett.* **247**, 224 (1990)

Parke, S. J., Walker, T. P., *Phys. Rev. Lett.* **57**, 2322 (1986)

Pauli, W., *Niels Bohr and the Development of Physics*, McGraw-Hill (1955)

Peccei, R. D., Quinn, H. R., *Phys. Rev. Lett.* **38**, 1440 (1977)

Peccei, R. D., Quinn, H. R., *Phys. Rev.* **D16**, 1791 (1977)

Perl, M. L., *et al.*, *Phys. Rev. Lett.* **35**, 1489 (1975); *Phys. Lett.* **63B**, 366 (1976)

Peterson, J. R., *Phys. Rev.* **105**, 693 (1957)

Pevsner, A., *et al.*, *Phys. Rev. Lett.* **7**, 421 (1961)

Pjerrou, G. M., *et al.*, *Phys. Rev. Lett.* **9**, 114 (1962)

Plano, R., *et al.*, *Phys. Rev. Lett.* **3**, 525 (1959)

Polyakov, A. M., *JETP Lett.* **20**, 194 (1974)

Powell, C., *et al.*, *The Study of Elementary Particles by the Photographic Method*, Pergamon (1959)

Rasetti, F., *Phys. Rev.* **59**, 706 (1941); **60**, 198 (1941)

Rochester, G. D., Butler, C. C., *Nature* **160**, 855 (1947)

Rubakov, V. A., *Nuc. Phys.* **B203**, 311 (1982); *JETP Lett.* **33**, 644 (1981)

Sakata, S., *Prog. Theo. Phys.* **16**, 686 (1956)

Sakharov, A. D., Pisma, Z., *Eksp. Teor. Fiz.* **5**, 32 (1967)

Sakurai, J. J., *Phys. Rev. Lett.* **9**, 472 (1962)

Salam, A., *Proc. 8th Nobel Symposium, Aspenasgarden 1968*, Almquist and Wilksell, Stockholm, p. 367.

Sancisi, R., Van Albada, T. S, *International Astronomical Union Symposium No. 117*, *67*, (1987)

Schwinger, J., *Phys. Rev.* **91**, 712; **94**, 1362 (1953)

Sikivie, P., *Phys. Rev. Lett.* **51**, 1415 (1983); **52**, 695 (1984)

Sikivie, P., *Phys. Rev.* **D32**, 2988 (1985)

Smith, P. F., *et al.*, *Phys. Lett.* **B197**, 447 (1987); **B181**, 407 (1986); **B171**, 129 (1986); **B153**, 188 (1985)

Söding, P., Wolf, G., *Ann. Rev. Nuc. Part. Sci.* **31**, 231 (1981)

Steigmann, G., *Ann. Rev. Astron. Astrophys.* **14**, 339 (1976)

Thompson, R. W., *et al.*, *Phys. Rev.* **83**, 175 (1951)

Van der Meer, S., *CERN Report ISR-PO/72–31* (1972); *Phys. Rep.* **58**, 73 (1980)

Walker, W. D., *et al.*, *Phys. Rev. Lett.* **18**, 630 (1967)

Watson, M. B., *et al.*, *Phys. Rev.* **131**, 2248 (1963)

Weinberg, S., *Phys. Rev. Lett.* **19**, 1264 (1967)

Williams, E. J., Roberts, G. E., *Nature* **145**, 102 (1940)

Wolfenstein, L., *Phys. Lett.* **13**, 562 (1964)

Wolfenstein, L., *Phys. Rev.* **D17**, 2369 (1978); **D20**, 2634 (1979)

Wu, C. S., *et al.*, *Phys. Rev.* **105**, 1413 (1957)

Yang, C. N., Mills, R. L., *Phys. Rev.* **96**, 191 (1954)

Yukawa, H., *Proc. Physico-Mathematical Soc. of Japan.* **17**, 48 (1935)

Zacek, G., *et al.*, *Phys. Rev.* **D34**, 2621 (1986)

Abbreviations used:

JETP Lett.: Journal of Experimental and Theoretical Physics Letters
Nuc. Inst. Meth.: Nuclear Instruments and Methods
Nuc. Phys.: Nuclear Physics
Nuovo Cim.: Nuovo Cimento
Phil. Mag.: Philosophical Magazine
Phys. Lett.: Physics Letters
Phys. Rev.: Physical Review
Phys. Rev. Lett.: Physical Review Letters
Phys. Rep.: Physics Reports
Proc. Roy. Soc.: Proceedings of the Royal Society
Prog. Theo. Phys.: Progress of Theoretical Physics
Rev. Mod. Phys.: Reviews of Modern Physics
Z. Physik: Zeitschrift für Physik

Index